Teaching and Education in Fracture and Fatigue

Teaching and Education in Fracture and Fatigue

Edited by

H.P. Rossmanith

Technical University of Vienna, Vienna, Austria
and Secretary-General, International Society for Technology, Law and Insurance

CRC Press
Taylor & Francis Group
Boca Raton London New York

CRC Press is an imprint of the
Taylor & Francis Group, an **informa** business
A TAYLOR & FRANCIS BOOK

CRC Press
Taylor & Francis Group
6000 Broken Sound Parkway NW, Suite 300
Boca Raton, FL 33487-2742

First issued in paperback 2019

ISBN-13: 978-0-419-20700-9 (hbk)
ISBN-13: 978-0-367-86560-3 (pbk)

A catalogue record for this book is available from the British Library

Publisher's Note This book has been prepared from camera ready copy provided by the individual contributors.
The publisher has gone to great lengths to ensure the quality of this reprint but points out that some imperfections in the original may be apparent.

Visit the Taylor & Francis Web site at
http://www.taylorandfrancis.com

and the CRC Press Web site at
http://www.crcpress.com

Dedicated to my great teacher,

George R. Irwin,

The Father of Fracture Mechanics,

who has shown engineers

how to put fracture to work

Prof. Dr. George Rankine Irwin, December 1977
taken at the University of Maryland

Property of H. P. Rossmanith, Vienna, Austria

Indeed, if further confirmation of the importance of cracks is needed, simply return to Kipling. A Chief Engineer named McPhee describes the defects in his ship and ends by saying,

> **"But the warst was at the last, She'd a great clumsy iron nineteen-foot Thresher propellor... and just on the tail of the shaft, before the boss, was a red weepin' crack ye could ha' put a penknife to. Man, it was an awful crack.."**

And presently, his colleague McRimmon, asks

> **"Whaur's the flaw, an what like?"**
> **"A seven-inch crack just behind the boss, There's no power on earth will fend it just jarrin' off"**
> **"When?"**
> **"That's beyon' my knowledge"**
> **"So it is, so it is"** said McRimmon, **"We've all oor leemitations".**

That was written in 1895. Perhaps now we could have helped McPhee a little in estimating the remaining life.

*Taken from R.Kipling's short story **'Bread upon the Waters'**. It is said to be based on his own account of what happened when one of the propellers of the Groukan dropped off in the Bay of Biscay because of a fatigue crack in the tailshaft*

Communicated with pleasure by C.Turner, Imperial College, London

Contents

The Subject

Requests and Demands

Regional Differences

Advanced Teaching Aids

Proposals and Improvements

1 INTRODUCTION

I. MILNE
President of the European Structural Integrity Society
Integrity Management Services, Falkland House, Youlgrave, Derbyshire, UK

The terms Fracture and Fatigue evoke strong images of failed and tired components. The image is one where a structure or component loses its integrity and fails to do its job. The consequences of such an event can be felt in a myriad of trivial everyday events which annoy us all at a domestic level, and in major one-off' disasters. When these involve large scale pollution or loss of live, they capture the headlines of the world press. Fortunately disasters with these consequences happen infrequently. Less fortunately, those failures which don't end in such dire consequences often go unreported as if they are of less significance. This leads to a level of complacency in the design, construction and operation of large scale structures which is not merited. Again, failures are often put down to operational factors, when even if true, a design modification could effectively either eliminate the potential for such a failure, or significantly reduce the consequences. A good example of this is in the vulnerability of Ro-Ro ferries to loss of stability when the front doors are breached.

When we think of structural integrity we tend to think in terms of safety issues. However, the economic effects of fracture are enormous. Recent studies [1,2] have demonstrated that, annually, the total economic costs of fracture related events to the economies of advanced countries is 4% of gross domestic product. This is comparable with the largest national budget items. In the UK, this equates to what is spent on defence, and is several times the expenditure on transport and education [3]. Crucially, it is approximately twice the total expenditure on R&D, yet it has been estimated [2] that this loss could be halved within 10 years by application of our current knowledge, without the need for major breakthroughs, and at only 1% of the cost per year. What chancellor of the exchequer or business manager could refuse such a cost benefit?

Teaching and Education in Fracture and Fatigue. Edited by H.P. Rossmanith. Published in 1996 by Taylor & Francis, 2 Park Square, Milton Park, Abingdon, Oxon, OX14 4RN. ISBN 0 419 20700 7.

ESIS, the European Structural Integrity Society, has the following statutory objectives: to foster research into the prevention of failure of engineering materials, components and structures by fracture or other physical phenomena via:

- Interdisciplinary research into physical behaviour of engineering materials components and structures,
- Development and assessment of testing methods, numerical methods and engineering methods,
- Improvements of engineering designs,
- Dissemination of knowledge, and
- Education.

Although education is implicit in all objectives, it has been separately identified to emphasise the importance which ESIS places upon it. This is because structural integrity cannot be identified as an independent discipline, nor can it be regarded as contained within a single discipline. It is truly interdisciplinary, and makes challenging demands on physicists, mathematicians, engineers, materials experts and others. It is also case oriented, and indeed has been driven by applications. The nuclear industry, the aircraft and transport industries, the offshore industry and the petrochemical and power generation industries have all made major contributions to the development of what we term fracture mechanics methodologies, the contributions often being specifically industry oriented. Consequently, although each industry can learn from each other, the educational base is fragmented.

The importance of a sound and coherent educational base in fracture mechanics was further underlined at a meeting held in Sheffield in April 1993 which discussed structural integrity research and training strategies across the expanding ESIS membership. For the first time we had delegates from all independent states in Central and Eastern Europe meeting with ESIS Technical Committee chairman, to identify how best we could meet future needs. One of the important aspects identified was teaching and education, and a committee was set up to cover this. ESIS sponsors the international conferences, TEFF, which have to date been held alternatively in Vienna (Austria) and Miskolc (Hungary). These conferences are not just confined to European issues, but are a major international forum for this subject. It is emphasised that this forum does not debate the science or application of fracture and fatigue, nor is it a course. It debates how the subject is taught, how best we can develop future practitioners to a professional and coherent level which might include independent recognition by learned societies. Structural integrity is too important a subject to be left as an add-on to syllabuses, or as a subject taught as minor issue of the major subjects.

As President of ESIS I would like to make a personal observation here. It is important to recognise that fracture mechanics methodologies are required to match the continuum mechanics of engineering assessments with the non-continuum behaviour of materials. Coming from the materials discipline myself, I find that there is a poor understanding amongst many fracture experts of the behaviour of real materials and of the wide variation which often exists in their property values. Too often analysts find it too easy to incorporate a continuum model of a material property into a computer program. The consequent results, naturally, only reflect the quality of the input data, and in many cases can be seriously misleading. In the future, I hope for much more emphasis on the development of knowledge of the properties, behaviour and variability of real materials, and how this can be more effectively modelled in our problem oriented methodologies.

It is to be hoped that this volume of papers, which contains papers from the first 4 of the TEFF conferences, will help stimulate further improvements in the education of this absorbing and essential sub-discipline.

References

[1] The Economic Effects of Fracture in The United States (1978) US Department of Commerce, National Bureau of Standards, Special Publications, 647-1 & 647-2, 1978.

[2] Faria, L.: The Economic Effects of Fracture in Europe. Final Report, Study Contract No. 320105 between the European Atomic Energy Community & Stichting voor Toepassing van Materialen (Delft), CEC, 1991.

[3] Milne, I.: The Importance of the Management of Structural Integrity, Engineering Failure Analysis, Vol.1., No.3, pp 171-181, 1994.

2 EDUCATION IN FATIGUE AND FRACTURE: A TRUE CHALLENGE FOR THE ENGINEERING WORLD

H.P. ROSSMANITH
Institute of Mechanics, Technical University Vienna, Vienna, Austria

Failure Engineering is the technical discipline which comprises the definition, evaluation, analysis, assessment and implications of technical failures. Fatigue and fracture are among the various physical processes which are associated with the loss of integrity of a structure or a structural component

During the past 20 years the world of engineering has been subjected to fundamental changes with totally new developments and concepts in science, technology, industry, administration, business, marketing and education. This continuous sequence of changes has fundamentally revolutionized society where an ever-increasing pressure towards productivity and improved quality, engineering design for profit while keeping safety at a highest possible level pushes the engineer into a difficult position. Optimal education and training of engineers is the key for success and, therefore, should be a mandatory prerequisite for every engineer [1].

This volume deals with methods of education in the technical fields of fracture and fatigue (F&F) and highlights some of the pertinent issues on innovative teaching, learning and training in these engineering disciplines.

Experience unveils that teachers in technical subjects in general do have very little theoretical knowledge in didactics and pedagogy but still do astonishingly well in transmitting practical skills and know-how. The key factors signaling preparedness for learning are motivation of the student and practicing engineer for basic, advanced and continuing education and recognizing the cash-value of increased competence and continuing professional development gained at the workplace or by attending a course at an institution of higher learning offering the availability of continuing education.

Teaching and Education in Fracture and Fatigue. Edited by H.P. Rossmanith. Published in 1996 by Taylor & Francis, 2 Park Square, Milton Park, Abingdon, Oxon, OX14 4RN. ISBN 0 419 20700 7.

Traditional teaching aids such as books (text books, reference books, special issues, etc.) are exposed to highly competitive means such as those based on audio-visual technology. Within the framework of fracture and fatigue (F&F) an ever-increasing number of special technical publications and a few text books on the subject of fracture research face the competition of a growing number of audio-visual teaching and training kits [2].

Basic training and education in F&F of materials and structures has become a major issue and while the importance of research in the field of F&F is recognized and the application of the theory to practical applications is honored with great success, contemporary methods of teaching, education and learning in F&F are not appropriate and prove to be highly insufficient. Unlike education in the classical engineering disciplines where abundant experience with respect to optimal teaching and learning has been converted into appropriate textbooks and other teaching aids, counterparts in F&F, except for a few noteworthy examples, seem to lag far behind.

This situation has been recognized some time ago and in order to focus the emphasis of a wider group of experts to the lack of appropriate means of education in F&F special sessions on teaching and education in F&F have been organized in association with several international and national conferences [3-4]. It was very quickly realized that round table discussions would not suffice to correspond to all requests and a more detailed attack on the various aspects of education in F&F had to be launched.

In April 1993, an ESIS patronized East-West Symposium on the State of the Art in Fracture and Fatigue was organized at the University of Sheffield which was attended by more than 60 delegates from Eastern and Western European countries. On this occasion an ESIS Technical Committee on Teaching and Education was established under the joint chairmanship of H.P.Rossmanith (Austria) and L.Toth (Hungary) [5].

During the period 1990-1995 the "Office of International Exchange Programs for Central and Eastern Europe" as part of the joint activity between Austria and Hungary has sponsored a binational effort for establishing a common curriculum for teaching and education in fracture and fatigue during the course of which four international workshops/symposia have been organized in Vienna (Austria) and in Miskolc (Hungary).

The aims, objectives and topics of these workshops/symposia included various tasks. The workshops were designed to:

- provide an *international forum* of educators and teachers in the field of F&F from East and West,
- foster the *exchange of information* on education and training in design and failure engineering with particular emphasis on F&F,
- develop the *human resources* to their fullest potential,
  discuss the *current issue* on teaching and education in F&F in the light of a systematic and comprehensive development of teaching, training and learning methodologies,

- develop and extend *knowledge and skills* in teaching methodologies, develop *effective and efficient* ways of satisfying the needs and demands,
- establish a common *curriculum/course strategy* with respect to various levels of teaching and education of the students,
- assess and wage the *cost* of education against cost of failures in engineering,
- inform about recent *advances* in F&F in industry,
- unveil *advantages and disadvantages* of both, Eastern and Western educational systems and "What can East and West learn from each other?",
- settle on: "Are *traditional* teaching and education methods really so bad?",
- discuss the role of *computer-related* educational technologies,
- survey the jungle of *literature, program packages* etc. on fracture and fatigue,
- identify the *needs of industry* in terms of teaching and education,
- establish *industry-university* cooperation in training and research,
- exhibit and clarify "What do contemporary *universities* in different countries and social environment offer?",
- assess and identify indicators of *quality* of and *manage* educational professional development,
- address the *social acceptance* of present teaching and education models in failure engineering,
- assess the influence of *new technology* on new educational strategies,
- evaluate the benefit of *team learning, multi-cultural* and *multi-disciplinary* learning environments,
- identify the limits of the *transferability* of advanced computer-and video-assisted teaching methodologies,
- highlight the needs of *developing countries* in terms of education and practical application of F&F to their engineering problems, as well as
- present an *outlook* to future possibilities and limitations.

Because of their intrinsic occupation with material and structural failures, engineers play a central role in the process of education and teaching as well as training in F&F. Not only has learning been recognized as the source of wealth, welfare, competitive advantage, and quality of life, for many companies continuing updating, upgrading as well as the acquisition of new knowledge, skills and competence is essential for survival on the highly competitive inter-national market-place [6-8].

This volume as well as the workshops on Teaching and Education of Fatigue and Fracture were prepared to bring together all actors involved in the educational spectrum associated with F&F, especially

- engineers in the various engineering disciplines where product failure due to F&F is of concern,
- scientists involved in basic and applied research in F&F,
- practitioners applying the techniques and methodologies of F&F in their daily professional work,

- professional educators and teachers at the various institutions engaged in medium and higher level engineering education as well as in post-mandatory school continuing education,
- continuing professional development and training,
- developers of educational media,

  decision makers and managers in business, commerce and industry, as well as professional bodies, local, national and international institutions, administrations and government advisors.

This volume contains 30 contributions authored by 46 experts from 17 countries from all over the world.

In the Introduction, the President of the European Structural Integrity Society, Dr. Ian Milne, via cost of failure, sets the scene by emphasizing the strong interrelation between economics and education. Key issues related to teaching of fatigue and fracture and reflections on forty years of teaching of the engineering aspects of fracture are the subject of the introductory papers. Then the need of teaching the physical basis of fracture mechanics is demonstrated and the necessity of combining simplicity and rationality is treated in detail.

The role of practice illustrated by the case study teaching method is highlighted by examining the needs of industry and comparing experience gained at several universities in East and West.

The case study method in conjunction with multidisciplinary teamwork and the group study method will be presented in several contributions. Achieving quality and the assessment, evaluation and control of quality of education in fracture and fatigue is addressed in two important contributions.

Experience in teaching fracture and fatigue gained in various representative countries is comm unicated in a sequence of five papers which address teaching strategies and methods for stimulating the students interest in this important subject.

The use of advanced teaching aids, particularly the role of the computer and advanced student-tailored software for teaching and learning of F&F is favored in a sequence of four papers. The contributions in the concluding section address proposals for new courses and improvements in F&F terminology.

The large majority of the contributions to this volume have been presented at the TEFF work-shops/symposia. A few papers have been specifically invited to address particular issues of teaching and learning in the fields of F&F. This volume does also reflect the efforts and activ ities of the members of the ESIS Technical Committee on Teaching and Education

The editor would like to express his thanks to all colleagues who attended the TEFF events and contributed to this volume. Their interaction and criticism in the course of this activity is kindly appreciated. It is hoped that the ideas expressed in this volume will be put to work and will turn out to be fruitful and advantageous in improving the quality of engineering education in the special fields of fracture and fatigue.

References

[1] R.A.Smith (Editor): Innovative Teaching in Engineering. Ellis Horwood, 1991.

[2] H.P.Rossmanith: How to teach fracture mechanics? Proc. ECF-8, Vol III, 1703-1717, 1990, EMAS, UK.

[3] Education in Fracture (Eds. J.Rama Rao et al) In: Proc of ICF-6, New Delhi, 3-30, Pergamon Press, 1987.

[4] Round Table Discussion TEFF (Chairman: H.P.Rossmanith), ICF International Conference on Fracture, June 10, 1993, Kiew, Ukraine.

[5] H.P.Rossmanith and L.Toth (Organizers): International Workshops/Symposia on Teaching and Education in Fracture and Fatigue (TEFF-1,2,3,4; 1992-1995) Vienna (Austria) and Miskolc (Hungary).

[6] F.M.Burdekin: Report on "Education in Fracture" issued by the Fracture Sub-committee of the Royal Society, UK, 1983.

[7] The Economic Effects of Fracture in the United States. NBS Special Publication 657-2, 1983.

[8] L.Faria: The Economic Effects of Fracture in Europe. EC Report 320105, 1991.

The Subject

Teaching and Education in Fracture and Fatigue. Edited by H.P. Rossmanith. Published in 1996 by Taylor & Francis, 2 Park Square, Milton Park, Abingdon, Oxon, OX14 4RN. ISBN 0 419 20700 7.

3 KEY ISSUES RELATED TO THE TEACHING OF FATIGUE AND FRACTURE

R.A. SMITH
Department of Mechanical and Process Engineering, University of Sheffield, UK

Abstract
This paper discusses the need to teach fatigue and fracture, but is cautious about making appeals for funds for research based on a perception that fracture is important because the economic consequences of failure can be large. The author's experience in teaching fatigue and fracture on introductory courses is discussed. A non-mathematical approach to Linear Elastic Fracture Mechanics is presented, together with a discussion on the uses of case studies, history and size-scale in fatigue and fracture courses.
Keywords: Teaching, fatigue, fracture, fracture mechanics, history, case studies, size-scale.

1 Justification by economics?

There can be no doubt of the economic significance of the topic of fatigue and fracture. Recent European studies have indicated that huge savings could be made by avoiding failures caused by fatigue and fracture, presumably by improving design. The orders of magnitude of savings to be made by avoiding fracture problems are estimated as several percent of GNP. These studies reinforce similar findings to other work conducted in the United States. However, many pressure groups for other subjects have made similar claims. The topics of tribology, corrosion and vibration readily spring to mind. If the savings of GNP in all these areas were added together one might imagine that the total savings would be phenomenal! However, in reality all topics in engineering design have a complex interrelationship and great care should be exercised in singling out any particular subject since a well rounded engineer needs a wide range of knowledge.

Teaching and Education in Fracture and Fatigue. Edited by H.P. Rossmanith. Published in 1996 by Taylor & Francis, 2 Park Square, Milton Park, Abingdon, Oxon, OX14 4RN. ISBN 0 419 20700 7.

If the only object of our researches into fatigue is the avoidance of fatigue failures, we have to compete with claims from other completely non-engineering topics such as cutting crime, improving the safety of transport systems, providing better health care, all of which could benefit GNP. In reality the financial system boundary for which these claims are made needs careful definition. We also need to understand the complexities of accidents: let me give two examples. The direct cause of the deaths of spectators at the accident at Hillsborough Football Ground in 1989 was the failure of a wrought iron crush barrier within the crowd, the failure being due to corrosion and over-loading. However, many complex interrelated causes came together to form the cocktail that was required to precipitate the accident. It would be quite wrong, and indeed nobody would make the claim, that the single cause of the accident was corrosion. The safety measures invoked by the police and stewards in marshalling people into the ground, the design changes in the ground that had occurred over a period of years, the testing of the crush barriers, the erection of fences to prevent crowd access on to the pitch, all played a major part in the disaster. Many human reactions played a part: for example the difficulties of communication between police and stewards at the immediate site of the accident and their superiors in the control room who were slow to direct emergency vehicles to the site of the accident.

I am just reading a newspaper cutting from the national press of last week in which it is reported that control of a Jumbo Jet being towed at Heathrow Airport was lost: it then rolled down a slight slope and crashed into another Jumbo leaving British Airways with a £½ million repair bill. The single event which precipitated this accident was the rocking of the towed aircraft by high winds which caused the towing bar to snap thus disabling a cable linked to the braking system. The aircraft then lost hydraulic power and braking power. The ground pilot lost control over the aircraft and was powerless to prevent its downhill run. So although this accident was caused by the breaking of a simple component, the consideration of the hazards resulting from such a fracture had not predicted the consequences, which in this case proved to be extremely expensive.

We should beware of overstating our claims that increased knowledge of research would create great savings. We make these claims principally to bolster our research funding. We are saying our problem is so important to you (the potential funding agency) that you should give us money so we can help to save you (society) money. Now in general, if we look at the output of such research in the international journals devoted to fatigue and fracture, we find it is far too esoteric to be applied by design engineers to reduce failures. It is therefore difficult not to come to the conclusion that the real problem is not the need for more and more complex research, but the need is for better and effective dissemination of our existing knowledge and it is to that problem that I return for the bulk of this presentation.

2 The UK situation for University engineering courses

Many of the things that I will say, of course, stem from my own experience which has been principally concerned with the teaching of undergraduates and postgraduates in British universities over the last 15 years. I readily acknowledge there are many differences between our system and the education systems in Europe, the United

States and Japan. We are facing a problem (challenge) in the United Kingdom of rapidly increasing the proportion of our young people who attend universities. The proportion has increased in the recent decade from about 4 or 5%, to something now approaching 15%. We have changed from a system based on educating a rather small elite to what might, with some justification, be called a mass education system. At the same time, the state of preparedness of our students in many ways is deficient. In the areas of maths and physics, although their knowledge appears to be wider and in some cases more exciting than the knowledge of entrants some 20 or more years ago, the routine application of elementary maths and physics to problems in engineering is beyond many of our students at the beginning of their course. We therefore have to spend a considerable amount of time imparting knowledge, methods and techniques which previously we would have assumed would have been well honed at schools.

Furthermore, the demands on the output level of our students when they leave university have multiplied, with an increased emphasis on what are called transferable personal skills - the ability to make presentations, to have a greater knowledge of management, to speak in foreign languages and so on. If I may add here, I am pleased to say that more than half of our first year Mechanical Engineering students are studying a European language and a small number are studying Japanese allied to their Mechanical Engineering studies. So the pressure has increased at both the input and output ends of our system. The ability of students to study the long hours that are necessary to absorb knowledge in engineering has been criticised by many of my colleagues. We have tried to switch our teaching methods from the passive acceptance of knowledge in large doses by the students to their active learning and understanding of knowledge. All this has meant that the hours available for detailed study of engineering topics have been put under pressure. The extent of syllabi have therefore been reduced, the amount of time to spend on any particular topic has therefore been reduced. How therefore do we manage to fit in detailed knowledge of teaching the principles of fatigue and fracture to these students?

3 Principles of fracture and fatigue on undergraduate courses

In essence we teach principles to undergraduates. Recall that our courses in the United Kingdom are a very short duration of three years which is much less than the courses of many of our European colleagues. So we use the undergraduate period to teach principles and we rely on postgraduate courses to teach details. At the University of Sheffield, many of you will know we have the SIRIUS Institute - the Structural Integrity Research Institute of the University of Sheffield - which teaches a postgraduate course in structural integrity. It is through this type of vehicle of delivery that our main detailed teaching will be performed.

As far as undergraduates are concerned, we have changed the order in which material is being taught such that, and this may surprise many of you, fracture mechanics is taught in the first year as part of the Materials Engineering course. The reason for doing this is that our Materials course has changed from being detailed consideration of scientific principles of structure to a starting point of consideration of properties, particularly mechanical properties, which are used in engineering design. The orders of magnitude of such properties for a wide range of materials are introduced and only then are the physical behaviours of materials which cause these properties and the structural microstructural inputs from the material which caused these properties, discussed. In this way it is rather easy, with the students having acquired an elementary knowledge of strength of materials, to introduce the topic of fracture mechanics in a conceptual sort of way without over much reliance on advanced mathematics. Thus, at the end of a 20 week Materials Engineering course of something like two lectures per week, the students can quite happily perform, inter alia, simple fracture calculations based on linear elastic fracture mechanics: they can integrate fatigue crack propagation laws, so called Paris-type laws, they can calculate guaranteed lifetimes of pressure vessels subject to proof testing. All this knowledge was not in existence 40 years ago. It is now being tackled by undergraduate in the first term of their course. So what has dropped out to enable this modern information to be taught? Largely speaking, detailed studies of the phenomenological behaviour and the empirical design correlations of fatigue and fracture.

4 Elementary approach to fracture mechanics concepts

Let me describe an intuitive method of approaching fracture mechanics. The concept of stress concentration is approached via the "fluid flow" and "bunching of stress lines" analogy. The pivotal role of notch route radius in determining the stress concentration factor of a notch is introduced via the classic Inglis formula:

$$K_T = 1 + 2\sqrt{\frac{D}{\rho}}$$

where D is the notch depth and ρ the notch root radius.

The special case of a circular hole of stress concentration factor 3 is mentioned. The observation is made that as the root radius tends to zero, the notch becomes a crack and the stress concentration factor tends to infinity. Then the limit of the product of the maximum notch root stress and the square root of the notch root radius is examined and shown to remain finite, and equal to the product of the nominal stress times the square route of the crack length, as the notch is transformed, by diminishing its root radius to zero, thus becoming a crack. It is then stated that a classic elastic analysis of a central crack in a uniformly stressed infinite plate leads to the result that all the near crack tip local stresses scale as the nominal stress square root of crack length product. Thus the physical meaning of the concept of an elastic stress intensity factor is established. The simple expression, involving the stress intensity factor, for the tensile stress ahead of a sharp crack tip is given without formal proof. The obvious point is made that the elastic stress distribution must be truncated by yielding and a simple quantitative estimate of the extent of the yield zone is derived.

The central credo of linear elastic fracture mechanics is then explained. That is, the elastic stress intensity factory controls the magnitude of the local near crack tip stresses which in turn control the extent of the yield zone which then controls the site of the fracture action, the so-called fracture process zone. The students then happily accept that the value of the remote elastic stress intensity factor controls local fracture events. Experimental evidence is then presented on how the stress intensity factor can be used as a fracture criteria by defining a "material constant", the fracture toughness.

An examination is then made of cyclic loading. By postulating that the irreversibly stretched material at a crack tip can only be accommodated on unloading by a geometry change leading to crack extension (backed up by electron micrographs of such deformation), a simple model of ductile fatigue crack growth is formulated. Consideration of strain compatibility at the elastic/plastic boundary, coupled with the previous estimate of the plastic zone size, lead to a surprisingly useful quantitative model of fatigue crack growth. Thus, without recourse to mathematics, the essential features of LEFM are presented to the students who are then in a position to apply this knowledge to a wide variety of fracture and fatigue problems. This approach has the benefit of maintaining the interest of the students through their ability to solve real engineering problems, rather than burdening them in the early stages with too much theory without visible end result. These simple concepts are reinforced by cartoon-like illustrations, examples of which are given in Figure 1. Note that in Figure 1, the applied mechanics inputs are clearly separated from the material property responses. This is an important clarification of 'cause and effect' which is often confused by beginners.

5 Historical case studies

Does the history of the development of our subject have an important bearing on the way it is taught? Can we find time in our crowded courses to teach aspects of history? I have used to good effect introductory material on failures which have been the catalyst for research into fatigue and fracture problems. Starting with the Versailles railway accident of 1842 caused by a broken railway axle, I introduced the mechanisms debate, that is, is fatigue the crystallisation of metal? Wöhler's work on the same topic introduces the concept of fatigue limits. The fracture of liberty ships in World War 2 and the Comet accidents of the 1950s serve to illustrate the rise in understanding of the stress analysis of cracks and the introduction of linear-elastic fracture mechanics. Many would choose to introduce the important work of Griffiths but as I explained above, in my introductory course I prefer not to use an energy based approach to fracture which is necessary, but not sufficient, and instead rely on a mechanics stress based approach, which has in-built necessity and sufficiency.

The use of case studies of failures is, of course, an important teaching tool. I am reminded of two sayings, the first from Ecclesiastes in the Bible:

> "What has happened before will happen again.
> What has been done before will be done again.
> There is nothing new in the whole world".

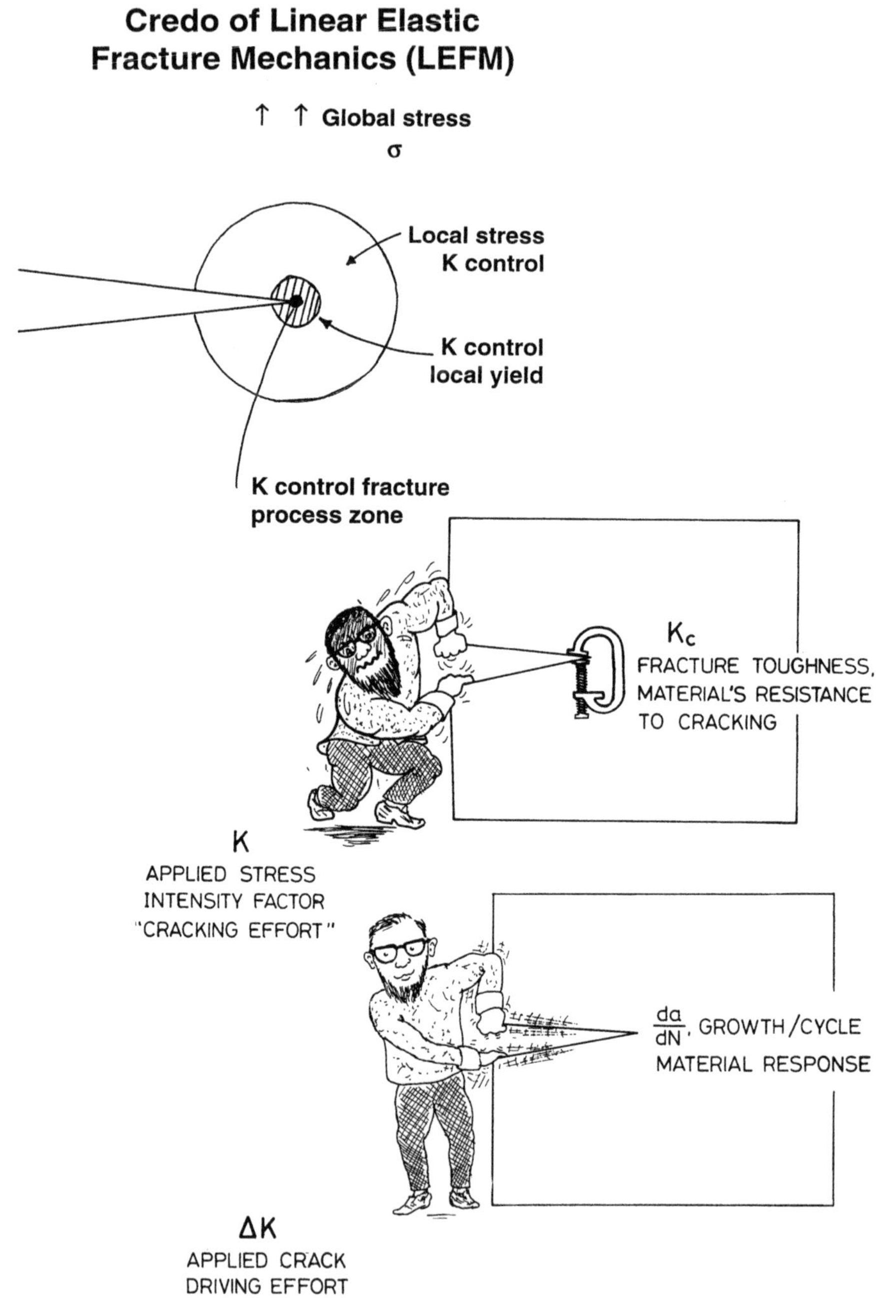

Fig. 1. Cartoons illustrating the concepts of fracture mechanics.

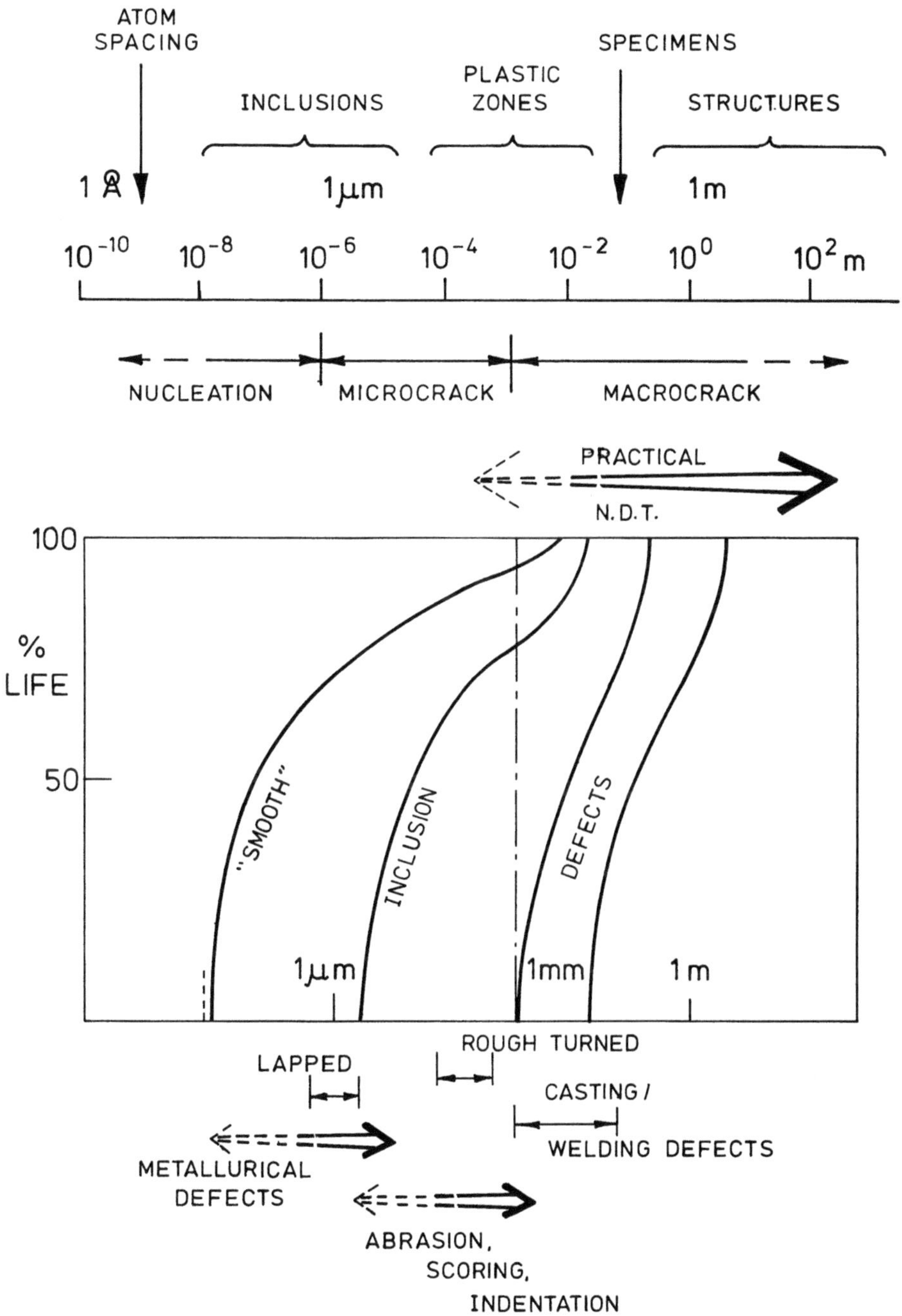

Fig. 2. The importance of size in fracture and fatigue.

Confucius said:

> "Man has three ways of learning.
> Firstly by meditation; this is the noblest.
> Secondly by imitation; this is the easiest.
> Thirdly, by experience; this is the bitterest".

I am delighted that a journal devoted to this topic has recently started publication (Engineering Failure Analysis edited by D R H Jones, published by Elsevier). I am sure that this will produce many fresh sources of teaching material in the future.

6 Importance of size scale

What I have said relates very much to a preliminary course in fatigue and fracture. First courses are important because they both lay the foundations for further study and can be used to fire the enthusiasm of students for the subject. If the latter is not accomplished then the former will never be undertaken. You may have criticisms of what appears to be a very simplified approach, but experience has shown that it does capture students' imagination at the foundation level. In order to dispel the notion that fatigue and fracture really is that simple, I end the course by spending some time in discussing the huge range of size scales involved in fracture problems (see Figure 2). We operate in the laboratory at a visible size scale ranging from millimetres to metres but we apply our theories to events on the sub micron scale in order to predict fracture in components or structures which might be tens or even hundreds of metres in size. This figure illustrates the near impossibility of detecting fatigue cracks grown from nominally smooth surfaces until a late stage in their development; at least by normal methods of non-destructive testing. Although we now know that all of the fatigue life of a component is spent in the initiation and development of a crack, or cracks, this invisibility caused fatigue to be regarded as a great mystery by our predecessors.

The quantification of this initiation and growth to failure in the design stage leads to both the fascination and difficulty of fracture avoidance. It both keeps us in the research business and frustrates our efforts to avoid fracture by teaching.

4 REFLECTIONS ON FORTY YEARS TEACHING OF THE ENGINEERING ASPECTS OF FRACTURE

C.E. TURNER
Mechanical Engineering Department, Imperial College, London, UK

Abstract
A brief history of the development of fracture mechanics is given to show the basis on which courses for undergraduate and graduate engineers were developed over the past many years. For undergraduates, fracture studies are not taught separately but integrated into a number of other courses. These start with Materials Science in the first year and lead to topics relevant to engineering applications in the final year. For graduates, a special course on fracture is given, with the aim of understanding the physical basis of topics such as constraint and temperature transition in steels as well as the use of K, CTOD and J for the assessment of the significance of defects through fracture analysis diagrams (FAD).
Keywords: Crack opening displacement (COD), elastic-plastic fracture mechanics (EPFM), fracture analysis diagrams (FAD), history of fracture studies, J-integral, linear elastic fracture mechanics (LEFM), teaching fracture.

1 Introduction

Surely no-one would propose a single ideal syllabus for teaching fracture studies. The course must be related closely to the nature of the students in both status, undergraduate (UG), post-graduate (PG) and post-experience (PE) of various types, and to their professional interest, mainly the branch of engineering or materials science. The present remarks can therefore apply only to the groups with which the writer is familiar: mainly mechanical engineers but with some aeronautical students at UG level; PG mechanical and civil engineers at Master and Ph.D level and PE courses for technical staff from industry.

For any such group, the key issue is seen as the amount of time that can be devoted to the subject for the course in question. That depends on the perceived importance of fracture studies vis-a-vis other calls on the time of the participant.

The normal undergratuate course in the U.K., leading to a Bachelor's degree, has for a long period been of three years' duration. Some four- or five- year courses now exist, perhaps leading directly to a Master's degree, but most still contain only a three-year component of conventional technical studies with the extra time devoted to management

Teaching and Education in Fracture and Fatigue. Edited by H.P. Rossmanith. Published in 1996 by Taylor & Francis, 2 Park Square, Milton Park, Abingdon, Oxon, OX14 4RN. ISBN 0 419 20700 7.

type subjects and/or industrial project work. To anyone who has ever tried to plan a new course, the accommodation of core technical subjects and then essentials such as at least a second language, economics, management, environmental and social studies, coupled with a perceived growing gap between University entry level and any agreed first degree, is itself a nightmare, without the introduction of a specialist topic such as fracture mechanics! Courses with four years' study of technical subjects have been introduced recently, and may become more widespread in the future, but in parallel there is a call from industry for more breadth and attention to the whole industrial background against which engineers have to practice, not conducive to the study of specialist subjects. Nevertheless, most modern courses are much more flexibile in structure than hitherto, leading to elective subjects, such as fracture mechanics, any one of which might be taken by only a small group of students in their penultimate or final year of study. The total time devoted to fracture related studies by the UG students in mechanical engineering with which the writer is familiar, might be from 6-10 hours spread through various courses and years as outlined later; an elective specialist course might contain 10 hours formal presentation, of which a variable fraction might be fracture related and onto which an additional rather indefinite 'own-time' effort of perhaps 20-50% of the formal hours is also implied.

Post-graduate work is mainly of one year's duration for a Master's degree or three years for a Ph.D. In the writer's Department, it is common practice for Ph.D students to study one or two subjects from the Master's courses, particularly since the majority of Ph.D students are from other Universities so that there is no common background to their prior knowledge. A course within a Master's programme might contain 20-30 hours presentation, again with a rather indefinite 'own-time' contribution. The outline of such a course therefore provides the best opportunity to discuss the teaching of fracture studies.

2 A brief historical review of fracture studies

This review may seem an indulgence in nostalgia but it is also meant to serve the purpose of outlining what seemed to be the essential features for understanding fracture, as the subject developed. At the end of each following sub-section the main advances and teaching difficulties created are summarised. Where possible, reference is made to books, rather than to original research articles, as being more suitable for taught courses.

2.1 Brittle fracture before Fracture Mechanics

Brittle fractures have been known throughout recorded history, in certain rocks and metals. These were indeed termed 'brittle' and it was accepted as a natural phenomenon (Cottrell and Kamminga, 1990). Mediaeval tool and weapon users experienced failures and some centres became known as producers of both better and more reliable artefacts than others, an early example of 'know-how' rather than 'know-why'. Some large scale failures were reported in the late 19th and early 20th century in structures made of notionally ductile material and were mainly seen as due to poor material or an 'act of Fate'. Brittle fracture come to the general attention of engineers as an identifiable problem through the failure of a welded bridge in the late 1930's and, notably, of the so-called Liberty ships of World War II. The relevance of, or indeed the recognition of, a transition in the fracture mode of structural steels was still appreciated in only two then seemingly unrelated circles --- some practical engineers in the field and some research workers. Fracture studies were first introduced into the Mechanical Engineering Department, Imperial College, about forty years ago. The first teaching of a transition with temperature that the writer is aware of was via laboratory demonstrations of the effect in V-notched Charpy pieces in the early 1950s. These were conducted on a 1913-vintage Charpy machine, which, with instrumentation added to the striker, still remains a current research tool! 'Fracture mechanics' did not then exist as a recognised subject in that Griffith's classical work was but little known and Irwin's seminal papers had not been written.

A PG course of ten lectures on Brittle Fracture soon emerged. Early work by Tipper

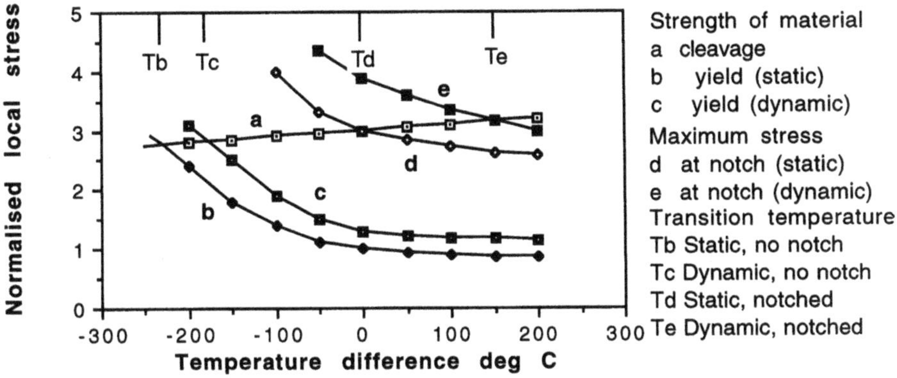

Figure 1. The Ludwig-Davidenkov diagram for the fracture-mode transition.

(1962) on the micro-change from ductile on the 'upper shelf' to cleavage on the 'lower shelf', allowed some explanation to be given and the Ludwig-Davidenkov diagram, Fig.1, became, and still remains, the introduction to the interplay between notch constraint, temperature, strain rate and the fracture mode transition. But the distinction between the the micro- and macro-meanings of ductile and brittle was very confusing even to fracture specialists, with post-yield cleavage appearing to be a contradiction in terms.

Laboratory tests on conventionally sized plate samples a few tens of millimetres wide, still with V or saw-cut notches, could not be made to give a below-yield failure except at extremely low temperatures. It seemed as though there were some barrier that prevented a cleavage crack growing under monotonic loading despite the low toughness with which it propagated, once started. The nature of such an 'initiation barrier' became clearer when Robertson circumvented it to produce low stress failures in pre-loaded wide plates tested at room temperature. The key was the introduction of a running crack caused by impact loading on a small region at very low temperature, adjacent to the test plate proper. Later, Wells produced brittle fractures in welded plates into which a deliberate flaw had been introduced in the heat affected zone (HAZ). Loading was parallel to the line of the weld and a crack ran into the parent plate, transverse to the weld, often completely severing the plate at low applied stress. In short, for a brittle (cleavage) fracture to occur in steel, the role of temperature and transition in micro-mode for propagation, coupled with some sort of unusual condition or 'damage' to allow a crack to start whilst the nominal applied stress was quite low (i.e in the region of a design stress) was becoming apparent, as summarised by Boyd (1970).

2.1.1 Teaching comments

The alternative micro-modes of fracture in structural steels and the role of temperature, notch constraint and strain rate had become well recognised. The lack of a coherent theory to relate these factors and the separate transitions in micro-mode and in macro-ductility for structural steels, was, and still is, a source of difficulty to many students, albeit alleviated if the behaviours are distinguished as in Fig.2. It seems a pity that the various bodies using a specialised fracture terminology still do not make this distinction clear, perhaps by introduction of suitably different terms for the micro- and macro- scales.

2.2 The coming of LEFM

In the late 1950s and early 1960s, news of Irwin's work on the K singularity concept first

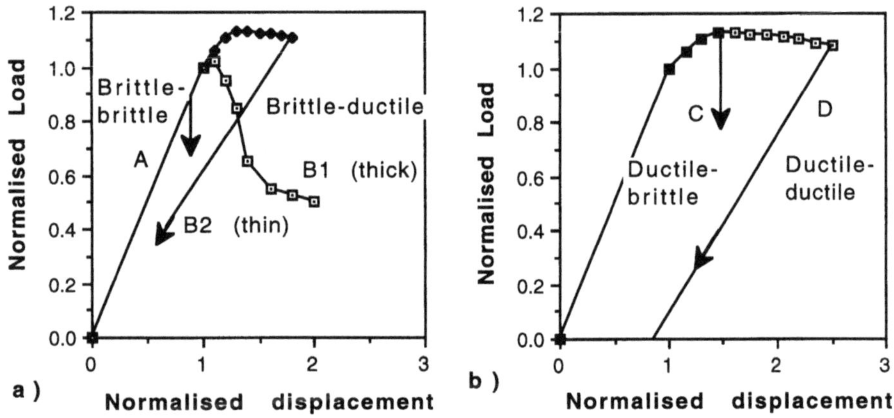

Figure 2. The Macro-micro description of fracture; a) LEFM cases; b) EPFM cases.

reached Europe and the era of Fracture Mechanics was inaugurated. There was a period of great puzzlement over the choice of the G or the K concept but that was overcome by the relationship $K^2 = E'G$, deduced by Irwin a few years later. Thereafter, solutions for K for many different geometries were a great aid to progress, using either concept.

In LEFM, the role of plane stress and plane strain and a transition in the overall appearance of the fracture surface from flat to oblique was emphasised with little or no reference to the micro-scale. The need was emphasised for pieces very thick and wide in relation to the local crack tip plastic zone (some 50 fold), if the minimum (i.e.plane strain) toughness, K_{Ic} was to be found, ideally associated with a diagram like Fig.2a case A but sometimes with case B1(thick). Having established this new concept of a material fracture property, K_{Ic}, the thin sheet (oblique fracture) behaviour was then presented as a resistance or R-curve of toughness versus crack growth, ending in an overall instability controlled by the second derivative of energy with a diagram like Fig.2a cases B2 (thin). These case B diagrams are macro-elastic, micro-ductile, possibly without complete separation, and had not in fact been seen in the tests on structural steels of the rather low strength C-Mn types in thicknesses then used for ships and bridges, referred to in Section 2.1. The case B2 (thin) diagram is not un-like Fig.2b case D in form, but elastic (in the macro-sense) with an unloading line that returns to near the origin. Fig.2b case C was not seen within the LEFM picture so that the fracture behaviour of the low and medium strength ferritic steels seemed an apparent anomaly, except in so far as, with damage, the brittle Fig.2a case A diagrams could be obtained.

These difficulties took several years to resolve, even amongst the leading research workers. Indeed, and for many years, a quite separate and more empirical approach, the so-called Pellini diagram, (see for example, Boyd 1970) was advocated, outside the patterns of LEFM. At first, it was not appreciated, at least within the U.K., that Irwin's studies were on high strength materials where low-stress failures (i.e. at say a nominal stress of one quarter or one third of yield) were fairly easy to induce provided a sharp notch was used but where there was little effect of temperature or strain rate and no transition in micro-mode, thus making the Davidenkov diagram irrelevant!

A plausible picture of the peculiarity of the C-Mn-type steels with regard to transition in micro-mode of fracture with temperature and strain rate, was given by Cottrell in 1958, see for example Knott (1973), so that the 'black-magic' of fracture both brittle and ductile at the same time (Fig.2, case C) started to look understandable at the micro-level. But also in 1958, the existence of the separate micro- and macro-transitions was perceived as a major stumbling block to the engineering application of the Griffith-Irwin model of LEFM

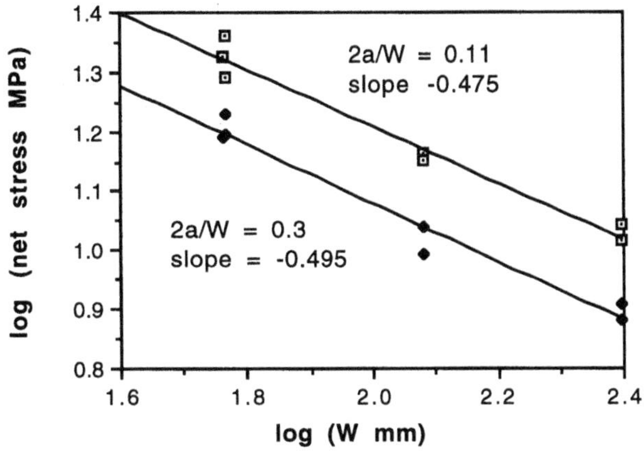

Figure 3. Early tests on cracked glassy plastics; a slope of -0.5 supports the LEFM law.

to structural steels. This was highlighted when Mylonas *et al* (1958) reported tests in which notched pieces, tested in tension after local damage caused at the notch root by pre-compression, failed well below yield (i.e. Fig.2 case A) but which for various widths, did not follow the Griffith type relationship of a square root dependence on size. The history of this hiatus, which lasted for nearly a decade, seems to have escaped the text books. Meanwhile, other advances were being made.

There is a separate history, which is not discussed here, of the development of a parallel version of fracture mechanics including large elastic displacements, by the rubber industry, some of which was being imported into polymer studies in the early 1960s. However, at about the same time that LEFM was being used for metals, it was applied to the fracture of glassy plastics, notably by Kies (whence the symbol K). Oddly, hardly any data was published at that time to show that K_{Ic} was indeed a material constant. Such early data as the writer is aware of shows the need for plastic zone correction (in relation to the Fig.2a case B diagrams), without evidence that the simple LEFM concept of fracture toughness was indeed meaningful. Direct application of LEFM to several glassy plastics, van den Boogaart & Turner (1963), Fig.3, gave results used by the writer to illustrate his courses of that time and quite adequate to support the broad concept against early doubters.

The lack of relevance of LEFM to the brittle fractures of steel as reported by Mylonas *et al* (1958), was later studied by Jones and Turner (1967) who showed that, when notched, pre-compressed and then tested, two distinct bands of data existed within a wide scatter in the fracture loads. The scatter was, no doubt, due to both rather coarse control of the damage by pre-compression and to the well observed variability in cleavage fractures but, within the scatter, a sharp borderline was found, Fig.4, between the stress above which a crack would propagate completely through the plate but below which it would arrest after a few millimetres growth. Tests were then made over a range of widths and this borderline stress was found to follow the LEFM relations quite closely, thus implying there was a width independent term, expressible as K_{Ic}, that might be viewed as a material fracture property for crack propagation.

2.2.1 Fatigue crack growth

As a rather separate thread that is of great importance, LEFM was applied to fatigue crack growth by Paris and Erdowan in 1964, see for example Broek (1986). That concept has perhaps had a wider effect than the use of fracture mechanics for brittle-fracture problems. Fatigue had, of course, been known and taught for many years through the S-N diagram.

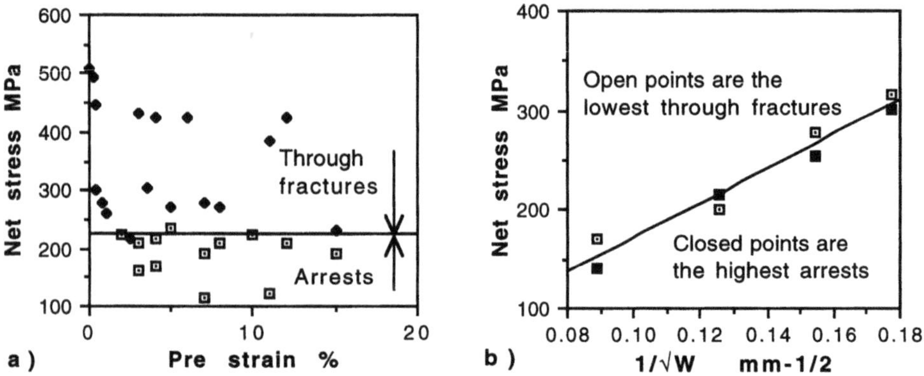

Figure 4. An early demonstration that LEFM could be applied to fracture in structural steel; a) the borderline between fracture and arrest; b) its variation with width follows LEFM.

It is a sobering thought that, according to Paris, no learned journal that he approached would publish the now familiar 'law' of crack growth, apparently because the well-known S-N treatment showed no connection with cracks and the connection now made was semi-empirical. Yet sixty years earlier, in the late 1890s, there was a short story written by the famous writer Rudyard Kipling (1895), in which a ship's Chief Engineer explained to his Captain that the tail-shaft of the ship was cracked nearly half way through and described all the signs of corrosion-assisted fatigue started at a stress concentration. Asked by the Captian when the shaft would actually break, Kipling even had the engineer (in those days naturally a Scotsman) reply, "That's beyon' my knowledge" which got the riposte, "So it is. We've all (got) our limitations". That seems to be the first recorded short course in fracture mechanics, based on what Kipling, as a journalist, had learned in conversation with ship's officers.

2.2.2 Teaching comments

Once K solutions became known, much teaching of fracture seemed to pay more attention to the stress analysis than to its consequences. The reason was perhaps that the concept of a singularity that was to be embraced rather than avoided was a novelty to most engineers. Spending some time on the related algebra was, and by some still is, seen as the easiest way to let the idea sink in.

The recognition that LEFM can often be applied to fractures in high strength materials (Fig.2a case A) but not to many practical cases of brittle fracture in structural steels (Fig.2b case C type of failure) is still a stumbling block to many students, particularly at the UG level when elastic-plastic mechanics has probably not been entered into. The fact that stable tearing in thin sheet can give a non-linear loading diagram resembling Fig.2b case D, yet be elastic, i.e.not the EPFM case D but the LEFM case B(thin) is still not realised by some unfamiliar with tests on high strength thin sheet.

Quite apart from the more recent insights into the growth of cracks too short to be treated by LEFM, the application of LEFM to fatigue crack growth was, and to some still is, a stumbling block in that the use of ΔK to describe the amplitude of severity at a crack is central but the material characteristic is no longer a material toughness. As a matter of teaching in a logical way, it is a pity that the half power singularity associated with an elastic crack, first introduced by Wieghardt (1907), nearly 20 years before Griffith's work, did not come to be well known as an example of stress analysis before the relevance of K to fracture was seen. Indeed, it is arguable that teaching K as a measure of crack severity and applying it via macro-scale fatigue crack growth would even now be a simpler starting point, and one relevant to nearly all materials, than via Griffith and brittle fracture with the material dependent complexities of micro- and macro-transitions in behaviour.

2.3 The coming of EPFM (but without R-curves)

Few structures are designed into the plastic regime although survival in that state may be called for if accidents happen. The first call was therefore to understand the effects of localised plasticity on cracks, such as at stress concentrations and in regions of residual stress at weldments. However, since as a matter of convenience and economy, most test pieces are small, tests on low and medium strength metals are usually taken into the fully plastic regime before failure occurs and a strong motivation for understanding the role of plasticity on fracture is the need to interpret such fully plastic test data, probably including a region of stable ductile growth after initiation.

Wells introduced the concept of crack opening displacement, (COD), or crack tip opening displacement, (CTOD), as it is now usually called, in the early '60s, with the brittle fracture of weldments in mind, see for example Knott (1973). CTOD was related to applied stress by the Dugdale strip yield model, essentially a plane stress model for 'a small crack in a wide plate'. At that time, finite element or other numerical procedures were in their infancy and played little role in the development of the ideas. The Dugdale model was however soon superseded by an empirical relationship to allow for the effects of plane strain. It is still relevant in that most of the fracture data from which the related method for assessing safe usage of weldments was developed, British Standard document PD6493, see for example Latzko *et al* (1984), were found from tests on welded plates with small cracks. Test plates of about 1m square were used, with an artificial defect of about 20mm in extent (unfortunately not of a shape readily analysed) inserted in the weld or heat affected zone (HAZ) region. With a small degree of work-hardening (wh), such that σ_{wh} (W-2a) > σ_0W, many of the fractures experienced general yield of the plate up to about two or three times the yield strain. However, to ensure capturing the worst risk of brittle fracture, the standard CTOD test method used high constraint deep notch bending pieces, albeit in the relevant full thickness if possible, which for many steels allowed full yield of the ligament to occur, but without yet more widespread general yield beyond that. Although at the time, that seemed but a reasonable conservatism, it has since led to the charge of undue conservatism because in the deep notch case all the plasticty is concentrated at the notched section whereas in the shallow notches relevant to most structures, notch plasticity may be alleviated by more general yielding away from the notch. In 1965, McClintock pointed out from slip-line field arguments that for non-hardening materials there could be no unique characterisation in the fully plastic regime for both deep notch bending, with curved slip-lines and high constraint, and centre-cracked tension with straight slip-lines and low constraint. This argument was later extended to shallow notches where constraint is also low. Nevertheless, for another 20 years or so, that point was largely ignored. A COD curve for the acceptance of defects in welded structures was introduced by Harrison and colleagues in 1968. With modifications, some appreciable, it is still the basis of today's COD assessment methods. These early developments of EPFM in terms of COD are summarised in Latzko *et al* (1984).

In the same year Rice introduced the now well-known J-integral concept, although both Eshelby and Cherepynov had independently made similar proposals rather earlier. The form of Rice's presentation, together with the supporting characterising stress and strain fields of Hutchinson and Rice and Rosengren (HRR), made the topic available to engineers in a way that the Eshelby and Cherypynov work did not. All these solutions were however based on non-linear elastic (NLE) behaviour that many found, and still find, difficult to accept as relevant to real elastic-plastic (REP) material. Finite element studies now played an important role by showing that, even when using the Prandtl-Reuss type laws for incremental plasticity (but with no un-loading), the J contour value was as near path independent in full yield as in the LEFM regime (where any path dependence is only due to numerical modelling errors). The first such demonstration known to the writer was by Hayes in 1970. Experimental work under monotonic loading by Landes and Begley in 1972 gave the final launch to what has now become the best known and most widely used fracture model, by showing an apparent invariant value of J_{Ic} for initiation. In the light of the ensuing explosion of interest in J it is ironic that they subsequently modified some of their interpretations because stable crack growth clouded the initiation event. The explosion was caused by the synergy between the bewitching simplicity of the contour integral formulation in relation to the then growing power of finite element methods. Such finite element studies validated the use of G with an Irwin-type plastic zone correction (G_{pzc}) up to stress levels well beyond that anticipated, approaching the limit load, Fig.5a, (where the normalised load exceeds unity because some work hardening is allowed for). Further studies on cracks at regions of stress concentrations, by Sumpter and Turner in 1973, made clear that the essential change, whereby J became very substantially larger than G, arose only when yield became uncontained by an outer elastic field, Fig.5b. These early developments of J-methods are also outlined in Latzko *et al* (1984).

One important aspect of the use of J was the acceptance of test piece sizes, for the purpose of measuring initiation toughness, J_{Ic}, much smaller (by some 10 fold or more) than required for valid LEFM data. Thus for a valid LEFM test in low strength steel, a thickness of a few hundred millimetres might be required, but J data was quoted from tests at perhaps only 25 mm thick. A difficulty in understanding then arose from the translation of the J_{Ic} value into a seemingly acceptable K_{Ic} value by writing $K^2 = E'J$. The argument was that a near common crack tip condition existed in the two sizes of piece advocated (it was pointed out that there was a lower limit of size below which that was not true) so that provided the correct analysis was applied (i.e. use of J when there was appreciable yield) then the value obtained was physically meaningful. This argument is now generally accepted with some reservations on how closely similar the conditions have to be. In particular, some believe that a quite small difference in severity between two sizes might just switch the fracture from micro-ductile in the smaller (less constrained) piece to cleavage in the larger (more highly constrained) piece. The difficulty of distinguishing between a 'true' initiation point and a small amount of Stable crack growth again emerged,

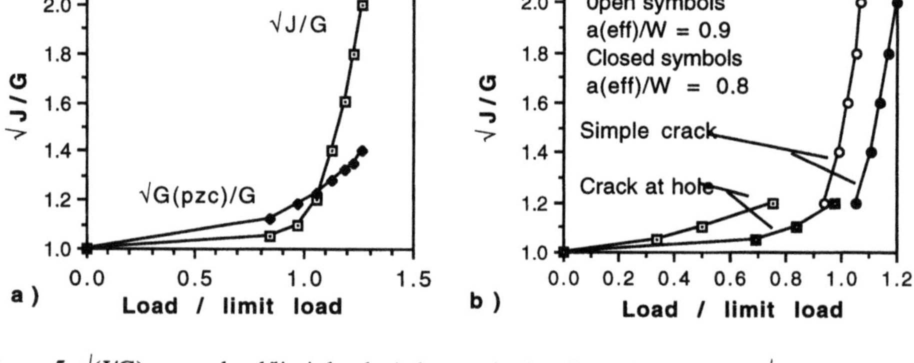

Figure 5. $\sqrt{(J/G)}$ versus load/limit load; a) three point bending, also showing $\sqrt{(G_{pzc}/G)}$; b) a crack in a region of stress concentration at a hole, radius R, and tensile plates of the same a(eff)/W where a(eff) is (a + R) for the crack at a hole or just a, for the plate.

despite refined experimental techniques such as the un-loading compliance test method, see for example Latzko *et al* (1984). The definition of initiation after an arbitrary but small amount of growth, such as 0.2mm, remains a point of contention.

A valuable offshoot to the J-concept was the application of what is now called C* (with dimensions of J/time) as a characterising parameter for creep crack growth, da/dt, Nikbin *et al* (1976), somewhat analagous to the use of ΔK to describe fatigue crack growth in the LEFM regime. The similarity in form between power law secondary creep rate in metals and NLE power law hardening models of the HRR field gives the underlying justification.

2.3.1 Teaching comments

A problem for all students not already familiar with plasticity is the incremental nature of the stress-strain laws, the understanding of plastic constraint at the crack tip and its dependence on configuration.

The first great merit of the J-integral method is its direct degeneration to G and hence the whole range of LEFM theory. The difficulty of explaining to non-mathematically minded students a meaning for the integral itself and to demonstrate to all, when first met, that J has a relevance to real elastic-plastic material despite being based on NLE behaviour, exercises the ingenuity of all who teach it, ameliorated only by avoidance of R-curves!

An apparent de-merit of the first experimental work on J, noted above, was the cumbersome nature of the experimental procedure. However, that was soon replaced by a simple relationship between J and area under the loading diagram so that, up to initiation, the test procedure became near trivial and a positive benefit to the method. The translation of a J_{Ic} value into a seemingly acceptable K_{Ic} remains a stumbling block for many and the lack of emphasis on micro-mode, indeed almost a lack of mention of it in many early uses of J, could lead to some students missing the real problem of brittle fracture in steels.

2.4 Fracture analysis diagrams, R-curves and other recent developments

As already noted, a CTOD-based method for assessment of defects, loosely called a 'design curve', was developed in 1968. The curve related only to fracture; failure by overload (i.e. reaching the limit load) was assessed as a separate exercise.

A J-based method, known as R-6, was published in 1976 by a team from the then CERL facilities of the then CEGB, the electric power generating authority in the U.K. With just under 20 years usage, it is perhaps borderline between 'history' and 'recent development'. R-6 introduced a so-called fracture analysis diagram (FAD) with an ordinate of K/K_{Ic} and an abscissa of Q/Q_L (where Q is applied load and sub L denotes the limit load). The ordinate thus gives a measure of the approach to brittle fracture whilst the abscissa shows the approach to plastic collapse. A semi-empirical curve, clearly derived from the Dugdale strip yield analysis, encloses a region in which operation is considered to be safe and outside which some form of failure is a distinct risk, Fig.6. The relationship of this diagram to an inverted form of Fig.5 is clear -- $\sqrt{G}$ is proportional to the LEFM severity that might cause brittle fracture and if a critical value were supposed for $\sqrt{J}$, that would be proportional to K_{Ic}, as outlined at the end of the previous Section. Several features are included to account for residual stresses and in a recent revision to the method, to allow for hardening where a clear-cut limit load does not occur, see for example Latzko *et al* (1984).

Although a formal relationship between J and CTOD in the Dugdale model, of the form $J = \sigma_0\delta$, was pointed out by Rice when J was first introduced, this had to be written as $J = m\sigma_0\delta$ for cases not based on that model. The value of m was a function of both hardening and geometry, not to mention the precise measure used for CTOD in other than the strip-

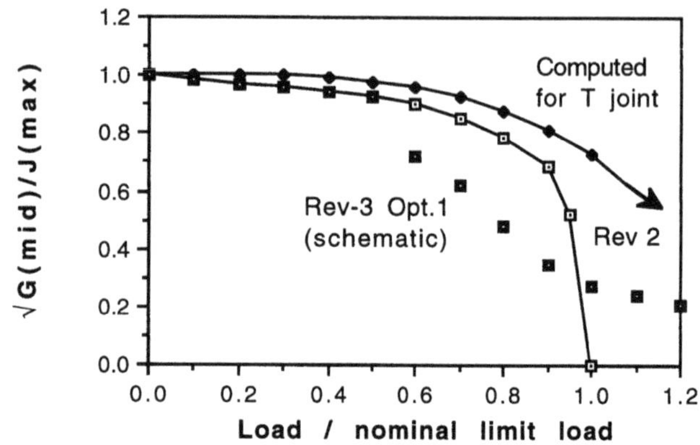

Figure 6. The R-6 type FAD; in original form (Rev.2) and revised form (Rev3 Opt.1). The application is to a nodal T joint in an offshore platform with the brace under tension.

yield formalism, notably in finite element analyses. For some years the duality of CTOD and J-methods of treating fracture caused much concern. An explanation can only be seen if the reasons for their development and the opinions of the time are studied in more detail than justified here. In brief, CTOD was originally developed to guard against the risk of cleavage fracture in weldments used in the transition regime. J was originally developed to allow initiation toughness to be measured in the presence of plasticity, thereby allowing the use of small pieces. There appears to be an implication that cleavage was not a risk, or at least that avoiding initiation (of ductile growth) would eliminate that risk. Both methods have recently moved towards each other in that certain test data can be interprted by either method, but the effect of size and constraint on fracture still causes differences of opinion where cleavage is a possible mode of failure.

The logic of this welcome coming together deserves some attention at the teaching level. The original COD 'design curve' had an abscissa of strain (normalied by yield strain) although the normalised strain induced by primary loads (as opposed to strains from secondary loads and residual stresses) was taken as equal to the normalised applied stress. This, at least in principle, allowed for strain-governed situations where plastic collapse was not an issue. On the contrary, R-6 was developed for pressure vessel uses where the dominant load (i.e. by pressure) might well cause failure by plastic collapse. It was realised at an early stage in the development of R-6 that fracture might occur from a defect as plasticity spread through the wall of a vessel, well before the collapse condition for the whole vessel was reached. Thus the normalising factor, Q_L, referred to above as 'collapse load', became the contradictory concept of a 'collapse load local to the crack being studied'. It was this treatment that allowed a single non-dimensional FAD to be proposed. In the early '80s a schematic relationship between the then COD design curve and the R-6 diagram was seen. The later attention to the effects of work hardening destroyed this single picture but in the most recent revisions the FAD type presentation, whereby both fracture and plastic behaviour are considered on the same diagram, is now used for both methods.

It is however of interest that the concept of the normalising load, Q_L, being the 'collapse load local to the crack' as defined in the R-6 method based on application to pressure vessels, is not suited for use with a highly statically indeterminate structure. An example of the very satisfactory use of the modified R-6 FAD, to predict the crack severity for a nodal joint in an offshore platform with a part through-thickness crack in a weld, is also shown in Fig.6, Kristiansen & Turner (1993), for a structural steel with some hardening allowed for. The point of interest is that the normalising load, Q_L, is there not the 'local collapse load' but the several fold higher value of the collapse load of the overall joint. The reason for this is that the 'local collapse', as defined in R-6, allows plasticity to spread through the wall thickness, implying an approach to uncontained yield (and thus a rapid increase in J) in a statically determinate system, whereas in the complex shape of the nodal joint, local plastic deformation allows the loads to 'by-pass' the crack, leaving it at a near elastic severity until overall yield occurs, somewhat akin to the stress concentration case of Fig.5b.

2.4.1 R-curves

The extension of J-methods into stable growth, whereby a so-called J-R-curve is developed as a plot of J versus Δa, at first arose as a help in defining an initiation point by backward extrapolation to some small extent of growth, zero or otherwise. Despite different views on what that small amount should be, the notion is surely acceptable. However, the seemingly obvious extension to using the R-curve itself as a measure of post-initiation toughness, i.e. an extension of the G-R-curve concept into the plastic regime, is highly contentious. It was soon realised that such R-curves were strongly geometry dependent and could not therefore represent a material property. Nevertheless, R-curves found from high constraint pieces were taken to give a lower bound and their use to predict a maximum load whilst stable growth occured, was implented as an extension to the R-6 FAD, see for example Latzko *et al* (1984).

It was then shown that some R-curves (notably for the higher strength lower hardening metals) were size dependent as well as configuration dependent. Recently, a different approach has been suggested by Turner & Kolednik (1994), based on the true energy dissipation rate for rep materials with crack tip opening angle (CTOA) assuming the role of the characterising parameter for stable growth. In short, the validity of the J-R-curve approach for the treatment of stable growth is not at all secure, at least in the opinions of some, including the writer.

2.4.2 Characterisation by two parameters

The concept of a unique characterisation of initiation by a single parameter, be it COD or J, has also come under attack recently since the degree of constraint with almost any degree of plasticity is dependent on the configuration. In particular, the constraint in the fully plastic state reached in most EPFM type tests made on compact tension (CT) or deep notch bend (DNB) pieces is high whereas in many (though not all) structural problems the constraint at a shallow or surface notch, particularly when under tensile loading, may be much lower. Two versions of a two parameter description have been proposed and developed, Hancock *et al* (1993), O'Dowd & Shih (1994), to characterise the crack tip state, implying a severity that is a function of triaxiality. These two-parameter arguments certainly describe the observable effects of constraint, thereby making applications of EPFM more realistic at the expense of yet more complicated algebra. It remains to be seen whether the concept and effects of constraint can now be presented more easily to students, since they are no longer contradicted by use of a one parameter model of the elastic-plastic stress field. Further semi-empirical, semi-theoretical arguments have also been used to describe the effects of such a picture on cleavage fracture by Dodds and colleagues, see the latest edition of Anderson, (1994).

2.4.3 Teaching comments

The early form of the R-6 FAD, (Fig.6, curve Rev.2) was easily presented in both courses on fracture mechanics, per se, or as part of design or elective application type courses where some students would have no specialist knowledge of fracture, perhaps just an acquaintance with LEFM and possibly with temperature transition for steels. The bringing together of CTOD and R-6 methods, though clearly welcome, has been at the expense of, or at least has coincided with, a considerable increase in complexity of the FAD treatment making a realistic coverage open only to the specialist course.

The position on R-curves and the two parameter models for initiation is even worse in that what seemed received wisdom 10 years ago is now stated (at least by some) to be incorrect. Meanwhile, the new proposals are still under development. The implications of these aspects on teaching are clearly debatable and seem better treated not as part of the historical developments of Section 2 but as current problems, mentioned further in the next Section.

3 Courses in fracture mechanics

3.1 Undergraduate courses for mechanical engineers

It has already been said in the Introduction that in UG years, fracture enters through elements of the core courses. Thus in a first course on material properties, the notion of avoiding failure by fracture, fatigue, creep and corrosion will be discussed and the concept of surface energy introduced. In a second course, introduction to LEFM, temperature transition in steel and crystal structure will be made, comparable in depth to the treatment of criteria of yielding, fatigue by S-N methods and creep deformation behaviour. In third year design or elective courses on topics such as Nuclear Power or Gas-turbine Technology, application might be made of items such as the R-6 FAD and fatigue or creep crack growth, as appropriate, but with little of the backgound beyond that of the second core course. A similar elective on Materials and Joining would treat the micro-aspect more fully, with the role of inclusions, brittle particles, grain boundaries and HAZ as initiation sources in metals and introduce fracture treatments for polymers, adhesives and composites. In so far as an elective course on Fracture Mechanics is available in some 4 or 5 year courses at Imperial College, it would be essentially part of the work prescribed in the Masters course so that the main discussion will be made in terms of that course, variously of 20 to 30 formal hours presentation.

At this depth of treatment, where the minimum is that of the core courses, with no elective subjects relevant to fracture or indeed materials at all, the message that is, hopefully, retained by all students is that not only are cracks potentially dangerous when loaded but that, for reasons of fabrication or in service deterioration, cracks are likely to be present in all real structures whereas they often are not present in conventional, carefully prepared, laboratory test pieces, nor allowed for in many routine design procedures.

3.2 A course for post-graduates

It is intended that a PG course of some 20-30 hours tuition would take a student up to the forefront of presently agreed knowledge wherefrom specific (and quantitative) application to such as fatigue in aircraft, fracture-safe procedures for welded structures, behaviour of polymeric pipelines and similar industrial problems could be approached later in one's career with understanding of the principles involved and a notion of some of the pit-falls.

The simple suggestion is made that the understanding required from such a course (for mechanical engineers) is that summarised in the foregoing historical review. The great difference, now, is that the facts and insights that emerged over a period of some 30 years (from the World War II ship failures to the R-6 method) are known before the course is written. In the period described, many of the most essential behaviours had to be left as not understood whereas that now applies mainly to some topics mentioned in Section 2.4. Athough the sequence of events may well be altered to ease presentation, most aspects of fracture involving both LEFM and EPFM behaviour can be presented as rational effects in a course of about 25 hours. It is unlikely that several of the uncertainties of the time, such the application of LEFM to structural steel, as illustrated in Fig.4, need be entered into. Indeed, once the climate has changed from uncertainty over, or downright disbelief of, a new theory, even illustration of relevance, such as Fig.3, may be waived. In short, once the matter is treated in a text book it is readily taken as relevant and correct by most students. The fear must be that with its entry into a computer system the item becomes 'gospel' and quite inerradicable, whether it is in fact correct or not! But as noted in the

Teaching Comments already made, many of the 'mysteries of the time' present difficulties in understanding, even with the present appreciation of their causes.

The key point in the approach used is to give understanding of the physical basis of practical engineering problems on the macro-level. Little time is spent (for engineers) on the *algebra* of the Westergaard or Williams singularity solutions or the *theory* of slip-line fields. Those types of analyses have virtually disappeared from core courses in favour of numerical solutions. The gain is that the latter are also central to almost any fracture problem for which a standard solution does not already exist. The loss is in understanding of the physics of the problems which was given more emphasis during an algebraic solution than for the numerical replacement. Presumably computer aided learning approaches to the physics of fracture will become available at both micro- and macro-level, if not already so, but the writer has no knowledge of them.

One of the difficulties experienced with the particular courses at Imperial College with which the writer is familiar, is that the students come from widely different backgrounds with a majority from overseas. The concept of the above work is to superimpose the understanding of crack behaviour onto the classical undergraduate level continuum understanding of stress analysis and mechanical properties, since that is exactly the pattern by which the subject developed. In fact, the teaching of what one Department would regard as mainstream to an undergraduate course may differ considerably from the work chosen by another. The largest discrepancies relevant here are at the borderline between what may loosely be termed the mathematically-based topics of stress analysis and the physically-based aspects of mechanical properties. Thus the prior knowledge by students starting M.Sc. or even Ph.D. work of topics such as criteria of yielding, the Charpy temperature transition behaviour, the S-N treatment of fatigue, the primary, secondary and tertiary stage of deformation creep and a simple picture of the micro-structure of metals and polymers, varies from good to non-existent. The immediate solution is to offer to postgraduates who need them, the notes on the relevant topics as issued for the core courses to undergraduates.

A number of other aspects of fracture studies, some already mentioned in passing are noted briefly as deserving more attention in courses than given here; micro-aspects of toughness and fracture; stress corrosion and corrosion fatigue; short crack growth in fatigue; crack growth direction; behaviour in modes II and III and in combination with mode I; dynamic initiation and growth; ductile instability; non-destructive examination; weldments; creep crack growth; other materials, notably polymers, composites and cement or concrete. Cases can be made for including any or all of these topics. At one time or another, most have featured on the fringes of the courses given. The one over which the writer had most doubts of relegating to that position is non-destructive examination. The reason for its marginalisation is pressure of time in relation to the topics already described in Section.2 (and perhaps for some, the descriptive nature of the subject matter with little simple algebraic theory in support). The reason for the regret at this treatment is that there can easily be no mention of it at all in a fracture course (as opposed to at least a passing reference to the other aspects listed) yet it is the absolute key to practical application of the subject. Without an appreciation of what defects can be detected and sized, a quantitative estimate of critical defect size, unless at one extreme of size or the other, cannot be made meaningful. It is easy to set an examination question by postulating a crack size--who has not done so ?-- but we academics might find as apposite the recipe from Mrs Beeton's famous 19th century book on ccokery; "Recipe for Jugged Hare. First catch your hare; then....."

3.3 Specialist post-experience courses

From about 1970 to 1980, courses were offered in conjunction with the Welding Institute (now TWI). That period spanned the time when those in industry who needed knowledge of fracture mechanics had heard nothing about it during their own undergraduate career. Thus courses giving a general introduction were offered followed by use of the COD method leading to the use of PD6493 (see Latzko *et al* (1984)), in about a one week course of 30 hours tuition. Later that was supplemented by more specialist courses introducing J, elastic-plastic finite element methods (at a time when they were but little known) but treatment of R-curves and ductile instability became too specialist to be viable. This reflected the small group of industries calling for such topics and also the arrival in industry of new Ph.D. recruits who had received advanced fracture study in their own training. It should be mentioned that TWI always ran complementary courses on topics such as non-destructive examination and quality control, independently of the joint courses mentioned here.

More recently, the College has run specialist courses on topics such as life enhancement of high temperature plant (mainly creep damage and fatigue crack growth, Nikbin *et al* (1986)) and adhesive technology (including adhesive fracture and fatigue, Kinloch (1987)), for which a good knowledge of conventional fracture mechanics is a pre-requisite. In summary, such courses have always been taken to near the frontier between agreed knowledge and research in some field of application but that frontier has changed rapidly over the period in question, in both level and topic.

3.4 Teaching and research

An important amout of teaching of new subject matter is done through research, including at appropriate levels, both final year UG and PG projects. The attitude adopted over the last forty years in 'strength of materials' has been to inculcate a balance of theory, experiment and application. Nowadays that invariably includes use of numerical procedures, but unless a student is pursuing a numerical analysis specialism (a quite legitimate role but not, by definition, a fracture project) a deliberate attempt is made retain a balance between the various aspects of the work.

One important tool has been the use of in-house developed finite element programs for which there is complete access to all the algorithms. Over the years these have included some of the first versions of elastic-plastic modelling, elasto-dynamic crack growth and stable tearing. The teaching point is not so much the economy of the program or even close accuracy of the data produced as the recognition by the student of the potential difficulties of the modelling and the fallibility of computer programs. The 'it's correct because it's computed' syndrome is very pernicious, the more so as new generations of students arrive who have received all their instructions during the computing era. A recent study of a quite straightforward elasto-plastic fracture problem gave three different answers from three different numerical techniques, finite element, boundary element and finite volume, all 'known' by the users to be correct and superior to any other. Fortunately experimental evidence was available to show that at least two of the solutions were incorrect.

Rather similar remarks can be made about laboratory hardware. It is not easy to get the balance right between in-house repetition of items industry has already developed or the acceptance of a unit designed with unknown objectives; for example, a few years ago a certain instrumented impact machine gave a readout of toughness that neglected the effect of inertia loading in deducing K from the striker load.

4 Discussion

An explanation, or even excuse, is first offered for the number of diagrams and references relating to the writer's own work with colleagues. This is a direct consequence of the PG course developing hand-in-hand with ongoing research work, itself primarily attempting to understand the principles with data on particular materials or development of standard techniques a rather minor issue. No doubt, results similar to those instanced could often be found elsewhere but the crucial understanding arose from 'doing it oneself' and attempting to put together one's own version of at least part of the jig-saw of fracture.

There is no doubt that with the texts available today including, as well as those already cited, Ewalds & Wanhill (1985) for general student use, Atkins & Mai (1985) for a deep-probing background to energy balances and plastic deformation and Williams (1987) for polymeric materials, a quite suitable general course at PG level could be given by staff without direct research activity in the fracture field. Such a course could hardly be enlivened by personal case material but by the nature of things, that must be so for many well respected courses not explicitly research related. The same remarks apply to PE courses, where developments current at the time would have to be handled by visiting staff. For the UG content, some would argue that a restriction imposed by no direct research contact was an advantage in avoiding over complexity at that stage.

Little has been said here on the use of computers as a teaching aid other than the obvious use of finite element programs for numerical analysis. In fact, programs were introduced in the late '80s to allow derivation of Paris' law constants from experimental data and application of the law to a limited range of fatigue life calculations. The simple R-6 FAD was also so treated and other topics were envisaged. Each contained a library of K and other relevant solutions for several types of configurations over a small range of crack size variables. Many of the niceties required for following the formal requirements of a particular standard document were omitted in order to leave clear the essential logic of the methods. The intention, tested only once during the writer's connection with that work, was to allow students on PG or elective courses (and therefore limited in number to the hardware facilities available simultaneously) to use such programs during examinations as well as during taught classes.

The inclusion in taught courses of concepts still under development is a matter of contention. It can be argued that at post-graduate level it is appropriate to do so whereas at undergraduate level it is not. In the writer's view, the opportunity to pass advanced and possibly still contentious work into courses should be taken very sparingly for reasons of both confusing and overloading the student, but not entirely eliminated. The real point is whether the matter is presented, at whatever level, as apparently known fact or openly shown to be unresolved, an issue that may be a matter of judgement. After how many years and papers is a matter known to be correct? Indeed how many engineering statements are known to be incomplete or approximate yet are both widely used and useful? It is interesting to recall the benefits and advances that the uses of J-methods have brought, yet chastening to observe recent proclamations of the inadequacy of the one-parameter HRR field as a characterising parameter and the controversy over the meaning of the J-R-curve, some 25 years after the concepts were first introduced.

5 Conclusions

Despite an unashamed enthusiasm for the subject, the writer has supported and participated in teaching that, at undergraduate level, has integrated fracture studies into a variety of more general courses. That still seems appropriate for fracture or other specialist subject that is but a minute part of the curriculum for a typical engineering degree.

For post-graduate students in mechanical and also structural engineering, the essential subject matter has been seen as the understanding of the physical processes of fracture at the macro-level with some support at the micro-level and access to the design methods available (PD 6493 and the R-6 FAD) rather than the ability to manipulate algebraic theory for stresses at singularities of make recommendations on material selection or welding processes for particular structures.

For teaching at research level, as distinct from the value of the research data produced, access to instrumentation and programs at a rather basic level has been pursued and is strongly advocated in order to show the crucial nature of the underlying assumptions and techniques, despite the attractions of elegant bought-in facilities.

Acknowledgements
The writer gladly acknowledges the assistance of many colleagues over the years through which the courses described were developed. The views expressed are however his own.

6 References

Anderson, T.L., 1994, *Fracture Mechanics; Fundamentals and Applications,* (2nd Ed), CRC Press, Boca Racon, Fla, U.S.A.

Atkins, A.G & Mai, Y-W., 1985, *Elastic and Plastic Fracture*, Ellis Horwood, Chichester, U.K.

v.d.Boogaart, A. & Turner, C.E., 1963, Fracture mechanics: a review of principles with special reference to applications for glassy plastics in sheet form, J. Plastics Inst., Aug., pp.109-117.

Broek, D., 1986, *Elementary Engineering Fracture Mechanics,* (4th Ed.) Martinus Nijhoff, Dordrecht.

Boyd, G.M.(Ed.), 1970, *Brittle Fracture in Steel Structures*, Butterworth, London.

Cotterell, B.C. & Kamminga, J., 1990, *Mechanics of Pre-industrial Technology*, Cambridge U.P.

Ewalds, H.L. & Wanhill. R.J.H., 1985, *Fracture Mechanics*, Edward Arnold, London, with Delftse U.M.

Hancock, J.W. *et al*, 1993, Constraint and toughness parameterised by T, in ASTM STP 1171, Am.Soc. Test. Mats., Philadelphia, pp. 21-40.

Jones, G.T. & Turner, C.E., 1967 A fracture mechanics interpretation of low-stress fractures in pre-compressed mild steel. Jour. Iron & Steel Institute, 205, pp.959-965.

Kinloch. A.J., 1987, Fracture Mechanics of Adhesive Joints, in *Adhesion and Adhesives: Science and Technology*, Chapman & Hall, Chap.7.

Kipling, R., 1895, Bread upon the water, in *The Day's Work*, Macmillan, London, (First printed 1898 with many reprints up to 1948).

Knott, J.F.1973, *Fundamentals of Fracture Mechanics*, Butterworths, London.

Kristiansen, N.O. & Turner, C.E., 1993, Defect assessment of tubular T-joints using the R6 failure analysis diagram, Jour. Press. Vess. Tech, ASME, New York, 115(4), pp.373-380.

Latzko, D.G.H.*et al.*, 1984, *Post-Yield Fracture Mechanics*, Elsevier-Applied Science, London.

Mylonas, C. *et al.*, 1958, Static brittle fracture initiation at net stress of 40% of yield, Welding Journal Research Supplement, 36(10), pp.473s-479s, Am. Weld. Soc., New York.

Nikbin, K.M. *et al.,* (1976) Relevance of non-linear fracture mechanics to creep cracking, in ASTM STP 601, Am.Soc.Test.Mats, Philadelphia, pp.47-62.

Nikbin, K.M. *et al.,* (1986) An engineerging approach to the prediction of creep crack growth, J.Eng. Mat. & Tech., ASME, New York, 108, pp.186-191.

O'Dowd, N.P. & Shih, C-F., 1994, Two-Parameter Fracture Mechanics: Theory and Applications, in ASTM STP 1207, Am. Soc. Test. Mats, Philadelphia, pp. 21-47.

Tipper, C., 1962, *The Brittle Fracture Story,* Cambridge U.P.

Turner, C.E & Kolednik, O., 1994, A micro and macro approach to the energy dissipation rate model of stable ductile crack growth, Fat. & Fract. Eng. Mats. & Struct., 17(9), pp.1089-1107.

Williams, J.G., 1987,*Fracture Mechanics of Polymers*, Ellis Horwood, Chichester, U.K.

Wieghhardt, K., 1907, Über das Spalten und Zerreissen elastischer Körper (On the splitting and rupturing of elastic bodies) Zeitschrift für Mathematik und Physik (Jour. of Mathematics & Physics), 55(1-2), pp. 60-103, (English translation by H.P.Rossmanith, to be published, Fat. & Fract. Eng. Mats. & Struct., Dec. 1995).

5 APPLICATION AND INSIGHT ARE THE KEYS TO LEARNING IN FRACTURE MECHANICS

D.A. FORBES and T.G.F. GRAY
University of Strathclyde, Glasgow, Scotland, UK

Abstract

A strategy for teaching sharp-crack fracture mechanics to engineers is described. The philosophy behind the approach is highlighted and key elements, such as the need to understand principles, the provision of solution compendia and the quality of graphics, are emphasised. Sample graphics are included in the text. The sequence of presentation is outlined, starting from simple linear elastic models and developing into non-linear models and the *J*-integral concept. At each stage, the factors which are believed by the authors to promote insight into the principles of fracture mechanics, are exemplified. Applications to toughness testing procedures are described (this is presented to the students through a specially made video programme) and three industrial case studies are outlined to exemplify the need to interest the student in applications. Various information compendia and texts which have been found useful by the authors are referenced in the paper.
Keywords: case studies, fracture mechanics, principles, teaching strategy

1 Introductory philosophy

The interest shown in the present Conference, together with a variety of comments from colleagues over the years, suggest that fracture mechanics teaching presents some special difficulties. This feeling is captured in a remark made to one of the authors by a recent graduate from another institution - "I tried to learn fracture mechanics in the same way as other engineering subjects, but it didn't work." What difficulties lie behind such comments?

What he probably meant was that in many topic areas in mechanics he can expect to

Teaching and Education in Fracture and Fatigue. Edited by H.P. Rossmanith. Published in 1996 by Taylor & Francis, 2 Park Square, Milton Park, Abingdon, Oxon, OX14 4RN. ISBN 0 419 20700 7.

build up a capability to solve problems of a certain type by mastering the analysis of a variety of cases in the topic area, almost as mathematical puzzles. Fracture mechanics does not seem to yield to that treatment, however, as real problems never seem to conform neatly to prototype cases. The critical need is to understand the basic characterising (descriptive) parameters for cracked bodies *eg* crack-tip stress-field-intensity, crack driving force, strain energy release rate, crack-tip-opening displacement, *J*-integral. These concepts, based on approximate and analytically complex models, are often found to be difficult to reconcile with more familiar strength of materials characterising parameters *eg stress* and *strength* which will already have been learned. Even the units of fracture mechanics parameters will be unfamiliar at first Also, the idea of failure by crack propagation does not seem to fit with the classical textbook 'theories of failure'. (These theories are actually misnamed, in that they characterise the breakdown of linear elastic behaviour and are not descriptive of failure behaviour in the broader sense.)

There is therefore a need to go back to the beginning to define what fracture mechanics is about and to build up a teaching and learning strategy appropriate to the task.

The task of Mechanics - according to Gustav Kirchhoff - *consists in describing in the completest and simplest way, the observable movements in nature* [1]. Such a definition is equally applicable to sharp-crack-fracture mechanics and, as a statement, it offers considerable insight to the purpose of the subject. In case the full import of his assertion would be missed, Kirchhoff continued - *Herewith I wish to say that what we are concerned with, is to state the phenomena occurring but not to determine the causes.*

However 'descriptions' can seem to be merely academic and arid to engineers, unless they have a use. In his autobiography [2] Stephen Timoshenko comments, somewhat bluntly, on those of his teachers of mechanics *who 'did not know what it was that future engineers needed to know, or in what form.'* On the other hand, he gratefully acknowledges elements of his own mechanics teaching philosophy to V.L. Kirpichev - '*a marvellous lecturer...*' who always had full, attentive classes. Kirpichev believed that '*once a principle is understood, technical details no longer cause difficulties.*' (As an aside to those who see the future of engineering teaching to be dependent on Computer Aided Learning or entirely independent study, Timoshenko's concluded that '*Kirpichev's lectures were a living proof that the lecture method of teaching is not dead.*')

Several other relevant admonitions for the teaching of fracture mechanics can be drawn from Professor Kirpichev's parting advice to his eminent pupil - '*a lecture achieves its purpose only when carefully prepared, with the necessary proofs adduced by the simplest method, the theory illustrated by good examples. It is preferable to start one's explanation of a subject with very simple cases, and only when one's listeners have thoroughly grasped these to go on to more general and complicated problems.*'

These thoughts from a great teacher and historian of strength of materials form the background to the present assertion that insight and application form the key to learning in fracture mechanics. It would be fascinating to know what Timoshenko would have made of the task of teaching fracture mechanics, but it follows from the points he regards as significant, that the best available tools of presentation would have been used to give insight, together with the most interesting examples of engineering application.

Referring back to Kirpichev's dicta, there are very few *'necessary proofs'* in fracture mechanics, but those that are relevant, are not easy to *'adduce simply'*. The impracticability of deriving solutions from first principles has led the authors to provide extended reference notes, which incorporate necessary solutions and data. These notes provide a resource for exercises and `open book'examinations. The order of presentation of topics is therefore very important and it is vital to devote time at each stage to exercising use of the material, so that the next step can be absorbed. The following discussion therefore parallels the order of presentation of topics in practice, in order to show the sequence.

2 Visual presentation

Many of the key concepts in fracture mechanics are difficult to visualise. This is not always appreciated by experienced practitioners for whom a simple, stylised representation of a cracked body may be adequate. Background knowledge of the literature and associated graphic conventions equips the expert to visualise steep stress gradients, the three-dimensional nature of stress patterns, the implications of 'semi-infinite' boundaries and a variety of other implied features, which are accepted without detailed description. Line drawings of part cracks, where the solid material is not distinguished from the crack cavity, are common in academic articles, but may be seen, on reflection, to be confusing.

The novice is at a disadvantage without such background familiarity and special attention is needed to the communicative quality of graphics used in fracture courses. The increasing availability of simple computer drawing packages in recent years has been of great benefit in this respect, not least because it is relatively easy to improve graphic representations over time, in the light of student response. In the author's experience, there is little doubt that incremental improvement of visual representation has been accompanied by improved and more rapid comprehension. Electronically generated graphics are also highly portable from one medium to another and therefore it is possible to reinforce images in printed notes by similar large-scale images on overhead slide projector. Judicious use of colour can also assist in interpreting a complex concept.

A more recent development, which is under trial by the authors, makes use of direct projection of computer images, thereby allowing gradual development of a model, beginning with the solid component and developing through overlaid addition of coordinate systems, boundary loads and textual keys. This technique can combine the best features of the live chalkboard creation of a figure and the superior accuracy and clarity of prepared slides. (The sample figures included in this paper represent the final form of the figures, rather than the sequentially overlaid versions.)

3 The context of fracture mechanics application

Fracture mechanics application in mainstream engineering is relatively recent, but it is important to emphasise for the learner that many important engineering components and load-carrying structures could not be operated safely, if at all, without the framework of assessment which fracture mechanics provides *eg* aircraft structures, nuclear plant, turbines, chemical process vessels, offshore structures, cranes, bridges etc. This is done in the author's teaching through slide presentations of celebrated catastrophic failures, including several from consultancy files covering machinery, fairground equipment, lifting equipment and the like. It is shown that the failures occurred at low values of calculated stress and that they were therefore unsatisfactorily characterised by conventional stress/strength or 'failure' criteria. It is also possible to highlight factors which are frequently implicated in crack-dependent failures *viz* low ambient temperature, cyclic loading, thick construction, stress concentrations, welds.

This discussion is sometimes seen by students as merely 'descriptive' and therefore 'ignorable.' To counter such a response, case study analysis exercises are set in terms of the failure mode categorisation of figure 1.

This contextual phase leads to a conclusion that the application of fracture mechanics consists in identifying the possibility for a crack to exist in a loaded structure, determining its geometry and establishing the capability of the material to resist crack growth.

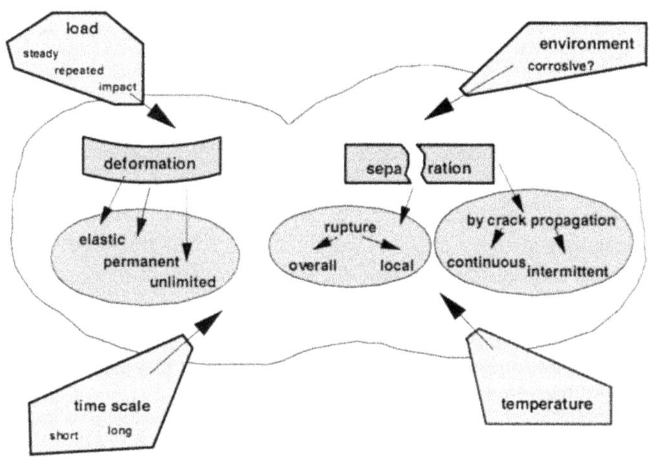

Fig. 1. Failure modes for materials

4 Historical approach to LEFM

The main benefit of using a historical framework is that it provides a natural sequence of '*simple cases*', leading to '*more general and complicated problems*'. (It also gives the opportunity to 'humanise' the subject through anecdotes about key personalities in the field.)

The analytical treatment begins with the 1913 Inglis-Kolosov solution for an elliptical hole in an infinite, elastic plate [3] (see figure 2).

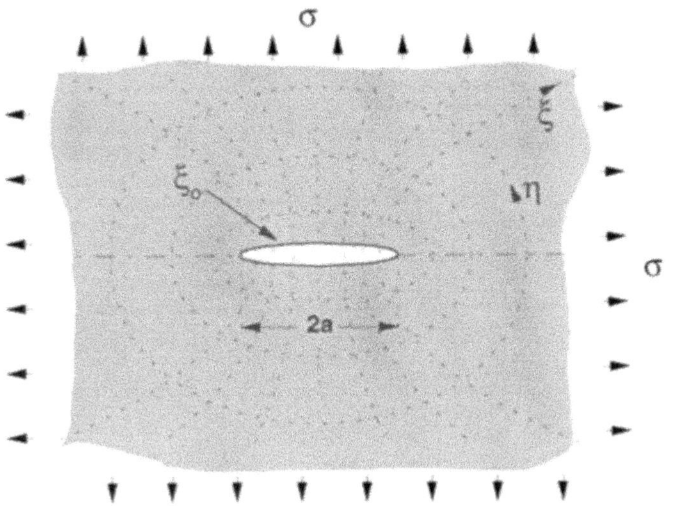

Fig. 2. Inglis-Kolosov elastic crack model

This treatment has the advantage that it is a stress-based characterisation, giving continuity with the stress-concentration concept, and it is complete and exact. Derivation of the stress pattern-description equations, shown below, is feasible from reference 3, but in the author's opinion, the time and intellectual effort necessary to deal with this solution is a distraction from the main points of interest.

(As an aside, Kolosov was Timoshenko's examiner in trigonometry when he applied from school for admission to the St Petersburg Institute of Ways and Communication in 1908).

$$\sigma_\eta + \sigma_\xi = 2\,\sigma \sinh 2\xi \left(\cosh 2\xi - \cos 2\eta\right) / \left(\cosh 2\xi - \cos 2\eta\right)^2$$
$$\sigma_\eta - \sigma_\xi = 2\,\sigma \sinh 2\xi \left(\cosh 2\xi_o - \cos 2\eta\right) / \left(\cosh 2\xi - \cos 2\eta\right)^2$$
$$\tau_{\xi\eta} = \sigma \sin 2\eta \left(\cosh 2\xi - \cos 2\xi_o\right) / \left(\cosh 2\xi - \cos 2\eta\right)^2$$

A more useful exercise which can be carried out by students at this stage, is to formulate and plot expressions for equivalent (Von Mises) stress from these equations. This gives some insight into the crack tip state (see figure 3). (Explanation of the plane strain idealisation will be necessary if this exercise is followed.) Another possibility, which is not beyond the capability of mathematical packages such as MATHCAD, is to plot the corresponding distribution for uniaxial loading. The point that the resulting pattern is insignificantly different from the biaxial tension case, will be useful in later discussion of the Sneddon-Irwin-Williams analysis.

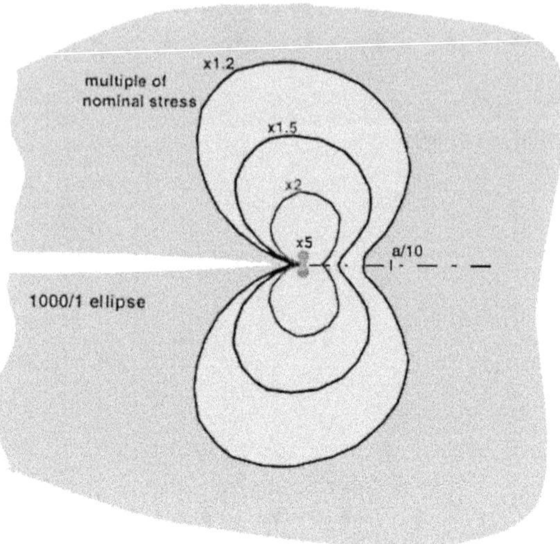

Fig. 3. Plane-strain equivalent stress plot for elliptical hole model

Although the equations above are complicated, the important conclusion is that at the tip of the sharp ellipse, where $\xi \rightarrow \xi_o$ and $\eta \rightarrow 0$, the maximum stress σ_η is given by:-

$\sigma_{\max} = 2\sigma\sqrt{\frac{a}{r_o}}$, where r_o is the asymptotic crack tip radius. Considering r_o to be a constant for typical cracks, leads to the finding that the maximum crack tip stress is proportional to $\sigma\sqrt{a}$.

The little known published discussion of Inglis' 1913 paper is useful at this stage [4] as it shows how difficult it was at that time to understand the relevance of an elastic solution which implied a very high, if not infinite stress at the crack tip. (Even more fascinating is the discussion contribution of a Professor B. Hopkinson, who suggested the small scale plastic zone size adjustment some 50 years earlier than is generally acknowledged in the literature.)

The natural historical progression would be to treat next the 1921/24 strain energy release rate analysis of Griffith [5],[6]. In fact many textbooks *start* at this point. However the authors have found that this introduces a new and unnecessary concept at this stage and therefore avoid discussion of the strain energy treatment. Griffith's crack model simply repeats Inglis' linear elastic model, complete with infinite stress. Moreover, neglecting geometric and material constants, Griffith's strain energy based characterisation reduces to the square of $\sigma\sqrt{a}$. Hence it is difficult to argue that time spent on the Griffith approach will increase insight at this stage.

The more important argument, which is certainly worth emphasising, is that Griffith made a major step forward by *verifying* the matter experimentally, at least for elastic failure of brittle material. One can show, therefore, that the model satisfactorily 'describes observable movements', despite the obvious lack of realism in the crack tip stress magnitude.

The forward link to the crack-tip stress-field-intensity approach of Irwin and, independently, Williams can be made in a variety of ways, but the authors highlight the analyses of Westergaard [7] and Sneddon [8] on the grounds that they revealed the uniqueness of near-crack-tip elastic patterns.

It is important to pause for explanation when introducing the well-known singular equations of Irwin [9]; (the principal stress formulation is given here, as this is convenient for simple student exercises.)

$$\begin{matrix}\sigma_1\\ \sigma_2\end{matrix} = \frac{K_I}{\sqrt{2\pi r}}\cos\left(\frac{\theta}{2}\right)\left[1 \pm \sin\left(\frac{\theta}{2}\right)\right]; \quad (r \ll a)$$

The main point emphasised by the authors, through variations on figure 4, relate to the use of the stress-field-intensity description to *compare* crack-tip states and, in particular to compare the behaviour of an observed fracture toughness specimen with a real structure containing a crack. The *absence* of a basis for comparison between different loading Modes is also important. The near-crack-tip basis of the formulation should also be emphasised by comparing the patterns with those of figure 3.

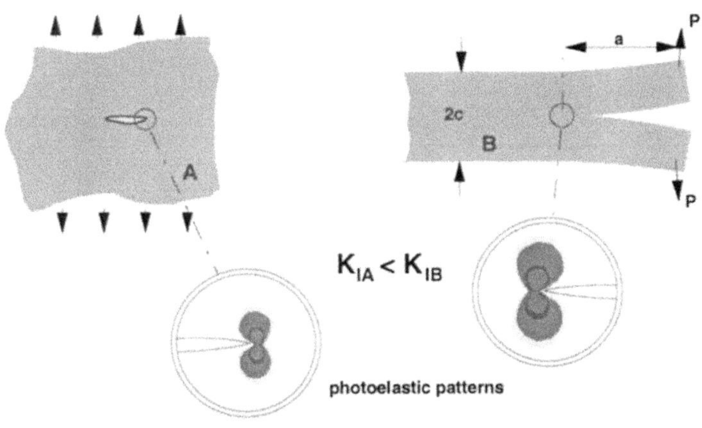

Fig. 4. Crack-tip stress-field-intensity concept

5 Application of stress intensity factor solutions

From this starting point of familiarity with the basic concept of stress field uniqueness, the student needs to be exercised in the principles of finding stress intensity factor magnitudes for realistic engineering problems. The basic requirement for such exercise is a convenient compendium of 'configuration factors', which is provided in 'open book' form in terms of equations and nomograms for 14 different standard cases (see, for example, figure 5). (Many of these solutions are original to the authors.)

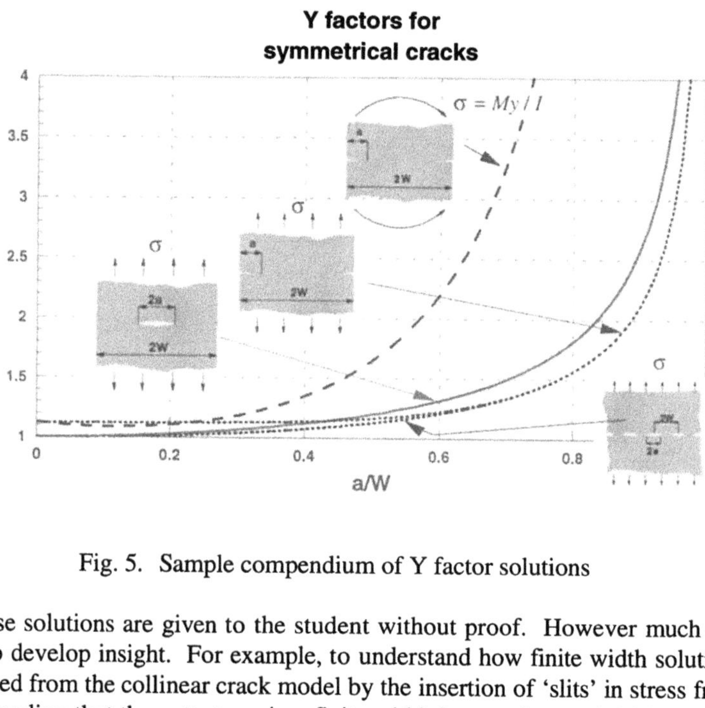

Fig. 5. Sample compendium of Y factor solutions

These solutions are given to the student without proof. However much can still be done to develop insight. For example, to understand how finite width solutions can be generated from the collinear crack model by the insertion of 'slits' in stress free regions; understanding that the nett stress in a finite width is not relevant (which is against most student's previous training); how to identify the 'effective' crack size in the case of cracks at notches or holes (see figure 6); comparisons between two-dimensional and three-dimensional cases; the reasons for higher stress intensities existing at the minor axis points on elliptical planform cracks and so on. Broek's textbook [10] is useful in this context.

When the student has developed a 'feel' for the basic geometries, there are many superpositional approaches to be taught - the use of linear elastic and 'null-crack' superposition; Green's function; weight functions; 'compounding' [11]. Some discussion

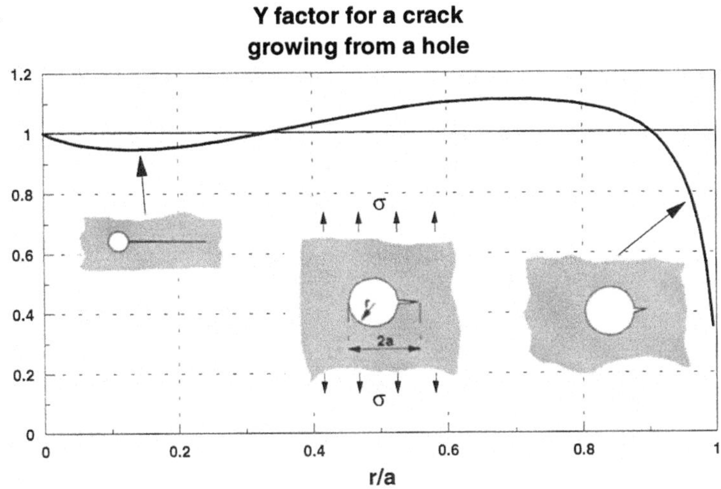

Fig. 6 'Effective' crack length concept for cracks growing from holes or notches

and exercise is required in relation to the basic idea that the stress intensity at a crack tip can be developed from a knowledge of the stress field normal to the crack plane in the uncracked state. Application of this principle also assists understanding of the standard cases *eg* stress intensity factors for bending loading and the results for short cracks at the edges of holes.

Figure 7 shows some sample exercises where the student is asked to develop and justify approximate elastic stress intensity factors.

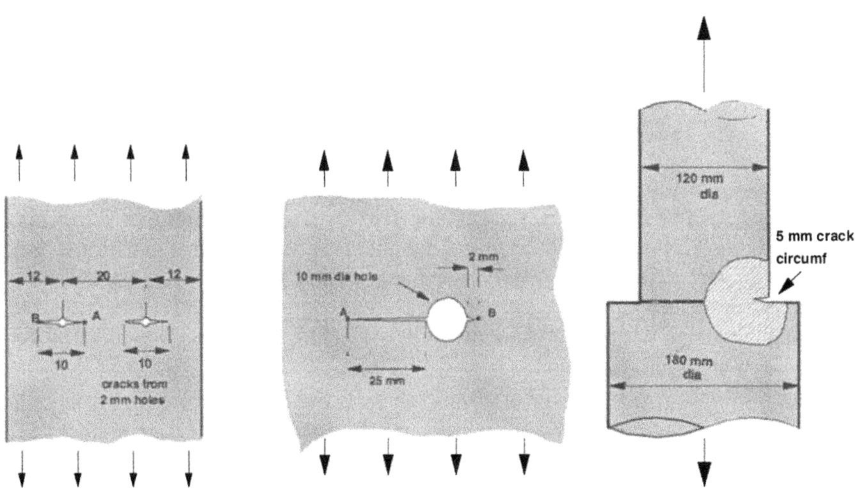

Fig. 7. Sample exercises in the application of Linear Elastic Fracture Mechanics

St. Venant's Principle also requires a subtly different interpretation from usual - 'a statically equivalent distribution of loads on a boundary can be employed, provided that the length over which the change in boundary loading is made is small in relation to the distance to the *crack-tip* where the stress intensity factor is sought.' This principle is also valuable for the generation of notch stress concentration factors from crack solutions and *vice-versa* [12] (see figure 8).

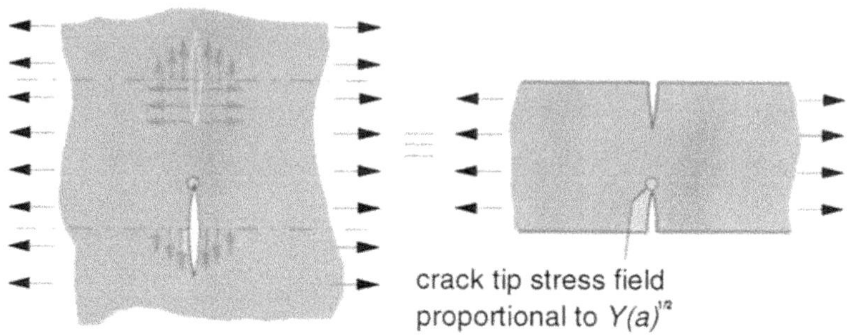

Fig. 8a Boundary forces required to transform colinear crack case to finite width

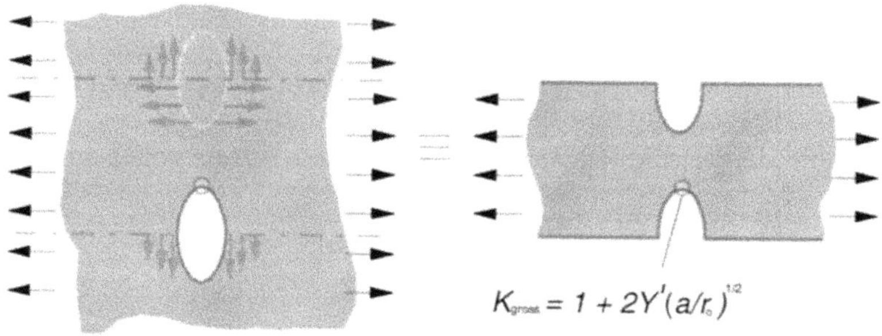

Fig. 8b Boundary forces required to transform infinite plate solution to equivalent finite width solution for notch

The basic cases, together with the combination techniques, make it possible to treat a large variety of practical cases, provided approximation is encouraged. (Students used to 'exact' answers are often disturbed by the lack of definitive solutions, but that probably is more revealing of conventional engineering exercises than a criticism of the present recommendations.)

Finally, students with knowledge of finite element stress analysis may be introduced to the use of singular elements [11].

Once a thorough grasp of linear elastic methods has been developed, the significance of crack tip yielding is introduced, initially via the small-scale yielding concept and the phenomenon of thickness-related triaxial transition (figure 9). Good graphics are essential. Earlier practice which the student may have had in plotting plane strain/stress fields will be invaluable at this stage. Some very clear three-dimensional finite element results quoted by Anderson [13] are especially helpful in tracking the cascading effect as plane strain constraint relaxes with increasing plastic zone size relative to thickness (see figure 10).

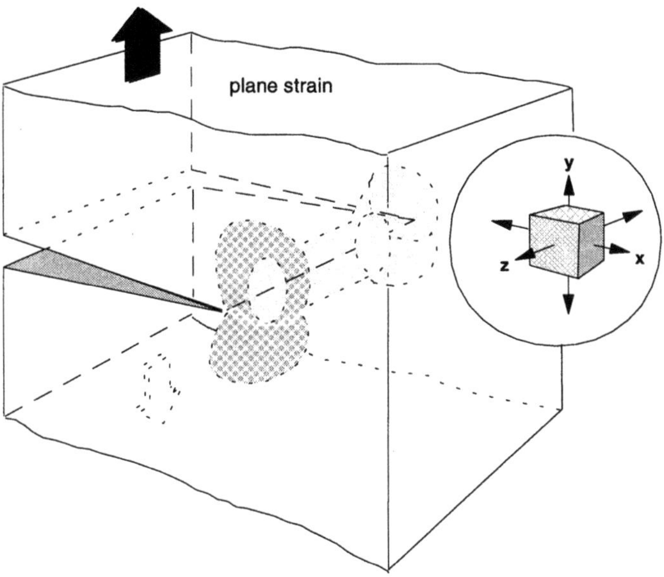

Fig. 9. Plane strain plastic zone representation

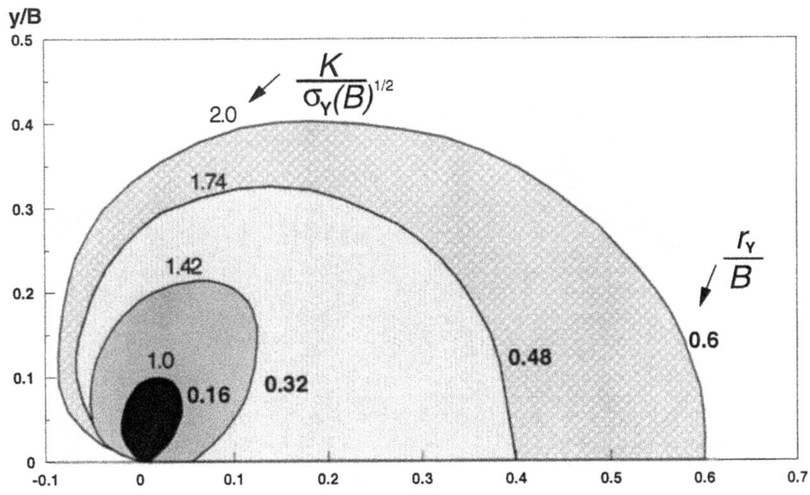

Fig.10. Three-dimensional, plane strain, elastic-plastic analysis - showing effect of constraint transition on mid-thickness plastic zone geometry (after [13])

6 Non-linear fracture mechanics

Provided that the near-crack-tip restriction on the use of LEFM has been clearly understood in the first place, the logic of restricting the Irwin approximate plastic zone size adjustment to a small proportion of the crack length follows. It is then clear that a different approach will be required for materials which are sufficiently tough to develop large relative plastic zones without crack growth. This difficulty then forms a natural introduction to the large plastic zone case, which is treated initially via the Dugdale model.

This description is interpreted broadly *ie* as a general model where the plastic zone is simulated by extension of the real crack. The yield strength of the perfectly-plastic material is represented by the forces in 'shoe-laces' which restrain opening of the crack faces. The force levels in the plastic zone will reflect the triaxial stress effect on the yield criterion (plane-stress or plane-strain), work-hardening etc.

If some exercises have been completed on the use of Green's function superposition of crack surface forces, it is relatively straightforward for the student to follow the standard proof of the relationship between plastic zone length and stress level. Interpretation as a series expansion of the small-scale plastic zone size expression [11] also provides a strong link to the LEFM model.

It is also fairly obvious that the mechanical state of the real crack tip will not have the unique character of the LEFM crack tip. Hence the comparisons between observable behaviour in a test and prediction of behaviour in a real situation will not be possible. This difficulty leads naturally to discussion of the CTOD concept (figure 11)

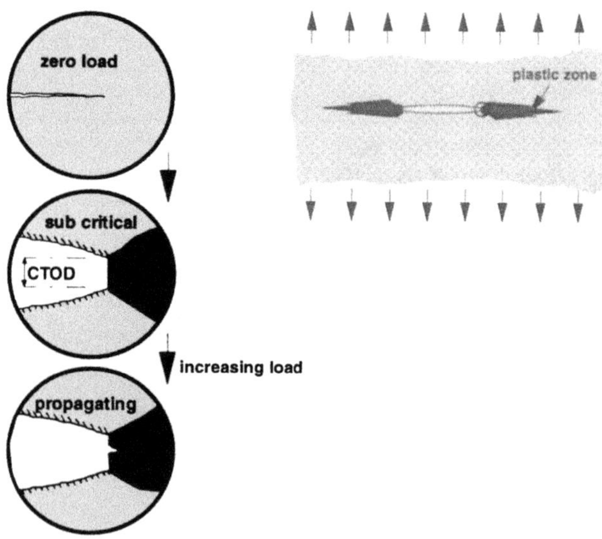

Fig. 11. Dugdale strip-yield model

The basic infinite plate solution for CTOD is extended via published finite element solutions [14] to provide a limited compendium of equations and nomograms covering six finite-geometries - see example in figure 12. These solutions provide descriptions which carry through from the small-plastic-zone approximations to limit load [15], [16]. Failure in ductile materials is more often than not governed by limit load criteria rather than by crack-tip toughness, especially where the cracks are small. This point is strongly emphasised in the exercise examples.

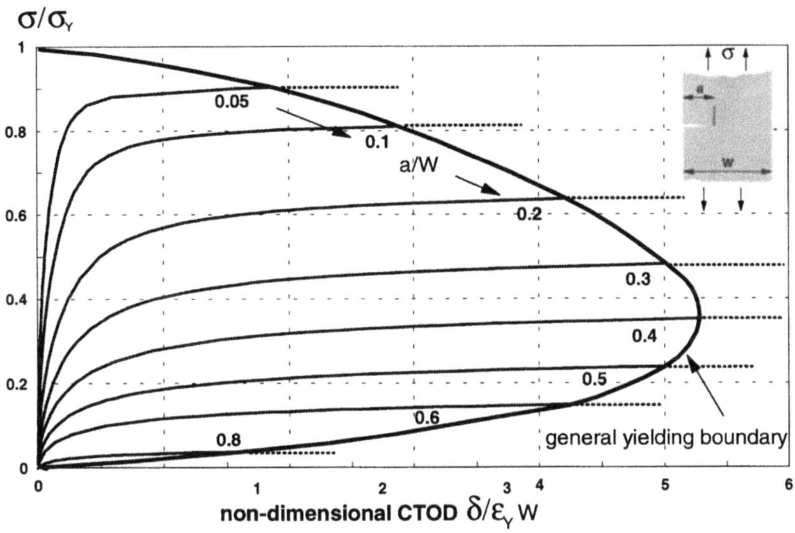

Fig. 12. Dugdale model result for single-edge-crack in tension

7 Variational energy approach to fracture mechanics

The energy approach is introduced primarily as a unifying concept, which provides a basis for comparing different cracked-body models. The reasons for including an energy discussion are therefore related more to *insight* than *application.*

Griffith's analysis is developed in terms of the potential energy associated with the introduction of a crack ($-\sigma^2 \pi a^2 B / E$) leading to the differential change in energy due to a virtual crack extension. The concept of an energy requirement associated with the creation of new crack surface is highlighted and the idea of crack instability is noted. (Analogy with the energy approach to Euler buckling may be helpful to students who are aware of this treatment.) However the main emphasis is placed on finding an approach which will localise description to the crack tip region.

The Griffith result is then derived alternatively via Irwin's asymptotic stress and displacement equations and the idea of virtual crack length *reduction.* (The history is interesting here, as Irwin was obliged to show compatibility with Griffith's analysis and this led to the redundant π term in stress intensity factor solutions.)

Given a fundamental definition in terms of total potential, it is then relatively straightforward to derive a linear compliance route to strain energy release rate, *viz* $G = \frac{1}{2} P^2 \frac{\partial c}{\partial a}$; and this in turn can be interpreted in terms of a practical compliance test or a finite element model of the same [11].

The potential energy definition also provides a reasonable basis for understanding the J parameter and the terms in Rice's contour integral. Application to non-linear material behaviour can then be shown. It is then possible to show how K_I can be related by path integration to J_I. Further, the proof of $J = \sigma_Y \delta$ can be demonstrated [17]. The definition of J in terms of non-linear compliance and limit load is also useful at this stage.

Figure 13 summarises the unifying nature of the J concept and the links between different models. The extent to which these relationships are dealt with depends on the time available. Detailed treatment is not necessary from the point of view of application, but it does develop insight into the literature, finite element applications, toughness testing procedures and application to non-homogeneous structures such as adhesive joints [18].

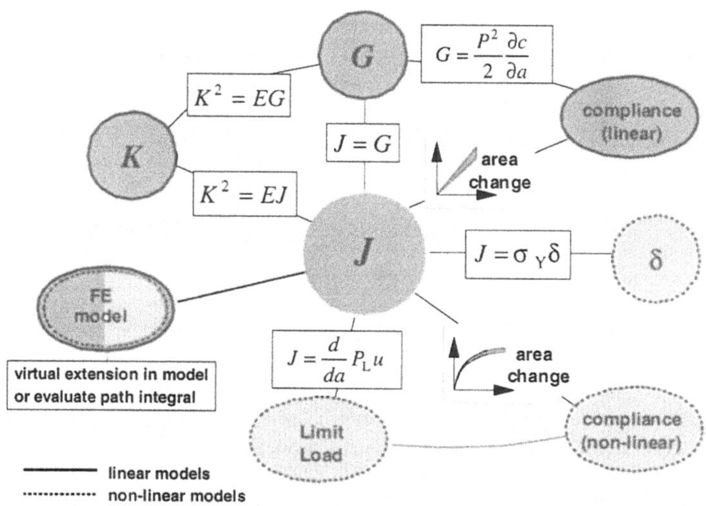

Fig 13 Interrelation between J and other fracture parameters

8 Fracture toughness

Deterministic application of *fracture mechanics* depends on material tests. Of course it is important for the student to know that various standard procedures exist and how to apply them. However it is much more important to understand how these procedures relate to the underlying mechanics models. The major point to be grasped is that, whereas the models deal with stationary (non-growing) cracks, practical toughness tests in other than ideally brittle materials have to deal with stable crack extension.

Some contact with real artefacts is indicated in this topic area and a specially made video programme is used to demonstrate a range of brittle and ductile tests on a single material. The effects of section size and temperature on brittle/ductile behaviour are shown and the actual specimen surfaces are available for direct examination by students in the class.

The simple step-by-step graphical procedures used to screen the test results for the intervention of stable crack growth and triaxial constraint are demonstrated by a laboratory technician as part of the lab programme. Standard K_{Ic} and CTOD test procedures are followed.

Other toughness test procedures are treated in the lecture programme in different ways and for different reasons. The idea of resistance curves is introduced through a computer-based exercise whereby the effect of initial crack length on the instability toughness is found through graphical comparison of constant load crack-driving-force curves with an invariant resistance curve.

Treatment of impact notch-bend tests in terms of limit load and critical CTOD is also included on the grounds that most engineers will have to deal with Charpy impact test results and be unable to relate them, even in a qualitative sense to proper fracture mechanics tests. It is instructive to show how impact energy requirement needs to be scaled up with respect to higher yield materials and thicker structures.

9 Case studies

Verbal accounts of real engineering problems and disasters lodge in the memory more securely and change attitudes more profoundly than text descriptions. In analytical terms, engineering case studies are often untidy and sometimes inconclusive but they are usually inspiring in terms of insight. It is most effective if such examples are interleaved between the theoretical treatments.

Fracture of a steam turbine LP disc rim

The details of this failure, including the analyses, were provided by GEC Alsthom as a special input to university courses in fracture mechanics and fatigue failure [19]. The approach used by the industrial analysts is maintained when presented by the authors to students, as this emphasises the realism of the case. It also demonstrates that the language of fracture mechanics is universal and not confined to 'ivory towers'.

The steam turbine was built in the late 1950's and was undergoing an overspeed test, following refurbishment, when it failed catastrophically in the works spin-test facility. The source of the failure was a sharp corner in the disc rim at an annular T-slot which retains the low-pressure blading (figure 14). Prior fatigue cracks of 0.2 - 0.6 mm depth were found.

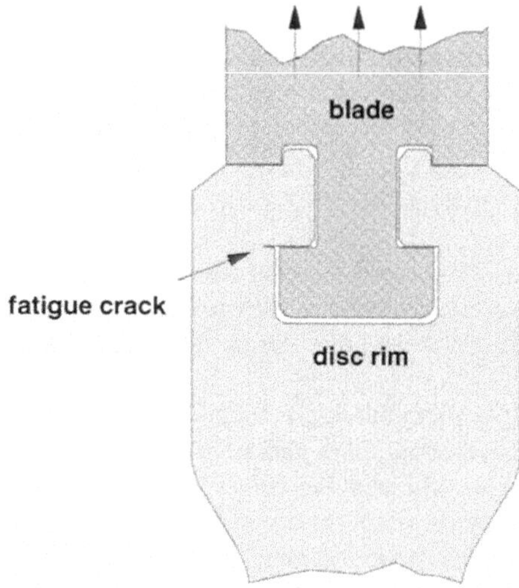

Fig. 14. Cross-section of steam turbine disc rim and blade

The cracked rim was modelled initially by interpolating between the two geometrical extremes, as shown in figure 15. Elastic superposition was used as indicated to combine stress intensity factors for different standard load cases. The upper curve in this presentation also highlights the principle whereby a crack growing from a step in dimension can be treated conservatively in terms of the total depth of the step plus the crack. The two finite element generated values agree well with the interpolation curve.

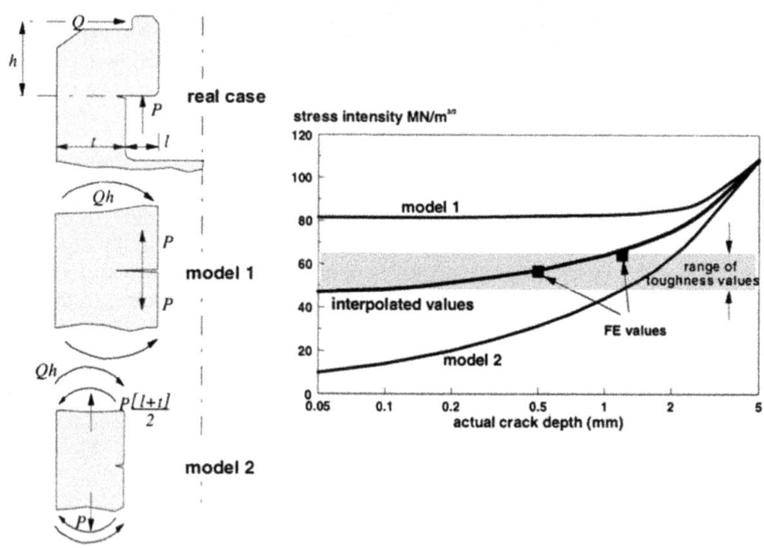

Fig. 15. Alternative models for disc rim stress-field-intensity factor

The toughness levels shown were obtained from samples cut from the disc and the assessment shown provides a credible explanation of the failure, despite the small size of the initiating cracks.

The full case study also covers related metallurgical examinations, low-cycle fatigue tests and calculations used to justify a redesign of the rim detail geometry, incorporating larger-radius corners.

Whisky Tank Failure

An example of a fracture mechanics *mis*-diagnosis was provided by the analysis of the failure of a vertical, cylindrical storage tank, used to contain waste products in a whisky distillery. A catastrophic failure occurred when a vertical seam weld ripped open from bottom to top and the tank unwrapped itself to form a flat plate on the ground. The failure actually occurred at the edge of the butt weld where there were toe cracks associated with corrosion and severe thinning of the shell.

Materials consultants carried out CTOD tests on the material and concluded that the critical CTOD levels exhibited by the material were unusually and unacceptably low for a standard pressure vessel steel. Lengthy debate concerned with the possible effects of hydrogen, whisky waste products and other ‘red herrings’ were cut short when it was shown that the condition in the thinned and cracked shell, under the head of product at the time of the failure, exceeded the limit load. Moreover, the critical CTOD for the material lay substantially outside the general yield envelope for the cracked geometry (for example as shown in figure 12). Hence, although the critical CTOD may have been low for the type of material, this factor was immaterial to the failure. (Novice fracture mechanics practitioners sometimes overlook the routine step of dividing the load by the nett cross-section to see if the limit load is exceeded.)

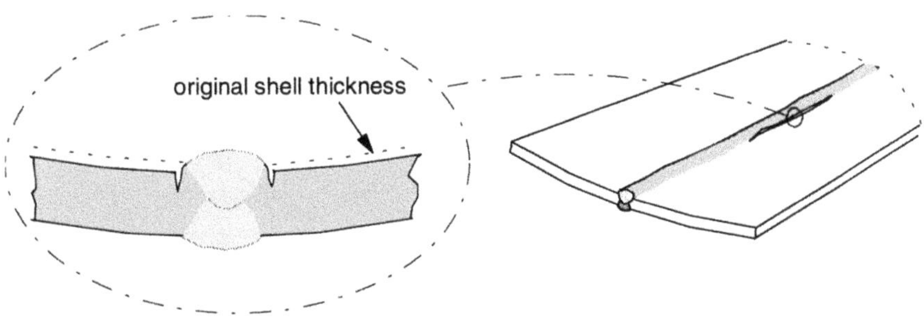

Fig. 16 Location of stress corrosion cracks in whisky waste tank

Chocolate Grinding Machine

The failure of heavy-duty gearboxes driving six chocolate-grinding machines revealed extensive cracking in solid and hollow drive shafts containing keyways. The machines were expected to have 20,000 hour lives, but lasted 800 hours in practice. The fracture paths provided classical examples of torsional and bending fatigue failure from stress concentrations (figure 17).

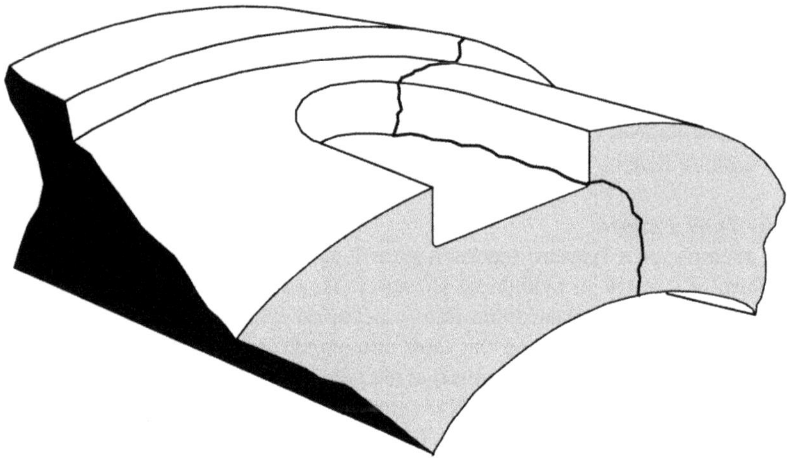

Fig. 17 Cracking from stress concentrations in machine shaft subject to torsion

The torsional fatigue cracks emanated from two locations, firstly the end-radius of the keyways at the point of maximum alternating principal stress on the surface of the shaft and secondly from the bottom sharp corner of the keyways (which can be seen as Mode II cracking). Figure 18 demonstrates the power of elliptical hole solutions in describing the non-dimensional dependence of stress concentration factor on notch depth and end-radius.

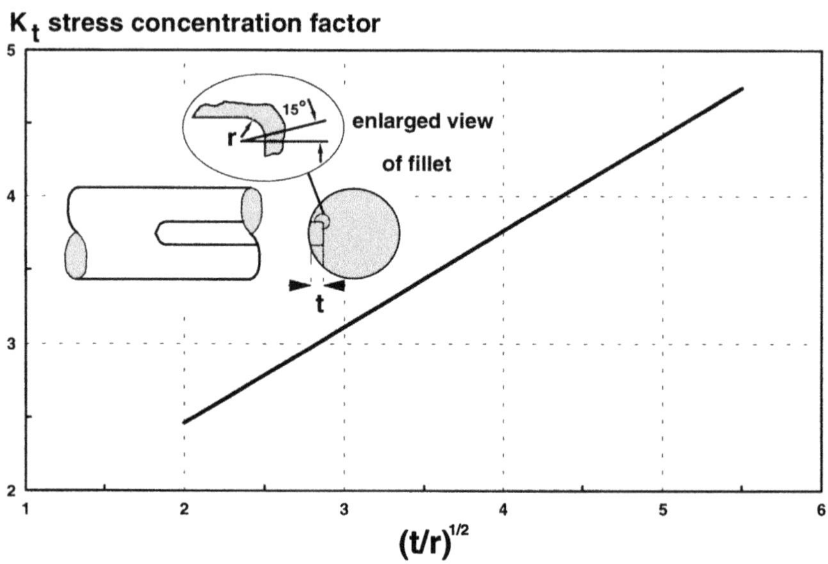

Fig. 18 SCF for torsion - dependence on radius and depth of keyway slot

In this case a major factor was torsional vibration, which was detected by condition monitoring instrumentation in all phases of operation of the mixer. Substantial alteration had therefore to be made to inertias and stiffnesses of the gear shafts, in order to disassociate gear meshing frequencies and natural frequencies of torsional vibration. However the keyways were also redesigned with larger root radii and gradual run-out of the semi-circular ends. The resulting design has given satisfactory performance ever since.

10 Summary and conclusion

A strategy for teaching the principles of fracture mechanics through the key concepts of *insight* and *application* has been described. Insight into the models and characterising parameters of fracture mechanics is essential if the material is to be applied from an understanding of principles, rather than through the practice of identifying a resemblance to standard 'worked examples'. Exercises and case studies in application are necessary to place the subject in a practical context and to provide the learner with experience of making the connections between real artefacts and continuum models.

Features which have been found to enhance insight include careful choice of learning path, beginning with simple cases and exercises, minimisation of analytical complexity through the provision of coherent compendia and an 'open book' exercise and test strategy. Good graphics are also important.

Three example case studies have been given which highlight the practical context of fracture mechanics and show the need to approximate and bound analyses in terms of theoretical models.

11 References

1. Timoshenko, Stephen P. (1953) *History of Strength of Materials,* McGraw-Hill.
2. Timoshenko, Stephen P. (1968) *As I remember,* D. Van Nostrand, Toronto.
3. Timoshenko, S. P., and Goodier, J.N. (1951) *Theory of Elasticity*, 2nd ed., McGraw-Hill, New York.
4. Inglis, C.E. (1913) 'Stresses in a Plate due to the presence of cracks and sharp corners', *Transactions of Institution of Naval Architects*, Vol. 55, No. 1, p 219.
5. Griffith, A.A. (1921) 'The phenomena of rupture and flow in solids', *Philosophical Transactions of the Royal Society*, Vol. A221, p163.
6. Griffith, A.A. (1924) 'The Theory of rupture', *Proceedings International Congress Applied Mechanics*, Delft.
7. Westergaard, H.M. (1939) 'Bearing Pressures and Cracks', *Journal Applied Mechanics.*

8. Sneddon, I.N. (1946) 'The distribution of stress in the neighbourhood of a crack in an elastic solid', *Proceedings of Royal Society*, Vol. A187, pp 229-260.

9. Irwin, G.R. (1957) 'Analysis of stresses and strains near the end of a crack traversing a plate', *Journal Applied Mechanics*, Vol. 24, p 361.

10. Broek, D. (1989) *The Practical Use of Fracture Mechanics,* Kluwer Academic Publishers, Dordrecht.

11. Parker, A. .P. (1981) *The Mechanics of Fracture and Fatigue,* E. & F.N. Spon, London.

12. Gray , T. G. F. (1995) Closed form functions for elastic stress concentration factors in notched bars, *Journal of Strain Analysis,* Vol. 30, No. 2, pp 143-154.

13. Anderson, T.L. (1991) *Fracture Mechanics Fundamentals and Applications,* CRC Press, Boca Raton.

14. Hayes, D.J. and Williams, J.G. (1972) 'A practical method for determining Dugdale Model solutions for cracked bodies of arbitrary shape', *International Journal Fracture Mechanics,* Vol. 8, p 239.

15. Gray, T. G. F and Spence, J. (1982) *Rational Welding Design*, 2nd edition, Butterworths, London.

16. Gray, T.G.F. (1977) 'A closed form approach to the assessment of practical crack propagation problems' in *Fracture Mechanics in Engineering Practice,* (ed. P. Stanley) Wiley Interscience, NY

17. Hellan, Kåre (1984) *Introduction to Fracture Mechanics,* McGraw Hill, Singapore.

18. Williams, J.G. (1984) *Fracture Mechanics of Polymers*, Ellis Horwood Ltd., Chichester, England.

19. Marriot, R. (1988) '*Investigation into failure of LP rotor disc rim'*, private communication, GEC Alsthom, Rugby, England.

6 TEACHING THE PHYSICAL BASIS OF FRACTURE MECHANICS

F. GUIU and R.N. STEVENS
Department of Materials, Queen Mary and Westfield College,
University of London, London, UK

Abstract
In this paper we emphasize the need for teaching the physical basis of fracture mechanics at both undergraduate and postgraduate level courses. It is recognised that much of fracture mechanics has developed a long way and is being applied without much reference to its basic physical principles. In a series of recent papers we have discussed with specific examples the dangers of this approach. False expectations can be easily be built upon arguments based on fallacious energy balance concepts and students should be made aware of these limitations.
In particular the notion should be dispelled that plastic deformation (or other dissipative processes) increase the resistance to crack propagation (or toughness) by an amount which can be calculated from the rate of energy dissipation. The effect of plastic deformation or any other dissipative process is that of modifying (in general relaxing) the stress field at the crack tip thus reducing the magnitude of the crack driving force, whether expressed in terms of energy release rate or by means of the stress intensity factor.

The difficulties encountered in trying to define a material parameter which characterises crack growth resistance in dissipative and structurally complex materials are discussed. A derivation of the J integral is given which has the advantage of physical clarity and allows easy deduction of its important physical properties. It is argued that crack tip plasticity cannot be modelled by non-linear elastic deformation and hence that caution has to be exercised in the use of fracture criteria derived from such modelling and from the critical value of the J integral.

1 Introduction

Linear elastic fracture mechanics (LEFM) has become an important engineering discipline and is taught in a systematic way as an introduction to any general course on fracture. This theory is now successfully established and it allows the fracture behaviour of relatively brittle materials in which plastic yielding takes place only on a

Teaching and Education in Fracture and Fatigue. Edited by H.P. Rossmanith. Published in 1996
by Taylor & Francis, 2 Park Square, Milton Park, Abingdon, Oxon, OX14 4RN. ISBN 0 419 20700 7.

small scale around the crack tip to be predicted accurately on the basis of a material constant (the fracture toughness) and stress calculations for the particular component and crack geometry. With the advent and development of much tougher materials it has become necessary to extend the theoretical basis of fracture mechanics to include materials which do not behave in a brittle manner, where plastic deformation plays a more significant role and where cracks can grow, with increasing resistance, in a stable manner. It is probably true to say that while considerable effort has been put into this exercise the success has been limited to date.

We are of the opinion that much of this effort has been partly misguided by an incomplete understanding of the basic physical principles on which fracture mechanics is based and we therefore believe that it is necessary to introduce the subject to students starting with a correct formulation of these principles and their correct application to the most simple cases. This is explained in the first part of this paper which is followed by a discussion of the difficulties which arise when fracture mechanics is extended to the elastic-plastic cases. Finally a derivation of the J integral is given which clarifies its physical significance, its use and limitations in elastic-plastic fracture mechanics.

2 Linear elastic fracture mechanics

LEFM is introduced in most text-books from two equivalent points of view. That which we might call the historical view starts with an exposition of the classical Griffith theory which is generalised into what has become known as the "energy balance approach". In this approach the relevant parameter characterising the behaviour or instability of a crack is the *crack extension force*, $\mathcal{G}$, also referred to as the *energy release rate*. The other approach starts with a development of the stresses at the tip of a crack and arrives at the concept of the stress intensity factor, K, as a characterising parameter of the state of the stress and strain at the tip of the crack. In this approach, K is used as the parameter which characterises the crack behaviour or crack instability and very often the energy balance is regarded of being of little more than historical interest. It is clear however, that in the LEFM case there is a simple connection between $\mathcal{G}$ and K and we have shown elsewhere that this connection is retained even in cases where different energy dissipation mechanisms occur simultaneously with crack extension[1,2]. The stress intensity factor should therefore be regarded in the fracture mechanics context as a convenient and very useful alternative parameter which is a measure of the *crack driving force* (or crack extension force) rather than a stress characterising parameter. The many advantages of using K rather than $\mathcal{G}$ in fracture mechanics calculations have undoubtedly helped the development of the subject considerably and later in the paper we shall make reference to some of these.

The Griffith energy balance formulation gives, however, a much clearer physical insight into the problem of crack instability and into the meaning of the critical parameters which determine this instability. It needs to be emphasised also that this energy balance is not based on an energy conservation law. Such a law is always satisfied for any crack length and boundary constraints and therefore no information about the crack behaviour can be obtained from it[2].

The Griffith theory [3] is a particular example of the theory of thermodynamic

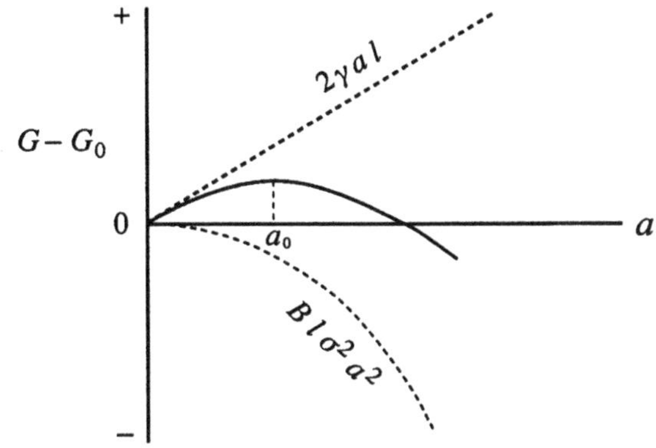

Figure 1
The variation of Gibbs free energy, G, with crack length, a, for a linear elastic material.

instability although Griffith's original formulation is in somewhat disguised form. Equilibrium conditions, stable or unstable, are expressed in terms of the thermodynamic potentials, of which there are two of particular interest in connection with fracture. These are the Gibbs free energy, G, and the Helmholtz free energy, F.

For a system or body constrained to be at constant temperature with constant tractions on its boundaries, the Gibbs free energy is a minimum when the system is in stable equilibrium and a maximum when it is in unstable equilibrium. On the other hand the Helmholtz free energy is a minimum for a system or body in stable equilibrium at a fixed temperature and having those parts of its outer boundary on which forces are exerted held fixed in position and a maximum for unstable equilibrium with the same constraints. It will be recognised that the constraints referred to above correspond to the fixed load and fixed grips conditions in fracture mechanics.

The Griffith energy balance can be illustrated by considering a tensile member of length L, thickness l and width w having a tensile force f exerted on its ends and a central through-thickness crack of length a. The Gibbs free energy of this body is

$$G = F - fL \tag{1}$$

The Helmholtz free energy has contributions from the elastic energy and the surface energy of the crack. The Gibbs free energy of the cracked body is

$$G_t = G_o + 2\gamma al - Bl\sigma^2a^2 \tag{2}$$

where G_o is the Gibbs free energy of the body without the crack, γ is the surface energy, σ is the tensile stress remote from the crack and B is a constant dependent on the geometry of the body and the elastic constants. The quantity G_t, which is the equivalent to the total energy of the body, U, in the more usual mechanical formulation of the energy balance, is plotted in Figure 1 where it can be seen that a maximum results from the subtraction of a term quadratic in a from a linear term. At crack lengths greater than that corresponding to the maximum the system is thermodynamically unstable.

The change in the total Gibbs free energy, $(\mathrm{d}G_t)_{T,f}$, at constant temperature and force resulting from a change in crack area is

$$(\mathrm{d}G_t)_{T,f} = [\mathrm{d}F_c - f\mathrm{d}L_c]_{T,f} + 2\gamma(\mathrm{d}A)_{T,f} \tag{3}$$

where dF_c and dL_c are the changes in strain energy and in length due to the change in compliance resulting from crack extension and A (= la) is the area of the crack. In the usual mechanical notation Equation 3 is written as

$$dU = d(E - fL_c) + 2\gamma dA \tag{3a}$$

where U is the total energy of the body, E is the elastic energy of the system and $(E - fL_c)$ is the potential energy of the system.

At the thermodynamic instability point $(dG_t)_{T,f} = 0$ and

$$-[dF_c - f\,dL_c]_{T,f} = 2\gamma(dA)_{T,f} \tag{4}$$

In Equations (3) and (4) the term $(dG_c)_{T,f} = [dF_c - f dL_c]_{T,f}$ is the energy (Gibbs free energy) available to drive crack growth (it is equal to $2Bl\sigma^2 a da$ in the Griffith theory) and $2\gamma l(da)_{T,f}$ is the energy opposing crack growth. The thermodynamic driving force (a virtual force) or crack extension force (also called the energy release rate), $\mathcal{G}$, is defined as

$$\mathcal{G} = -\left(\frac{\partial G_c}{\partial A}\right)_{T,f} = f\left(\frac{\partial L_c}{\partial A}\right)_{T,f} - \left(\frac{\partial F_c}{\partial A}\right)_{T,f} \tag{5}$$

If the analysis had been carried out under fixed grips conditions then it would have been found that

$$\mathcal{G} = -\left(\frac{\partial G_c}{\partial A}\right)_{T,f} = -\left(\frac{\partial F_c}{\partial A}\right)_{T,L} \tag{6}$$

the driving force and the instability condition being independent of the constraints applied.

It is very important to note that this energy balance does not tell us whether a crack will grow or not but it provides a lower bound, or a necessary condition, for crack propagation. It also allows us to identify the thermodynamic driving force on the crack or for crack growth i.e. the negative of the rate of change of the potential energy with crack extension. Here it is important to note that no real process will take place with zero net driving force (i.e. at the instability point in the case of crack growth when $\mathcal{G} = 2\gamma$) and it is to be expected that when crack propagation occurs $\mathcal{G}$ is greater, and perhaps much greater, than 2γ, the difference being dissipated as heat.

Whether the crack extension force can provide a criterion for crack growth is a matter for experiment to decide. This criterion states that the crack will grow when the crack extension force reaches a critical value, $\mathcal{G}_c$, or its equivalent, K_c. This critical value has to be determined by experiment, and in materials which obey LEFM it is found to be a characteristic material property (the fracture toughness) which is independent of the method used for its determination provided that some precautions are taken to limit the extent of of plastic deformation to a very small zone confined to the crack tip (the small scale yielding condition). The conditions which need to be satisfied for the determinations of $\mathcal{G}_c$ or K_c are specified by testing standards. It is generally found that the value of $\mathcal{G}_c$ experimentally determined is much greater than 2γ

and in view of our previous discussion this is not surprising. There is no need to make the incorrect assumption that $\mathcal{G}_c >> 2\gamma$ because the term 2γ is, in real materials, increased by the the work of plastic deformation associated with the growth of the crack, or by the energy dissipated by other processes which may accompany crack growth. The authors have discussed this point at length in some recent publications.[1,2,4]

Here we are faced with an important educational point. Should one impart to the students the commonly accepted notion that energy dissipation is responsible for, or contributes to, the increase in fracture toughness? Or on the contrary, should one dispel this idea and warn the students that this may be an unhelpful prop on which to base the understanding of fracture mechanics?

We prefer to adopt the second view and accordingly develop the physical basis of fracture mechanics on the idea that plastic deformation, as well as other energy dissipating mechanisms, produce an increase in toughness by virtue of the fact that they relax, or reduce, the magnitude of the stress at the crack tip and thus reduce the driving for the crack, i.e. the term $\mathcal{G}$ in Equation 6.

The extension of fracture mechanics concepts from the LEFM case to the elastic-plastic case is a logical development of this idea. First of all there is the need to identify a parameter which can be used to characterise the crack behaviour. It is to be expected that the behaviour of the crack will be determined by the force acting upon it and therefore we understand that the crack extension force, $\mathcal{G}$, or its equivalents, K and J, are the best candidates in this context.

3 Generalised crack extension force[1]

If a criterion for fracture based on the magnitude of the crack extension force is to be found for materials exhibiting considerable plastic deformation or for tough dissipative materials with complex microstructures, a complication arises which is not present in LEFM. This is the fact that the crack extension force cannot be calculated from the magnitude of the applied external tractions or loads alone; it must include contributions arising from the stress fields set up by the plastic deformation, or other microstructural material features (particles, voids, fibres, etc.) As a result it becomes more difficult to describe the behaviour of a crack, or to define a crack growth criterion based on a simple characterising parameter, such as $\mathcal{G}$ or K, based on the magnitude of the external loads and geometry of the body alone. In these cases it appears that the apparent toughness (as it would be obtained from the response of the body to known external tractions alone) or the so called fracture resistance is also a function of crack length.

All the entities which give rise to internal stress fields contributing to the value of the crack extension force can be modelled by elastic singularities and this makes it possible to extend the concept of a crack extension force to the general case of a loaded elastic body containing internal stress fields.

It is worth considering in detail the definition of an elastic singularity before proceeding further. An elastic singularity is subtly different from the mathematical singularity which is usually referred to in connection with cracks. The mathematical singularity is the infinity in the stress field at the crack tip in a linear elastic body. Infinite stresses are, of course, physically impossible and no real material can ever experience them. Real materials are always non-linear elastic when the strains are large and this non-linearity ensures that the stresses are always bounded and finite. The definition of an elastic singularity does not depend on the existence of fictitious infinite stresses.

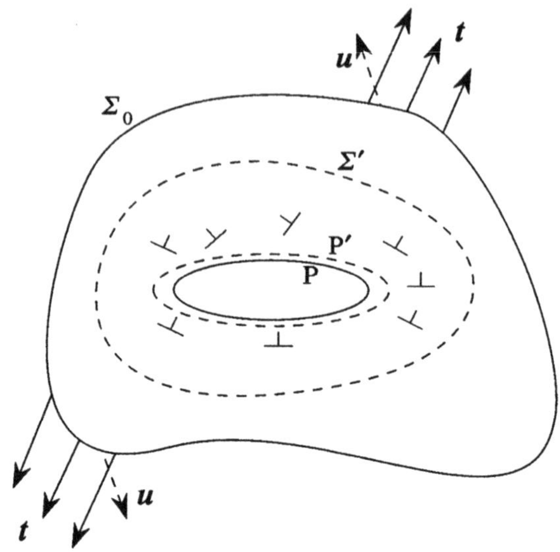

Figure 2
A crack of initial perimeter P in an elastic body with tractions, t, and displacements, u, on its outer boundary, Σ_0. It is surrounded by singularities, represented by edge dislocations.

If a closed curve is drawn in an un-deformed body the compatibility relations guarantee that the curve will also be closed in the deformed body. However, if the curve does fail to close in the deformed body one or more elastic singularities are present. Elastic singularities are therefore manifest by a failure of the compatibility relations. It can therefore be seen that cracks and dislocations cause closure failures and are hence particular examples of elastic singularities.

Consider an elastic body, linear or non-linear, with an outer boundary Σ_0 with tractions arbitrarily distributed on it which hold it in equilibrium (Figure 2). The body contains a crack of perimeter P and area A and at the same time randomly distributed elastic singularities, such as dislocations or force dipoles which generate the stress fields produced by the microstructural features.

If the crack area grows by dA as the crack perimeter changes from P to P' in an arbitrary manner the change in Gibbs free energy of the body is

$$(\mathrm{d}G)_{T,\sigma} = (\mathrm{d}F)_{T,\sigma} + 2\gamma(\mathrm{d}A)_{T,\sigma} - \int_{\Sigma_0} \delta\boldsymbol{u}.\boldsymbol{t}\,\mathrm{d}S \qquad (7)$$

where $(\mathrm{d}F)_{T,\sigma}$ is the change in strain energy inside Σ_0 and the integral is the work done by the surface tractions, $\boldsymbol{t}$, as each element dS of the bounding area of the body is displaced by $\delta\boldsymbol{u}$.

Consider now a reversible process allowing the transition between the states of Gibbs free energy G and G + dG with crack areas of A and A + dA respectively. A cut is made between the old position of the crack front and the new, and tractions applied to keep the cut faces closed. The tractions are then reversibly changed so that they become zero (i.e. the crack is allowed to open). The work done by the body against the tractions on the cut faces is $\delta\omega_c$. There will also be work, $\delta\omega_0$, done against the tractions applied to the surface Σ_0. The change in Helmholtz free energy is therefore, at constant temperature and tractions applied to Σ_0,

$$\delta F_e = -(\delta\omega_c + \delta\omega_0) \tag{8}$$

or

$$-\delta\omega_c = \delta F_e + \delta\omega_0 = \delta F_e - \int_{\Sigma_0} \delta \boldsymbol{u}.\boldsymbol{t}\, dS \tag{9}$$

Hence, by comparison with Equation 1

$$-\delta\omega_c = \delta G \tag{10}$$

Now δF_e can be expressed as a volume integral of the change in strain energy density, W, as the crack extends so that

$$-\delta\omega_c = \delta G = \int_{V_0} \delta W\, dV - \int_{\Sigma_0} \delta \boldsymbol{u}.\boldsymbol{t}\, dS \tag{11}$$

Suppose we draw a new surface inside Σ_0, say Σ' (see Figure 2). We can repeat the derivation of Equation 11 using Σ' instead of Σ_0. The part of the body inside Σ' is regarded as a new thermodynamic system, the part outside as belonging to the mechanical system applying tractions to Σ'. The result is clearly

$$-\delta\omega_c = \delta G = \int_{V'} \delta W\, dV - \int_{\Sigma'} \delta \boldsymbol{u}.\boldsymbol{t}\, dS \tag{12}$$

where V' is the volume contained within Σ'.

Equation 12 does not depend on any assumption about linear elasticity and non-linearity at large strains will cause the first integral to have a term $2\gamma dA$. Similarly the work, $\delta\omega_c$, done at the cut surface will have a term representing the work done against the cohesive forces and this will be equal to $-2\gamma dA$. It is conventional to subtract out this term from both sides of the equation leaving on the left-hand side $-\delta\omega_{ce}$, the work done against the *elastic* tractions. Thus we have

$$-\delta\omega_{ce} = \delta G = \int_{V'} \delta W\, dV - \int_{\Sigma'} \delta \boldsymbol{u}.\boldsymbol{t}\, dS \tag{13}$$

Clearly each of the integrals on the right hand side are different from those in Equation 11. However, providing the new surface, Σ', wholly encloses the strip of area dA which was cut to extend the crack, it is clear that $-\delta\omega_c$ and hence δG are the same as in Equation 11. Hence the right hand side of Equation 13 is invariant and is independent of the choice of the surface, Σ', provided it encloses the crack tip.

This proof of invariance is important. From it derives the path independence of the J integral. Note that although the right hand side of Equation 13 involves integrating over the volume V' and its bounding surface Σ' the left hand side is always the Gibbs free energy change in the volume V_0 bounded by Σ_0.

Inside the surface Σ' there are other singularities such as dislocations which have been assumed to remain fixed in postition. The work $-\delta\omega_{ce}$ is the virtual work of the elastic tractions acting on the crack faces as this grows from A to $A + \mathrm{d}A$. This work includes therefore contributions both from the stresses arising from the external tractions and those arising from the stress fields of the elastic singularities.

If the body is of uniform thickness, l, with a through thickness crack of length a having a straight front, then by definition the crack extension force is

$$\mathcal{G} = -\frac{1}{l}\left(\frac{\partial G_e}{\partial a}\right)_{T,\sigma} \tag{14}$$

but if the crack is of arbitrary perimeter, P, and its area increases by $\mathrm{d}A$ in an arbitrary manner the rate of change

$$\bar{\mathcal{G}} = -\left(\frac{\partial G_e}{\partial A}\right)_{T,\sigma} \tag{15}$$

is the average value of the component of $\mathcal{G}$ in the direction of movement of the crack front, and the crack extension force has different local values, $\mathcal{G}(\mathbf{r})$ at points with position vector $\boldsymbol{r}$ along the crack front[1].

Based on these three different parameters the following criteria could be considered for crack growth:

i) For a through-thickness crack with a straight front

$$\mathcal{G} = \mathcal{G}_c \tag{16}$$

ii) For an arbitrary crack shape moving in any arbitrary manner

$$\bar{\mathcal{G}} = \bar{\mathcal{G}}_c \tag{17}$$

iii) For a given crack front segment to become *locally* unstable

$$\mathcal{G}(\boldsymbol{r}) = \mathcal{G}_c \tag{18}$$

Although this is a criterion for local instability it is often used for total instability in cracks with curved fronts in the form

$$\mathcal{G}_{max} = \mathcal{G}_c \tag{19}$$

where $\mathcal{G}_{max}$ is the largest value of the local crack extension force around the crack front.

It is quite clear that it is a matter for experiment to decide whether any of the above criteria are satisfied and can be reliably used. From a conceptual point of view it is however very important to note that the values of $\mathcal{G}$ used here have been obtained taking the effect of the internal stresses into account. It can be recognised that this is the source of what is generally known as shielding of the crack because if these internal stresses are of opposite sign to those due to the external load, they will reduce the magnitude of $\mathcal{G}$ producing an increase in fracture toughness. Calculations of shielding can be carried out more easily in terms of K than in terms of $\mathcal{G}$ because according to the relations

$$K = K_1 + K_2 + \ldots K_i \tag{20}$$

and

$$\mathcal{G} = (\mathcal{G}_1^{1/2} + \mathcal{G}_2^{1/2} + \ldots \mathcal{G}_i^{1/2})^2 \tag{21}$$

for the superposition of the K's and the $\mathcal{G}$'s arising from loading systems 1 to i, it follows that K's can be added but $\mathcal{G}$'s cannot. Furthermore, there are mathematical methods available to calculate K from a knowledge of the loading configuration and the geometry of the body whilst calculation of $\mathcal{G}$ generally requires the cumbersome integration of stress times strain over the volume of the body.

We have to so far considered the extension of a crack alone with no other dissipative processes accompanying crack growth. Now we shall discuss the important case where plastic deformation, or any other dissipative process, takes place as the crack grows.

With reference to Figure 2 again let both the crack and some of the other singularities move in any arbitrary but conservative manner. The corresponding Gibbs free energy change will be

$$(\delta G)^{\text{tot}} = (\delta F_{\text{e}})^{\text{tot}} + 2\gamma \text{d}A - \int_{\Sigma'} \delta \boldsymbol{u}.\boldsymbol{t}\, \text{d}S \tag{22}$$

and by a similar virtual work argument as before, it can be shown that

$$(\delta G_{\text{e}})^{\text{tot}} = (\delta F_{\text{e}})^{\text{tot}} - \int_{\Sigma'} \delta \boldsymbol{u}.\boldsymbol{t}\, \text{d}S = -(\delta\omega_{\text{ce}} + \delta\omega_{\text{de}}) \tag{23}$$

The term $\delta\omega_{\text{ce}}$ has the same meaning as before while $\delta\omega_{\text{de}}$ is the work done by the elastic tractions on the surfaces of all the singularities that have been arbitrarily displaced. If these singularities are dislocations then $\delta\omega_{\text{de}}$ is simply the work of plastic deformation. The integral is again invariant with respect to Σ' *provided that it encloses all the moving singularities.*

We cannot obtain a crack extension force from the rate of change $(\partial G_{\text{e}}^{\text{tot}} / \partial A)_{T,\sigma}$. This is the total change in the Gibbs free energy due to both crack extension and the movement of other singularities divided by the change in area of the crack alone. This is obviously not a crack extension force. In the case of plastic deformation it would be the total energy release rate including that due to plastic deformation and it would be difficult to argue that this parameter, having little or no physical meaning, can be used to describe the behaviour of the crack. The true crack extension force has to be obtained from the energy changes evaluated using a surface Σ' which encloses the crack and no other moving singularities. Neither the work done in moving the singularities nor the energy dissipated in their movement enters the calculations of $\mathcal{G}$. Fracture criteria which are based on the total rate of energy dissipation seem to have little justification and they should be critically considered.

The only physically meaningful parameter that can be obtained from the total Gibbs free energy change $\text{d}G_{\text{e}}^{\text{tot}}$ is the quantity $(\partial G_{\text{e}}^{\text{tot}} / \partial A^{\text{tot}})_{T,\sigma}$ where $\text{d}A^{\text{tot}}$ is the sum of the areas swept out by the crack and all the singularities inside Σ'. This quantity is the average of the component of the force per unit length in the direction of displacement of both the crack and the singularities inside Σ' [1]. As far as we are aware such a parameter has never been used in the formulation of a fracture criterion.

4 The *J* integral and its use in fracture mechanics

The *J* integral is an invariant line integral evaluated over a path which encloses the crack. It can be derived after imposing some restrictive conditions from the general Gibbs free energy expression given by Equation 13. The *J* integral is identical to the crack extension force, $\mathcal{G}$, for an elastic material, linear or non-linear, and it has been shown that it characterises the deformation fields at the crack tip in an manner analogous to that of *K* in the linear elastic case[5]. In elastic-plastic fracture mechanics the notion of crack tip characterization has been taken further. If the elastic-plastic material is modelled by a fictitious non-linear elastic material with matching stress strain curve on loading, then the value of *J* from a contour enclosing the crack in the fictitious material is deemed to characterise the deformation fields in the real elastic-plastic material. Hence *J* is used in the formulation of a fracture criterion in materials which undergo significant plastic deformation, as a parameter which reaches a critical value when the crack grows. In most text books the derivation of the *J* integral and its path independence is done in a way that obscures it physical significance and although well established standards exist for the determination of *J* in elastic-plastic materials serious doubts about the validity of this approach arise when the physical meaning of the *J* integral is carefully examined.

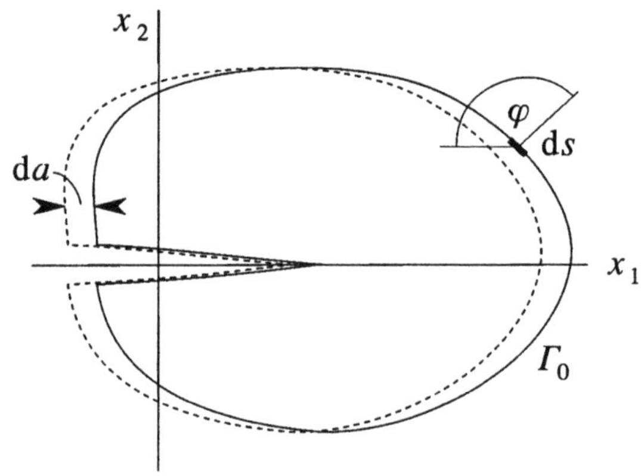

Figure 3
Illustrating the transformation of equation 13 to give the *J* integral. The body is embedded in an infinite body contain singularities outside Γ_0 such that the tractions on Γ_0 are the same as in the free body. The boundary is then displaced in the negative x_1 direction and the tractions readjusted.

To derive the *J* integral we consider a cylindrical body of arbitrary cross-section with surface tractions giving rise to a state of plain strain (Figure 3). The body contains a crack in the x_1x_2 plane with the crack front parallel to x_3. The tractions on the surface, Σ_0, of the body are **t**. The curve bounding the normal cross-section of the cylinder is Γ_0 and in what follows unit length of the cylinder is taken. Equation 13 applies to this particular case and *J* is defined to be

$$J = -\frac{1}{l}\left(\frac{\partial G}{\partial a}\right)_{T,\mathbf{t}} \tag{24}$$

where *a* the crack length and *l* is the thickness of the body.

To obtain the *J* integral from Equation 13 we transform the right hand side to a single integral over the surface Σ' or equivalently, since the problem is essentially two dimensional, along the contour Γ'. To do this we imagine the space outside Σ_0 to be replaced by matter which contains sources of internal stress (singularities) so arranged as to produce the same tractions on the surface Σ_0 as before. We now calculate the right hand side of Equation 13 by a two stage process.

Stage I. Instead of displacing the crack tip by δa we displace Σ_0 and Σ' by $-\delta a$ (δa it will be recalled, is in the x_1 direction). This will of course change the tractions on both surfaces and this will have to be taken account of in stage two.

Each element ds of Σ' will sweep out a volume per unit length of $-ds\,\delta a \cos\varphi$ where φ is defined in Figure 3. This is positive for elements with $\varphi > \pi/2$ and negative for elements with $\varphi < \pi/2$. The strain energy within Σ' changes by

$$\delta F_e^{\mathrm{I}} = -\delta a \int_{\Gamma'} W \cos\varphi \, ds \tag{25}$$

Stage II. The sources of internal stress outside Σ_0 are now adjusted so that the tractions on Σ_0 in its new position return to the prescribed values. During this process both the tractions and the displacements on Σ' change. Before stage I the surface Σ' had displacements $\boldsymbol{u}$ on it. At the end of stage I these had become $\boldsymbol{u}-\delta a\,(\partial \boldsymbol{u}/\partial x_1)$.

In stage II these are changed to final values, $\boldsymbol{u}_f$. The change in strain energy in stage II is therefore

$$\delta F_e^{\mathrm{II}} = \int_{\Gamma'} \left(\boldsymbol{u}_f - \boldsymbol{u} - \delta a \frac{\partial \boldsymbol{u}}{\partial x_1}\right) . \boldsymbol{t} \, ds \tag{26}$$

where the positive sign arises because the displacements are being imposed on the body inside Σ' by the external system. The total change in strain energy is therefore

$$\delta F_e = \delta F_e^{\mathrm{I}} + \delta F_e^{\mathrm{II}} = -\delta a \int_{\Gamma'} W \cos\varphi \, ds + \int_{\Gamma'} \left(\boldsymbol{u}_f - \boldsymbol{u} - \delta a \frac{\partial \boldsymbol{u}}{\partial x_1}\right) . \boldsymbol{t} \, ds \tag{27}$$

This can be written

$$\delta F_e - \int_{\Gamma'} (\boldsymbol{u}_f - \boldsymbol{u}) . \boldsymbol{t} \, ds = -\delta a \int_{\Gamma'} W \cos\varphi \, ds + \delta a \int_{\Gamma'} \frac{\partial \boldsymbol{u}}{\partial x_1} . \boldsymbol{t} \, ds \tag{28}$$

Comparing Equations 28 and 13 we have

$$\delta G = -\delta a \int_{\Gamma'} W \cos\varphi \, ds + \delta a \int_{\Gamma'} \frac{\partial \boldsymbol{u}}{\partial x_1} . \boldsymbol{t} \, ds \tag{29}$$

since $\delta \mathbf{u} = \mathbf{u}_f - \mathbf{u}$.

For an anti-clockwise circuit around the contour Γ'

$$\cos\varphi \, ds = dx_2 \tag{30}$$

and hence using Equation 24 we have

$$J = \int_{\Gamma'} W \, dx_2 - \int_{\Gamma'} \frac{\partial \boldsymbol{u}}{\partial x_1} . \boldsymbol{t} \, ds \tag{31}$$

which is the J integral in its usual form. It should be noted that the value of the J

integral, as well as the value of δG in Equation 13 is determined by the singularities inside Γ' or Σ'. A plastic deformation field will always contain singularities (other than the crack) whilst a non-linear elastic field without plasticity never will. Thus the usefulness of non-linear elasticity to model plastic behaviour must be extremely limited. Plasticity cannot be modelled by non-linear elasticity even if the mathematical relations between stress and strain are identical on loading in both cases, and even if the material is never un-loaded. Further input into this problem to reinforce the conclusion is provided by noting that there are no dissipative terms in the integrals of Equations 13 and 31. The strain energy density appearing in these equations is elastic strain energy of the body. In the elastic case all the work done on the boundaries of a region is stored as strain energy in this region, but in the plastic case only a fraction of the work done at the boundaries is stored even if the stress-strain relations on loading are perfectly matched. Thus the first integral of Equation 13 will yield a different result in each of the cases although the second integral will be identical.

5 Properties of the *J* integral

The properties of the *J* integral are now relatively simple to understand. They are as follows:

- The derivation does not appeal to a linear stress-strain relation in any way hence the *J* can be evaluated and has the same meaning for a non-linear elastic solid. Note the work "elastic". This conclusion does not apply to plasticity as will be seen below.

- If there is no singularity inside Σ' then *J* is zero. This is because there will be no surface inside Σ' on which tractions are exerted and on which work is done. With $\delta\omega_{ce}$ zero, Equation 13 simply states that the change in strain energy in a region is equal to the work done by the tractions on the boundary of the region.

- The *J* integral gives the x_1 component of the force on the crack when this is the only singularity inside Σ'.

- If there are other singularities such as dislocations inside Σ' then *J* is the net x_1 component of the force on all of them.

- *J* is only path independent if the paths always enclose the same singularities. Hence if we have a crack whose tip is surrounded by dislocations then *J* evaluated on a remote contour will be the net force on crack and dislocations. As the contour is shrunk towards the tip *J* remains constant until it passes through the dislocations when it is no longer path independent. This results from the fact that there are different numbers of singularities inside each contour.

- Plastic deformation is due to the creation and movement of singularities (dislocations). Because singularities affect the value of the J integral while non-linear elastic strains *per se* do not, the J integral for a material in which plastic deformation accompanies crack extension cannot in principle be calculated using non-linear elasticity to model plasticity. The J integral cannot be "fooled" into confusing plastic deformation with non-linear elastic deformation even if no unloading has taken place.

Using these properties we can comment on the philosophy of elastic-plastic mechanics. The methods used for the determination of the J integral, both experimental and numerical, give the value of J for a fictitious non-linear elastic material with the same loading stress-strain relations as the real material. Since this is in principle different from the value of J in the real material it is doubtful whether it can characterise the stress and strain fields around the crack tip in the real elastic-plastic material. A judgement of this question must ultimately rest on experiment. In the meantime it is important that students should be acquainted with the difficulties in the philosophy of elastic-plastic fracture mechanics as presently formulated.

References

1. F. Guiu and R. N. Stevens, Int. J. of Fracture, **52**, 1, (1991)
2. R. N. Stevens and F. Guiu, Proc. Roy. Soc. Lond., **A 435**, 169, (1991)
3. A. A. Griffith, Phil. Trans. Roy. Soc. Lond. **A 221**, 163, (1920)
4. F. Guiu and R. N. Stevens, J. Mater. Sci., **26**, 4375, (1991)
5. J. R. Rice and G. F. Rosengren, J. Mech. Phys. of Solids, **16**, 1, (1968)

7 TEACHING AND APPLICATION OF ENGINEERING FRACTURE MECHANICS: THE NECESSITY OF UNITING SIMPLICITY AND RATIONALITY

H.J. SCHINDLER
Swiss Federal Laboratories for Materials Testing and Research (EMPA), Dübendorf, Switzerland

Abstract

The situation in the field of engineering fracture mechanics is characterized by a wide and still increasing gap between the general knowledge in research institutes and what is actually applied in engineering practice. One of the sources of this unsatisfying situation might be the way this subject is taught to students of engineering: Since fracture mechanics is naturally a relatively difficult subject, many lecturers do not find the balance between a too superficial a treatment on one side and becoming lost in boring theoretical complications on the other. Many universities even do not offer courses in fracture mechanics at all. Therefore, in teaching as well as in application, one should look for convincing ways to unite simplicity and rationality. The key to this goal is introducing only as few as possible parameters, concentrating on their physical meaning, developing simple physical crack models that can be easily treated mathematically, and using them to establish a transparent theoretical framework.
Keywords: Teaching, application, engineering fracture mechanics, physical models.

1 Introduction

Fracture is the ultimate phase of a structure, where the transition from the original to the broken state occurs. Engineering fracture mechanics (EFM) provides the rational basis to treat this phase theoretically and experimentally. Since imperfections in the material, like cracks or flaws, play an important role in fracture processes, EFM concentrates on the theoretical description of the mechanical behaviour of structures that contain such imperfections. Their mechanical behaviour is mainly controlled by several inherent "nonlinearities" like large deformations associated with crack-tip

Teaching and Education in Fracture and Fatigue. Edited by H.P. Rossmanith. Published in 1996 by Taylor & Francis, 2 Park Square, Milton Park, Abingdon, Oxon, OX14 4RN. ISBN 0 419 20700 7.

blunting and the formation of zones of non-linear material behaviour. Correspondingly, the theoretical treatment of flawed structures includes some inherent difficulties, which are the reason for the complexity of analytical treatment of fracture problems and for the development of EFM as a special engineering and scientific discipline.

Disciplines of engineering science should in general stand on the three legs research, teaching and practical application, as shown in Fig. 1. . In today's EFM, the leg of practical application appears to be the weakest of the three: there is a large gap between what can be considered to be the state-of-the-art of the subject and what is actually used in engineering practice. The majority of the mechanical or civil engineers are not familiar with EFM. Even in large engineering companies it sometimes is hard to find engineers who are able to apply EFM correctly. If an EFM-analysis has to be performed - e.g. because of requirements of design standards -, then it is often done without the adequate expertise, thus often leading to questionable or meaningless results. EFM seems to be still an academic rather than a practical engineering subject. However, there is no doubt that there are many problems in engineering practice where EFM would be the adequate means or where some basic knowledge of this subject would be helpful. After all, EFM is the only engineering discipline by which fracture processes can be theoretically treated, predicted or explained.

As shown in Fig. 1, there are interactions between the three legs. Thus, teaching might be one of the reasons for the unsatisfying situation outlined above. Many universities do not offer courses or lectures in EFM. In other cases, the subject is often not taught in a fruitful way. Because of the above mentioned inherent theoretical difficulties of the subject, one of the main problems of teaching EFM is to find the balance between on one handa too superficial presentation, lacking solid physical and mathematical grounds, and on the other becoming lost in boring mathematical derivations, which can overshadow the fascinating physics of fracture processes and cause the initial interest of the students to fade away. Obviously, in teaching EFM, one is confronted with a certain dilemma.

In the present essay, the situation in the field of fracture is analysed and, by reflecting on the role of EFM in practical engineering, ways to improve it are sought, . Therefrom follow the aims of lectures in EFM and the reasons for teaching. As a key to overcome the above mentioned dilemma of teaching, it is recommended to use physical crack and fracture models that are capable to describe the behaviour of cracks in solids correctly, but also simple enough to allow for a simple analytical treatment.

2 Why teach fracture mechanics?

As the amount of knowledge in engineering is growing, the part of it that can be taught at universities is decreasing. In general, the role of universities in engineering is to provide the students with the fundamentals, whereas the specific skills have to be learned "on the job". According to the above description of the situation in practical engineering, only a minority of students need, in their later professional life, to apply EFM directly: for example to perform safe life calculations or failure assessments of structures by means of EFM. So the questions whether and why EFM shall be taught at universities should be answered before the one of what and how to teach. In fact, there are several good reasons, the main ones being the following:

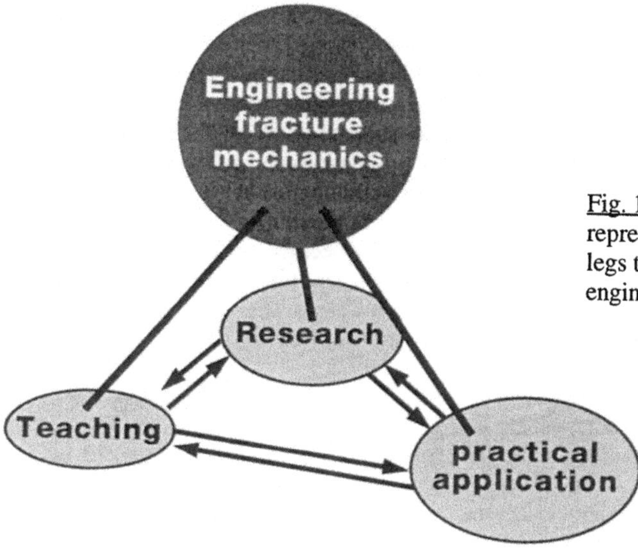

Fig. 1: Schematic representation of the three legs that support an engineering subject

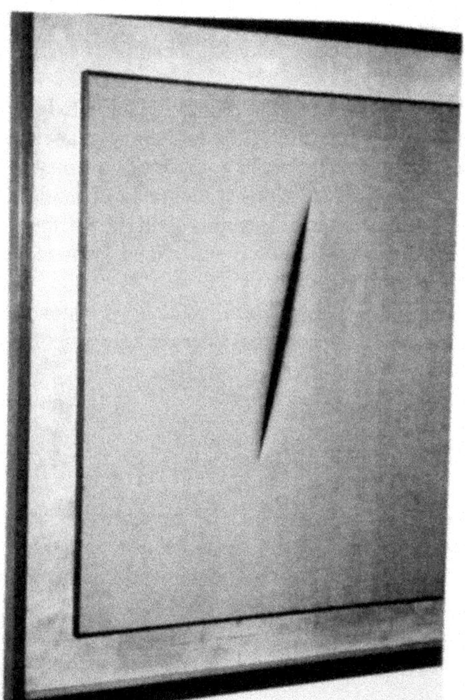

Fig. 2: Example of a crack as a piece of art (L. Fontana; photograph taken at Tate Gallery, Liverpool, UK)

- EFM enables the student to get additional physical insight in the mechanical behaviour of materials and structures. It is capable of explaining why in many cases the simpler procedures of the classical theory of strength of materials work well, despite of their disregarding the effects of defects and stress-raisers. Such questions are left open by the traditional stress analysis and strength of materials theories, leading to a latent feeling of doubt and uncertainty. Thus, EFM enables the students and engineers to feel safer when using engineering stress analysis, and to increase their self-reliance as structural engineers. In this sense, EFM belongs to the fundamental subjects of mechanics of materials and structural engineering. This means that the basic ideas, problems and theories are at least as important as the direct practical application.
- Students (as well as people in general) seem to have a natural interest in the subject. Fracture, cracks, notches, etc. are phenomena which everyone has his own physical every-day experiences with. To some extent, cracks are surrounded by a certain mystery, which is manifested, e.g., by its use in arts (see [1] and Fig. 2 as examples), as a symbol in languages ("a broken heart", etc.), or even as the name of a drug. A theory which is able to deal with cracks and fractures and to explain some of their features and predict their behaviour, is principally interesting and attractive. Keeping this natural interest alive and let it grow during the course, is one of the main challenges of teaching EFM.
- Besides of the knowledge of how to tackle the special class of crack problems, EFM offers some deeper insight in the world of physics and mechanical behaviour of materials, shows ways of how to treat complicated nonlinear problems by simple physical models, answering questions related to physical behaviour of materials in the presence of cracks or defects in general.
- Being a relatively young scientific discipline, the body of knowledge and the number of publications have grown very fast during the last thirty years and a large part of them is not yet well consolidated. The perfect textbook is not yet written. Learning it solely by textbooks or other publications is relatively difficult. Thus the students need some support to find the entrance into this field. Learning it "on the job" may be very time consuming or may lead to a superficial knowledge which often is not sufficient to really solve problems.
- Although slowly, concepts of EFM seem to gain attention in design standards (e.g. [2]). Thus, some basic knowledge of EFM becomes increasingly important when using these standards.

3 How to teach fracture mechanics?

According to the previous chapter, EFM belongs to the fundamental engineering subjects. Correspondingly, it should be taught in the same fashion as a fundamental subject. This means: all the presented theories and the fundamental relations that are necessary therein should be developed comprehensively and - first of all - comprehensibly from the very bottom, which means, from the basic laws of continuum mechanics. In general, dealing with formulas that are not strictly derived in the lectures is often not satisfying.

However, teaching EFM according to these general rules is not an easy task, since there are, on one hand, the inherent difficulties mentioned in the introduction, and, on the other, the poor mathematical tools engineering students are familiar with. One difficulty and probably one source of confusion about EFM is that many approaches are possible. Accordingly, many parameters can be used to characterize the loading state of a crack. It is preferable to concentrate on a few basic parameters and relations than to deal with the whole variety of them in a superficial way. Many of the parameters used in EFM are rather abstract, which means that their physical meaning is not obvious and that they need careful physical justification. Consider, for example, the stress intensity factor: it is not easy to understand why the intensity of a theoretical singular stress-field, which does not occur in reality by far, is able to characterize the loading state of a crack - additional rationales are necessary to show that.

For these reasons, emphasis must be laid first on developing a transparent rational framework that is able to give to the students a general guidance (see next chapter), and second on using simple physical models (e.g., dimensional analysis, weight functions, the Dugdale and Barenblatt crack models, global energy considerations and approximate methods) and their mathematical descriptions. In fact, one of the most amazing and even fascinating features of EFM are its capability to describe complicated physical processes by surprisingly simple models. These models are very powerful not only in reducing mathematical difficulties but also in promoting the physical understanding of the behaviour of cracks subjected to mechanical loading and to get a deeper insight in the subject. In EFM, like in other complex subjects, there is a pronounced synergy acting between physics and mathematics: the one helps to understand the other.

4 The rational framework of fracture mechanics

The rational framework of EFM as outlined above can be represented as a 4-story-building (Fig. 3). The basement is formed by the hypothesis of autonomy of the fracture process zone at the crack tip. Its loading and deformation state is characterized by local parameters like crack tip opening displacement, crack tip opening angle, plastic zone size, process zone size, local energy density, etc. These parameters form the first floor. The second floor is represented by the global parameters like energy release rate, J-Integral or stress intensity factor. Their role is to act as the link between the local parameters on one side and the loading and size parameters of the global mechanical system on the other. These floors are connected to each other and stabilized by the mathematical equations and physical relations between the parameters of the different levels.

As stated in the previous chapter, there is a variety of parameters that can be used to represent the different levels of the framework and correspondingly a large number of relations between them. Such a framework is therefore not unique, which means that several equivalent approaches are possible. In order to avoid confusion, only as few as possible parameters should be introduced. The question of which are the most suitable ones is, to some extent, up to the personal taste. According to the author's experience, the most suitable and convincing parameters to characterize the loading state of a crack are the geometrical ones, i.e. the crack-tip opening displacement CTOD and, when dealing with problems of crack extension, the crack-tip opening angle CTOA. Their capabilities to characterize the local loading state at a

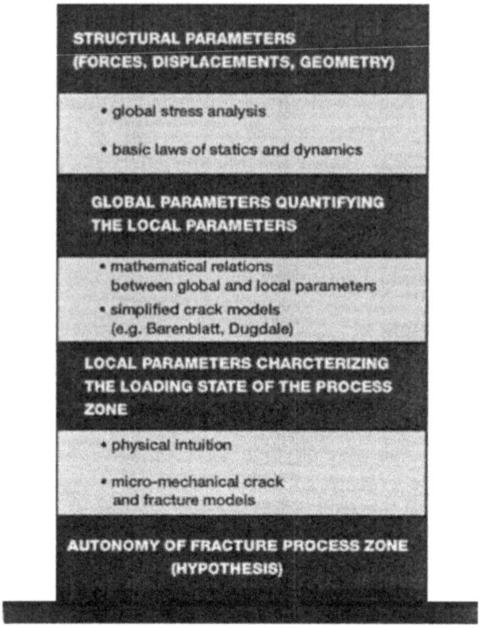

Fig 3: General structure of the theoretical framework of EFM

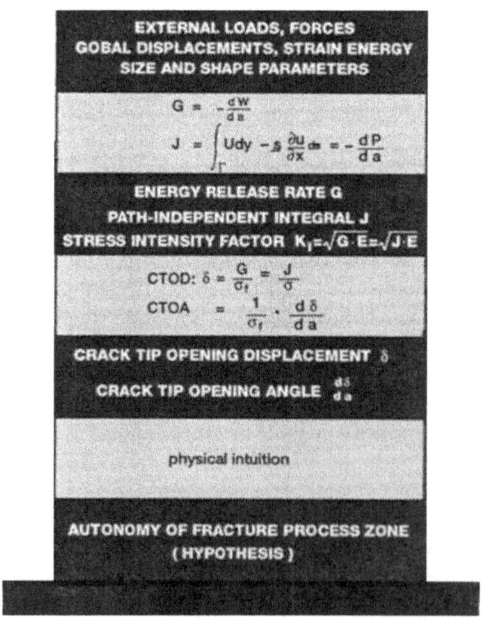

Fig. 4: Example of a theoretical framework based on CTOD and CTOA (see Fig. 3)

crack-tip are quite obvious, and they are applicable in linear-elastic (LEFM) as well as in elastic-plastic fracture mechanics (EPFM). One of their main advantages in teaching is their ability to be easily sketched on the blackboard, thereby linking convincingly the theory of EFM and the "real (physical) world". The corresponding theoretical framework according to Fig. 3 is shown in Fig. 4.

Concerning the required relations between the local parameters and the upper levels of the building according to Fig. 2 and 3, one has to look for appropriate ways to derive them straightforwardly and convincingly on the blackboard, preferably in less than about half an hour. Being displacement-like parameters, CTOD and CTOA can in general be easily related to the external forces by using simple crack models like the Dugdale model and applying well known principles like Castigliano's or Betty's theorem. It is interesting to note that for all the fundamental relations which form the basis of EFM, there are indeed theoretical models which allow for an easy derivation [3].

5 On Practical Application

As mentioned in the introduction, EFM is still not yet well established in practical engineering design of structures. The main reasons for this unsatisfying situation probably are the following:

- It is obviously possible to design proper mechanical structures without knowing anything about EFM. The benefit of EFM in terms of the additional degree of safety seems not to be high enough to justify the additional effort of applying EFM.
- Most of the design standards and material specifications are still not based on concepts of EFM, although EFM is known for more than thirty years. E.g., the toughness requirements of structural metals are still preferably specified in terms of Charpy fracture energy, not in terms of fracture toughness - although it is well known that the latter is physically and theoretically much better founded.
- From the viewpoint of practical application, the concepts of fracture mechanics seem to be not convincing enough. The source of this skepticism, altough sometimes justified, is often a lack of profound understanding of EFM.

One of the aims of teaching EFM should be to improve this situation and to promote its application in structural engineering. To achieve this, the tools for solving practicle problems have to be introduced. Again, preferably the most simple ones should be chosen. These include: Analytical estimation methods of stress-intensity factors, estimation of energy release rates by strength-of-material considerations, methods of numerical calculations of stress intensity factors, energy release rates and J-integrals. Furthermore, it should be shown how the theories of EFM can be simplified to practical concepts. The latter should be simple as well as basically convincing, and shall cover the most important fields of practical applications. These are the prediction of the component life time on the basis of subcritical crack growth due to fatigue, stress corrosion or creep and the failure assessment in the context of a safety analysis.

Concerning failure assessment procedures, simplicity is required. Even today's best known practical concepts like the R6-method [4] or the EPRI engineering approach [5] are probably too sophisticated for practical application. For these reasons, simpler methods like the concept of required toughness [6] have been suggested and should be explained to students.

Concerning the practically important topic of subcritical crack-growth, the established practice mainly relies on empirical laws. However, in a theoretical course on EFM, it is important that these empirical relations are supported by theoretical considerations to reduce the natural skepticism surrounding the empirical relations. By means of simple Dugdale-like models it is possible to calculate the behaviour of cracks under conditions of subcritical crack-growth [3].

6 Conclusions

Engineering fracture mechanics fills the gaps that are left open by the classical theories of strength of materials, gaps which are important to be filled in order to provide a solid ground for working self-reliable and creatively in structural engineering. However, EFM is not an easy subject to be taught and to be used in practice. Care has to be taken to do it rationally and fundamentally as well as comprehensibly and physically transparently. These partly contradictory goals can be achieved by using the simple physical models, which exist for most of the fundamental relations and formulas of EFM. Simple models that can be entirely mathematically described are always better than more sophisticated models that are limited in this respect.

There are three main challenges in teaching EFM: keeping the initial interest of the students in the subject alive, finding appropriately simple physical models and also convincing examples of practical application. These are not easy tasks, since topics like simplicity and practical applicability seem to be somewhat neglected in the research. Fracture mechanics as it is known today appears to be not yet fully consolidated with these respects, so there is still a considerable potential for further research.

7 References

1. Stedelijk Museum Amsterdam (1988) "Lucio Fontana"
2. CEN European committee for standardization (1992), Eurocode 3: Design of steel structures
3. H.J. Schindler, lecture notes "Grundlagen der Bruchmechanik", ETH Zürich (in german)
4. Central Electricity Generation Board (1986), Assessment of the Integrity of Structures Containing Defects, CEGB R/HR6 - Rev. 3
5. EPRI Electric Power Research Institute (1981), An Engineering Approach for Elastic-Plastic Fracture Analysis, EPRI NP-1931
6. Schindler, H.J. (1994), On the required toughness of structural steel, Proc. Int Conf., on Engineering Integrity Assessment, EMAS, Glasgow, 1994, pp. 397-405

Requests and Demands

Teaching and Education in Fracture and Fatigue. Edited by H.P. Rossmanith. Published in 1996 by Taylor & Francis, 2 Park Square, Milton Park, Abingdon, Oxon, OX14 4RN. ISBN 0 419 20700 7.

8 THE TEACHING AND FAILURE ANALYSIS AND ACCIDENT RECONSTRUCTION: AN OVERVIEW

A.A. JOHNSON
University of Louisville, Department of Mechanical Engineering,
Louisville, Kentucky, USA

An extended version of this paper at pages 331–351 includes fuller versions of the case studies.

Abstract
A course on failure analysis and accident reconstruction for graduate engineers has been developed at the University of Louisville. So far it has been taught three times and each time has been well received by students. The core of the course is a series of case histories selected from the instructor's consulting files. Each student submits a formal written report on each case and collaborates with another student in presenting one case orally to the class. Formal instruction is provided on topics such as fatigue, brittle fracture and stress corrosion cracking which frequently crop up in failure analysis. An attorney who has a degree in engineering teaches product liability law, contract law, civil procedures and rules of evidence. The students study videotapes of a major jury trial to familiarize themselves with courtroom procedures.
Keywords: Accident reconstruction, failure analysis, graduate course, product liability

1 Introduction

Several years ago the Chairman of the Department of Mechanical Engineering at the University of Louisville asked the writer to develop a graduate course on failure analysis and accident reconstruction. He agreed to do so because he had had many years of experience in these areas. At the outset, however, it was apparent that this was a project with formidable problems. Failure analysis and accident reconstruction are not well defined techniques such that one can teach a student a series of steps applicable to every case which he or she takes on. Rather, each case is different and requires a different plan of attack. Experience and a capacity for creative thinking are of great importance. Also, failures and accidents frequently lead to litigation and a

Teaching and Education in Fracture and Fatigue. Edited by H.P. Rossmanith. Published in 1996 by Taylor & Francis, 2 Park Square, Milton Park, Abingdon, Oxon, OX14 4RN. ISBN 0 419 20700 7.

failure analysis or accident reconstruction must be conducted from the onset with the possibility in mind that it may have to be defended in court.

2 Philosophy of the course

With these problems in mind the writer moved ahead with the development of the course. He set two broad objectives for it. First, it would give students an understanding of the diversity of failure analyses and accident reconstructions and of the techniques employed in them. Secondly, it would give students some understanding of state and federal law relating to product liability and of the role of an expert witness in litigation.

It was decided at the outset to base the core instruction in failure analysis and accident reconstruction on a series of case histories taken from the writer's consulting files. It was hoped that by using this approach the students would learn to appreciate that there is no one correct way to conduct a failure analysis or accident reconstruction and that two equally competent investigators can, in good faith, reach different conclusions in a given case.

Because of the importance of ferrous metallurgy and mechanical properties in failure analysis, formal reviews of undergraduate work in these areas were scheduled. New material aimed at deepening the student's knowledge in areas such as fatigue, brittle fracture and stress corrosion cracking, which are of particular importance in failure analysis, were added. Course notes summarizing some of these topics were prepared for distribution to the students. Also incorporated in the schedule were formal lectures on topics which are specific to accident reconstruction. These included accident vehicle examination and the interpretation of tire marks.

As he developed the course, the writer was fortunate to have the assistance of Mr. Gerald Stovall, a practicing attorney who holds degrees in both law and engineering. Mr. Stovall developed lectures and course notes dealing with product liability law, contract law, civil procedures and rules of evidence. Also, a complete videotape of a week-long product liability jury trial was obtained so that students could be shown the structure of a trial and the role of expert witnesses in it.

3 The case histories

The writer's consulting practice, from which the case histories have been drawn, is diverse. He has been involved in a wide variety of cases going back many years. His clients for failure analysis and accident reconstruction cases have included:

(i) Industrial corporations experiencing failures of manufacturing equipment.
(ii) Attorneys representing industrial corporations being sued because of death or injury caused by an alleged product failure.
(iii) Attorneys representing plaintiffs who have been injured or bereaved by an alleged product failure.
(iv) Insurance adjusters trying to asses the cause of accidents in which policyholders have been killed or injured.
(v) Attorneys representing insurance companies attempting to subrogate claims

by pinning liability on other insurance companies or industrial corporations.

(vi) Attorneys representing defendants in criminal proceedings following accidents.

(vii) Prosecuting attorneys preparing cases against criminal defendants following accidents.

The case histories chosen for use when the course was first taught were as follows:

Case History #1 - A tractor-trailer combination traveling north in downtown Louisville went out of control and grazed and hit the concrete median barrier. It was bearing fifteen junked and flattened cars. The six wire ropes holding the cars in place all failed. The cars slid over the median barrier into oncoming traffic causing three deaths and two serious injuries.

Case History #2 - The exit channel for the hot gases from the final stage of a gas turbine generator was constructed from 304 stainless steel sheet. An outer sheathing of plain carbon steel was joined to this inner wall by 304 stainless steel bolts. After a few weeks of operation numerous bolt failures occurred.

Case History #3 - a tractor-trailer combination being driven at a moderate speed suffered a spring hanger failure. It rolled on its side and the driver was severely injured. The spring hanger, which was made of malleable cast iron, was found to have been repair welded.

Case History #4 - A Charnley total hip prosthesis made of Vitallium failed after several years of service. It was found to have an extremely large grain size. X-ray diffraction showed that it had a hexagonal close packed crystal structure. Examination of the fracture surface revealed mainly cleavage.

Case History #5 - A Chevrolet Blazer ran out of gas and was being pushed by three youths. Another vehicle crashed into the rear of the Blazer killing one of the youths. Lamp filaments of both vehicles were examined to determine which lamps were lit at the time of the accident.

Case History #6 - A 1984 Nissan Sentra was involved in a violent rear end collision. Its gas tank exploded causing the deaths of both occupants. Examination of the gas tank weld revealed a 16 inch tear through its heat affected zone. The heat affected zone was softer than the weld metal or the base metal.

Case History #7 - A lineman was climbing a southern pine utility pole when it collapsed. At the time he was tightening a steel strand between it and another pole. Examination of the pole revealed that numerous stress concentrations contributed to the failure. These included a woodpecker hole, three bolt holes, decay, checks, etc. The lineman was injured both by the pole and by electrocution and was permanently impaired.

Case History #8 - A young man was killed when a 7200 power line made of ACSR (aluminum conductor, steel reinforced) failed. Examination of the steel reinforcing wire revealed that it had failed by fatigue. The span which failed was unusually long and the terrain was open. It was concluded that the failure was due to aeolian vibration.

Case History #9 - A grinder was being used to convert blocks of artificial rubber to granules. The screens were subjected to repeated impacts by the blocks of rubber as the blocks rotated in the grinder. Screen failures occurred after about 3 months of continuous service.

Case History #10 - The driver of a convertible lost control of the vehicle while driving at high speed on a highway. The vehicle left the highway, hit a grassy bank, turned over, and slid along the shoulder with the driver and passenger still in place. Both were badly injured. Examination of the vehicle revealed a broken tie rod in the front suspension. The vehicle left yaw marks on the surface of the highway after it went out of control.

The files for these cases are all very different. Case histories, #3, #4, #6, #7, and #8 were all major product liability cases. The files are extensive and include technical reports, interrogatories, depositions, and many other documents. The students were given a selection of documents and photographs from each one. Case history #1 is a criminal case which is complex, but the writer wrote a 62 page report on it in which the state's case was assembled. This was chosen to start the course because the existence of the report made it relatively easy for the students to grasp the facts of the case. Case histories #2 and #9 involve equipment failures and there was no litigation. Case history #5 is restricted to lamp filament examinations carried out for the Louisville Division of Police as a supplement to an accident reconstruction which they carried out. Finally, case history #10 involves an accident caused by driver error rather than a product defect. This was included to demonstrate that it is sometimes necessary to tell a client that there is no basis for a lawsuit alleging a product defect.

As the course evolved, the American Society for Metals International published a two volume Handbook of Failure Analysis Case Histories [1]. The Handbook uses a standard format for its case histories and that format was adapted for use in the course. Each student was given a copy of one of the two case histories published by the writer in the Handbook [2] [3] as a model. Case histories were limited to 1000 words each plus figures and photographs. Xerox copies of figures and photographs were allowed.

Students were paired and each pair made an oral presentation on one of the case histories to the rest of the class. Since the class size was set at about 20 there were enough students to present all ten case histories during the semester. A series of 35 mm slides was prepared for each case history and these were made available to the students.

4 The course schedule

In the terminology of American higher education, the course was designed to be 3 credit hours at the 600 level. This means that the class is supposed to meet for 3 50 minute periods each week and that all students taking it should have at least an undergraduate engineering degree. It was decided to teach the course in the early evening so that students employed locally full-time could participate. The class therefore met once a week from 5:30 until 8:00 p.m. for fifteen weeks.

The week by week schedule the first time the course was taught was as follows:

Week #1:	5:30 - 6:30	Lecture - Ferrous metallurgy I
	6:30 - 7:00	Introduction of case history #1
	7:00 - 7:30	Introduction of trial videotape
	7:30 - 8:00	Trial videotape
Week #2	5:30 - 6:45	Lecture - Ferrous metallurgy II
	6:45 - 7:00	Introduction of case history #2
	7:00 - 7:30	Student presentation of case history #1
	7:30 - 8:00	Trial videotape
Week #3	5:30 - 6:45	Lecture - Dislocations and Mechanical Properties I
	6:45 - 7:00	Introduction of case history #3
	7:00 - 7:35	Student presentation of case history #2
	7:30 - 8:00	Trial videotape
Week #4	5:30 - 6:15	Lecture - Dislocations and Mechanical Properties II
	6:15 - 6:30	Introduction of case history #4
	6:30 - 7:30	Lecture -Rules of civil procedure I
	7:30 - 8:00	Trial videotape
Week #5	5:30 - 6:00	Lecture - Recovery, recrystallization and grain size effects
	6:00 - 6:30	Student presentation of case history #3
	6:30 - 7:30	Lecture -Rules of civil procedure II
	7:30 - 8:00	Trial videotape
Week #6	5:30 - 6:00	Lecture - Brittle fracture
	6:00 - 6:30	Student presentation of case history #4
	6:30 - 7:30	Lecture -Rules of evidence I
	7:30 - 8:00	Trial videotape
Week #7	5:30 - 6:15	Review for first test
	6:15 - 6:30	Introduction of case history #5
	6:30 - 7:30	Lecture -Rules of evidence II
	7:30 - 8:00	Trial videotape
Week #8	5:30 - 6:30	First test
	6:30 - 7:00	Introduction of case histories #6 and #7
	7:00 - 7:30	Student presentation of case history #5
	7:30 - 8:00	Trial videotape

Week #9	5:30 - 6:15	Lecture - Fatigue I
	6:15 - 7:00	Lecture - Product liability
	7:00 - 7:30	Student presentation of case history #6
	7:30 - 8:00	Trial videotape
Week #10	5:30 - 6:45	Lecture - Fatigue II
	6:45 - 7:00	Introduction of case history #8
	7:00 - 7:30	Student presentation of case history #7
	7:30 - 8:00	Trial videotape
Week #11	5:30 - 6:15	Lecture - Fractography
	6:15 - 6:45	Lecture - Environmental effects on mechanical properties
	6:45 - 7:00	Introduction of case history #9
	7:00 - 7:30	Student presentation of case history #8
	7:30 - 8:00	Trial videotape
Week #12	5:30 - 6:45	Lecture - Metallurgy of welding
	6:45 - 7:00	Introduction of case history #10
	7:00 - 7:30	Student presentation of case history #9
	7:30 - 8:00	Trial videotape
Week #13	5:30 - 7:00	Lecture -Vehicular accident reconstruction I
	7:00 - 7:30	Student presentation of case history #10
	7:30 - 8:00	Trial videotape
Week #14	5:30 - 6:30	Lecture - Vehicular accident reconstruction II
	6:30 - 7:30	Lecture - Creep
	7:30 - 8:00	Trial videotape
Week #15	5:30 - 8:00	Review for second test

5 The grading system

The American system usually requires that in a graduate course each student be given a grade of A, B, C, or F. This was the case for this course. Since a graduate student must maintain a B average, a C, as well as an F, is a failing grade. Provided a student does a creditable job, therefore, he or she expects to receive a B. Good or excellent performance is rewarded with an A.

For an instructor to decide which students get A's and which get B's he needs some quantitative measure of each student's performance. In the present case each student was assigned a percentage score based on the 10 case histories submitted, the oral presentation, and performance in the two tests. The maximum scores possible in each area were as follows:

Case histories	50%
Oral presentation	10%
First test	20%
Second test	20%

6 Student evaluation of the course

Being very democratic, American Universities usually encourage students to evaluate their instructors. At the University of Louisville standard questionnaires are filled out by each student at the end of each course. This course has now been taught three times, once each in 1992, 1993, and 1994. There are, therefore, three sets of student evaluations available. These are shown in Table I which shows average scores in each of seven areas.

Table 1. The results of student evaluations of the course

	1992	1993	1994
(i) Visual presentation in class	4.55	4.13	4.13
(ii) Answers to questions asked in class	4.52	4.13	4.00
(iii) Effectiveness of classroom activity	4.15	3.81	3.88
(iv) Effectiveness of homework problems	4.33	4.00	4.00
(v) Effectiveness of assigned reading materials	4.30	3.67	4.00
(vi) Instructor's ability to stimulate interest	4.70	3.94	4.13
(vii) Instructor's overall effectiveness	4.57	4.07	4.38

They are on a scale in which 1=poor, 2=fair, 3=average, 4=good, and 5=excellent. These scores compare very favorably with scores achieved by instructors in other engineering courses at the University of Louisville.

7 Resources for the course

Most of the case history files contain print-outs from the searches of computer data bases made as the cases were pursued. Students therefore become very familiar with such searches even though they are not expected to carry them out themselves. The files also contain technical papers and reports obtained by copying material identified in the computer searches. Some of this material will have been found on the shelves of campus libraries. Other parts of it will have been obtained from other libraries through the efforts of library staff. Thus, the students are not expected to do a great deal of independent library work.

Reference works to which the students are introduced include the new two volume ASM Handbook of Failure Analysis Case Histories referred to above [1] and three volumes of the 17 volume ASM Metals Handbook [4-6]. In the field of accident reconstruction the students learn to use the Traffic Accident Investigation Manual by J. Stannard Baker [7].

8 References

1. American Society for Metals International (1992 and 1993), Handbook of Failure Analysis Case Histories, Vols. I and II, Materials Park, Ohio, USA.
2. Johnson, A.A. and Lattimer, C.A. (1992), Examination of broken lamp filaments following a rear end vehicular collision, Ibid, Vol. I, pp. 79-82.
3. Johnson, A.A. and von Fraunhofer, J.A., (1993), Mechanical Failure of a repair welded ferritic malleable cast iron spring hanger, Ibid, Vol. II, pp.76-78.
4. American Society for Metals International (1978), Metals Handbook, Ninth Edition, Vol. I, Properties and Selection: Irons and Steels, Materials Park, Ohio, USA.
5. American Society for Metals International (1986), Metals Handbook, Ninth Edition, Vol. XI, Failure Analysis and Prevention, Materials Park, Ohio, USA.
6. American Society for Metals International (1987), Metals Handbook, Ninth Edition, Vol. XI, Fractography, Materials Park, Ohio, USA.
7. Baker, J.S. (1976), Traffic Accident Investigations Manual, The Traffic Institute, Northwestern University, Evanston, Illinois, USA.

Acknowledgments

The writer is indebted to his secretary, Patricia Lumley, who typed the manuscript, and his wife, Barbara Johnson, who proofread it.

9 CASE STUDIES FOR TEACHING FAILURE ANALYSIS AND MATERIALS ENGINEERING

L.J. POWER
School of Mechanical and Offshore Engineering, The Robert Gordon University, Aberdeen, UK

Abstract
This paper describes the development, over a number of years, of the use of case studies to support the teaching of materials engineering and mechanical engineering at the Robert Gordon University, Aberdeen. These developments have been used successfully in technician engineer and professional engineer courses on full and part-time study modes. The case studies have been particularly successful in integrating a wide range of disparate parts of the syllabus, and providing students with deep learning. Motivation has been very good.
Key Words: Case studies, problem based learning, failure analysis.

1 Introduction

The Robert Gordon University began as a technical school in the latter part of the nineteenth century, although it can trace links back to the middle of the eighteenth century. Engineering courses were taught from the beginning, and, by the 1970s, the institution was awarding engineering degrees and postgraduate diplomas under the auspices of the UK Council for National Academic Awards. The institution was awarded University status in 1992, and continues to teach vocationally relevant courses. Currently there are some six thousand students and nearly five hundred academic staff in the University on a very wide range of courses.

The School of Mechanical and Offshore Engineering (SMOE) has developed to a state where it has about 450 students and over 27 academic staff. The School runs a range of courses. At undergraduate level it offers a technician Higher National Diploma (HND) and Bachelor of Science and (for professional engineers) three Bachelor of Engineering – Honours (BEng) degrees which are all professionally accredited with the relevant Engineering Institutions. There is also a more specialised

Teaching and Education in Fracture and Fatigue. Edited by H.P. Rossmanith. Published in 1996 by Taylor & Francis, 2 Park Square, Milton Park, Abingdon, Oxon, OX14 4RN. ISBN 0 419 20700 7.

Master of Science in Offshore Engineering course with 35 full-time and 25 part-time students. Ten research students are registered for MPhil/PhD degrees.

SMOE has tried to design the undergraduate courses so as to give a sound, general mechanical engineering core in order that graduates have flexibility in their professional careers. (If, for example, prospects in the offshore industry were to decline.)

2 Materials engineering and failure analysis

Tables 1 and 2 show the general structure of the BEng and MSc courses. The undergraduate course is a traditional Scottish four year course to honours. The first two years are common, and there is overlap in years three and four.

Table 1. General structure of BEng (Honours) degree courses

Year	Course	
	Engineering Technology	Mechanical and Offshore Engineering
4	Engineering Analysis Control and Instrumentation Plant and Project Management Failure Analysis Plant Performance Project	Engineering Analysis Control and Instrumentation Project and Operations Management Offshore Engineering Design and Selection of Plant Systems Project
3	Mathematics Engineering Analysis Control and Instrumentation Design Plant and Project Management Industrial Plant Engineering Applications	Mathematics Engineering Analysis Control and Instrumentation Design Project and Operations Management Offshore Engineering Design and Selection of Plant Systems Engineering Applications
2	Common Year 2 Mathematics Materials Manufacture and Design Business Administration Studies Engineering Applications	 Computer Applications Electrical Engineering Mechanical Engineering
1	Common Year 1 Mathematics Engineering Science Business Administration Studies Engineering Drawing and Design	 Computer Applications Mechanical Engineering Engineering Applications

Materials engineering is taught in the first year as part of Engineering Science, where the syllabus has a materials science bias. Some additional materials topics are taught in the second year as part of Materials, Manufacture and Design.

Failure Analysis features as a distinct subject in the fourth year; and is interpreted in a very wide sense: covering electrical systems breakdown and condition monitoring, as well as corrosion engineering and mechanical failure. Fracture mechanics forms part of Engineering Analysis in the third year, and of Failure Analysis in the fourth year.

Table 2. General structure of MSc course (lectures only)

Term		
1	Petroleum Technology Drilling Technology Materials Engineering	Subsea Technology Project and Operations Management
2	<u>Option A</u> Petroleum Technology Drilling Technology	 Subsea Technology Project and Operations Management
	<u>Option B</u> Failure Analysis Corrosion Engineering	 Welding and Non-Destructive Testing Project and Operations Management
3	Secondary Recovery Systems Drilling Technology Information Technology (Students choose three)	Underwater Support Systems Project and Operations Management Advanced Materials

The MSc course has a Materials and Corrosion Engineering option which also contains failure analysis - in the sense of fracture and failure. Other failure modes, such as corrosion, are addressed in the other subjects of this option.

The (Linked) HND/BSc in Mechanical Engineering contains a limited amount of materials and failure study in the first two years. Unfortunately, these years of the course are presented under the influence of a national, competence-based, system, and the approach described in this paper does not fit well with such a system. More will be said of this later.

Finally, there is an interfaculty course in Technology and Business. As the title suggests, the students on this course are less concerned with the details of mechanical engineering, but they do require to have a general grasp of engineering knowledge and skills, as well as economics and management.

This paper is concerned with case studies which are used, in different forms, across the full range of courses, with the exception (currently) of the HND and BSc in mechanical engineering.

3 Teaching approach

The formal syllabus content of many courses in materials engineering often contains the 'usual' structure, where a review of the relevant engineering properties of materials is developed from an atomic basis. Thus, the course would normally begin with a description of atomic structure, move through an examination of crystalline and amorphous solids and proceed to detail the relevant (for mechanical engineers) physical, chemical and mechanical properties. This approach has many advantages and it is very important that students have an understanding of the links between the microscopic, macroscopic and bulk features of materials. However, there is a risk that students may fail to perceive the links if they do not accept the relevance of this approach.

The staff of the School (including the author) have attempted to remedy this by adopting a reverse approach. Applications of materials are considered first, and the relevant properties identified, examined and developed (in terms of the atomic, microstructural and features) as necessary. Further, wherever possible, the applications, which are used, are drawn from documented failures of materials, components and structures. Consequently, the applications carry a sense of reality which is transmitted to the students, and prepares them for dealing with the problems they are likely to encounter in their work.

It has to be admitted that our approach was developed long before we were aware that other workers had formalised the educational methodologies and defined the terms. Our approach evolved partly by instinct, partly because the results were good and partly because the approach seemed to be enjoyable for staff and for students.

4 Problem based learning

This methodology, which would seem to have evolved in medical education, has been described before [1] [2]. The following list is an abridged version of the more general aims of such an approach:

- Develop skills of modelling, analysing and proposing.
- Develop skills of criticising proposed solutions.
- Develop independent study skills.
- Develop oral and written presentation skills.

Experience in other institutions has suggested that a gradual introduction of this approach, alongside standard lecture courses, was more likely to be successful. With careful planning, objectives related to knowledge, attitude and skills could be addressed. Student motivation was found to be good.

Other courses [3] have adopted this approach and it was found to promote active student centred learning, and was challenging and rewarding for staff and students. However, to be successful, group skills training was required.

Similar approaches [4], adopted to foster deep learning, have also been useful as a response to reduced staffing levels and low progression rates.

Interestingly, our approach has been gradual, too, with each case study being carefully structured and planned.

5 Deep learning and competence-based syllabuses

5.1 Deep, elaborate or contextual learning?

The term deep processing is used to indicate that students understand the meaning of their learning [5], and was associated with high scores on tests of their knowledge. However, reservations have been expressed about its real effectiveness. Elaboration of knowledge has, therefore, been suggested as a more successful strategy. These issues have been reviewed more extensively elsewhere [6], and the process of contextual learning has been identified. This is based on:

1. A concrete context, based on first hand experiences.
2. Related theoretical information, e.g. from notes and courses.
3. An opportunity to relate the abstract information to their experiences.

The differences between these forms of learning do not seem to be totally clear and they could, at least, be seen to be overlapping.

5.2 Competence-based syllabuses

As part of a major national programme to integrate academic and vocational qualifications - at all levels from craft to professional - syllabuses are being described in the form of learning outcomes and performance criteria. The methods of assessment are then specifically aimed at determining whether a student has met the relevant performance criteria.

In a university course there is a consequent risk that the experience of students becomes one of simple, mechanistic, instruction and assessment; the educational process (such as it is) is assessment driven. It can be seen that this poses a serious risk to deep learning. The approach can also be seen to be incompatible with programmes needed to identify such 'soft-skill' competencies such as initiative. [7]

The trend is worrying. Competence-based syllabuses fail to address the issue of: "What do I do if I don't know the answer to a problem I've never seen before?" Thus, while it is clear that engineers need to be 'competent', the system being imposed does not actually have anything to do with real competence in the workplace.

6 Teaching from failures

The staff of the School have a wide experience of failure analysis through consultancy work for local (and national) industry. This work is, of course, supplemented for teaching as necessary by information from the literature. The use of examples drawn from consultancy allows the wider engineering implications of proper design to be identified at the start of the course. Such implications would include economic and safety aspects, organisational and communications features, as well as more specific mechanical engineering matters such as design, thermofluids, strength of materials and materials engineering.

This approach also allows the use of role playing, where students can adopt the guise of failure analysis consultants. Information can be supplied to them in a limited way, through a formal request system, so that they learn to ask directed questions and sieve the data to eliminate contradictions. Students react well to this approach, and can work co-operatively when in small groups. They often integrate information from other subjects with little, or no, direct encouragement from staff.

The use of case studies for problem based learning was first tried in SMOE on postgraduate students in 1983. For such students it was possible to assume that they had some skills (perhaps limited) in group working. Success resulted, and encouraged staff to develop the approach. We have, deliberately, been fairly cautious. The approach was tried on undergraduate students on the School's engineering courses.

More recently case studies were used with students on the Technology and Business course: the 'engineering' content was diluted by staff - but was replaced by the students during the exercise. The students decided for themselves that certain technical information, omitted by staff, was relevant and researched the matter for themselves.

Earlier this year, one of the studies was used with postgraduate students on an Integrated Graduate Development Scheme. These schemes are modular Masters courses for professional engineers already in employment

Without exception, student reaction has been very good, and all the benefits noted by other workers have appeared. We are also detemined to ensure that there is as much emphasis on 'soft' skills, such as initiative and team working, as there is on the (perhaps) more detailed engineering knowledge.

7 Case studies

To date, case studies have covered such failure modes as corrosion, brittle fracture and fatigue. Time and space have limited a more detailed description of these to two examples.

7.1 Corrosion failure

In order to increase the proportion of crude oil which can be recovered from a reservoir, the technique of seawater injection is used. On offshore platforms, seawater is drawn from the ocean, filtered and deaerated on the platform, and then pumped (injected) underground into the bottom of the oil-bearing formation. This requires the appropriate handling systems - for large volumes of seawater - subsea, topsides and in the wells.

In this case, part of the system within the platform was constructed of grade 316L stainless steel, with some sections from higher carbon grades. After final assembly on the platform, the system was pressure tested with seawater and left, flooded but idle, for some five months. Inevitably, a combination of weld decay and pitting corrosion set in and the system was found to be leaking. The economic consequences were considerable: some £2m for replacement of the system, and lost production costs which would be of the order of £1m per day.

In technical terms this is a relatively straightforward problem to solve, if all the relevant data is available, but the case study is run so as to force the students to determine the information they need to know. Normally, a scenario is set where a group of four or five students act as specialist failure investigation consultants and are required to produce a report for the board of directors. The sequence is usually:

1. Initial briefing of students.
2. Meeting with company representative (a member of staff in an aggressive mode).
3. Investigation by students. Further company meetings are available by appointment, and, if specifically requested, technical details are supplied by a second member of staff - sometimes a technician. Students are at liberty to undertake whatever additional work they think fit, but, since most of the technical work has already been done by staff, the results of these are supplied if this saves time.
4. Interim oral report by students. (Where a group has not identified the essence of the problem, then some guidance would be given at this stage by staff.)
5. Formal written report.

The process is usually timed to take place for two or three hours per week for three or four weeks, but can be compressed into a day. Students are assessed on their reports and their enthusiasm and involvement. They also have an opportunity to assess each other.

7.2 Brittle fracture

This case concerns the sudden failure of a drill-pipe elevator during a critical lifting operation prior to a storm in the North Sea. The consequential damage was extensive and, like the first example, the lost production and recovery costs were large. Fortunately there was minimal injury to people. Matters were made worse by a sequence of similar failures.

Technically, this was a simple matter of poor heat treatment of the steel elevator; and there is a temptation to make a limited recommendation and finish the study. In practice, the problem was set about by other factors such as commercial pressures, design restrictions and legal implications.

We run this study with the same sequence and procedure as the corrosion problem.

8 The future

There are still some important extensions of these studies which can be made. Principally the fracture problems need to have more quantitative aspects of fracture mechanics built into the assessment. This should be relatively easy as the toughness is very poor and the defect size very small. There is some concern that an over-emphasis on calculation may detract from the actual processes which were used when the failure took place. These processes were related more to negotiation than to calculation.

Furthermore, discussion with industrial practitioners would suggest that the process of investigation is just as important as the details of each individual step.

9 Conclusion

Perhaps the discussion of failures, and the associated damage to property and people, appeals to the worst instincts of students: the idea that, as a result of the failure under consideration, life may have been at risk, or that large sums of money were lost, seems to attract a high level of interest. This is not really surprising.

More probably, students respond well to the responsibility involved in investigating real failures and learn well to analyse and synthesise the problem. They seem to enjoy following in the footsteps of a real investigation, whatever the theoretical educational merits seem to be.

Acknowledgements

I am grateful to the Robert Gordon University, the Carnegie Fund, my wife, and to my colleague NS Edward for their support in the preparation of this paper.

References

1 Cawley, P. (1989) The Introduction of a Problem-based Option into a Conventional Engineering Degree Course. *Studies in Higher Education*, Vol. 14, No. 1. pp 83-95.

2 Cawley, P. (1991) A Problem-based Module in Mechanical Engineering, in *The Challenge of Problem Based Learning*, (ed. D Boud and G Feletti), Kogan Page, London, pp. 177-185.

3 Matthew, R.G.S. and Hughes, D.C. (1991) Problem based learning – a case study in civil engineering, in *Innovative Teaching in Engineering*, (ed. R.A. Smith), Ellis Horwood Ltd, London, pp. 330 - 335.

4 Rogers, G.T. (1991) The student in control – an active approach to learning, in *Innovative Teaching in Engineering*, (ed. R.A. Smith), Ellis Horwood Ltd, London, pp. 348 - 353.

5 Marton, F. and Säljö, R. (1976) On qualitative differences in learning I - outcome and process, *British Journal of Educational Psychology*, Vol. 46, pp 4-11.

6 Coles, C. (1991) Is problem-based learning the only way? In *The Challenge of Problem Based Learning*, (ed. D Boud and G Feletti), Kogan Page, London, pp. 295-305.

7 Raven, R. (1988) The assessment of competencies. *In New Developments in Educational Assessment*, (British Journal of Educational Psychology Monograph Series; No. 3. ed. H.D.Black and W.B.Dockrell), Scottish Academic Press, Edinburgh, pp. 98-126.

10 DIFFERENCES BETWEEN SCIENTIFIC EDUCATION AND FURTHER RE-EDUCATIONAL INFORMATION TRANSFER TO PRACTITIONERS

C.O. BAUER
Wuppertal, Germany

Abstract
Educational tasks of universities differ from the needs industry is requesting in terms of re-education of practising engineers. These differences derive from the different age, experiences and the daily work of practitioners. This results in different independent training centers set up by the various branches of industry for re-educational training and lack of sufficient acceptable offers from the universities. The method of presentation as well as appropriate tools should be modified. The special consequences thereof are demonstrated.
Keywords: University education, practical needs of engineers in practice, methods for adult training, rules for presentation.

1 Scope and goal of university education

The task of the universities is to educate *untrained, inexperienced young people*. They lack of both, theoretical knowledge which has to be transferred on them and experience of practical work with its various influences. On the other hand they are very eager to learn and usually they are of "un-worn-out intelligence". Young people are ready to accept new facts and figures on an astonishingly large scale and to a surprising depth of scientific foundation. Because they normally do not have any or only little experience in practical work, they rarely ask for practical evaluation and transformation and their lack of knowledge does not allow them to compare with their own engineering work experience what is presented theoretically to them.

2 The sceptical experienced engineer

Transformation of knowledge to engineers with substantial experience in practical

Teaching and Education in Fracture and Fatigue. Edited by H.P. Rossmanith. Published in 1996
by Taylor & Francis, 2 Park Square, Milton Park, Abingdon, Oxon, OX14 4RN. ISBN 0 419 20700 7.

evaluation is of a quite different structure. Experience in their day-to-day work - be it good or bad, successful or with failures or even damages - has created a sound scepticism towards all theories and abstract models which immediately will mentally be checked directly against their own practical experience and their own necessities in the various sections of industrial life. Experienced engineers are normally well trained, and - as the expression states - experienced, they have confronted with a number of theoretical hypothesis and many of these theoretical assumptions have been abandoned when tested by practical necessity.

Many theoretical models with all their various possibilities in a broad system are accepted but above all workable means and tools for better solutions of actual problems. The number of new facts and figures duly remembered for further evaluation and transformation in day-to-day work is limited and reduced against student's intellectual potential. What is expected and needed is always the practical evaluation of the results and the proof of new scientific findings under the hard situations of working conditions. Here the dominant questions are:

- What does it help me in my daily work?
- What is its value for my work?
- How can I transform these findings and results to match my practical obligations, and
- How can they be utilised to generate better products, processes and safer procedures?

3 The need for special tools in adult training

These conditions need other approaches, methods and tools than the education of inexperienced, untrained students. They do not need long theories and theoretical deductions. What is wanted and expected is less theory, less theoretical deductions, but more specific examples to show the practicability of these findings, methods with applications as close as possible to their actual field of working.

They are not interested at all in the theoretical inclusions or contradictions to other theories or the including of abstract models in high flying engineering science. Mathematics and other theoretic tools will be accepted only when they are necessary for a better and quicker solution of practical problems. This should be proved before presentation and its value ought to be shown in practice immediately together with the explanation.

4 Consequences for adult training

These different expectations will lead to different solutions. Courses for adult engineers at universities are rare. They will rarely be accepted and often are far from remarkable positive results as long as they are organised and carried out by those lectors, professors, who normally train inexperienced young students.

This is the only reason why all over Europe various branches of industries have formed many independent training centres which concentrate on training practising engineers in new techniques where predominantly experts from the industry transfer their experience to their colleagues. Several statistics show that more than 60% - sometimes even up to 80% - of the speakers in such institutions are recruited from companies working in that field, a fact which reduces the contribution of professors or assistants of universities to less than 10%. A recent statement of the Association of German Engineers (VDI) on "necessary engineering qualification for the future" states - in simple words -, that for the requalification or further qualification of engineers in new procedures or processes for various branches of technology, no acceptable offer has been made by the universities.

When training experienced engineers a very remarkable necessity is the concentration on a limited time of such courses. If it is not a completely new process/procedure which needs to be understood and to be practised through training, for such a course span of a five days is the most one may consider. Experiences of the last years have shown that even with very interesting topics three and more-day courses are rarely accepted by employers. Intensive courses of one or two days form the bulk of activities even if the trained engineer recognises that such a training must necessarily be incomplete and the participants need a repetition of some kind as a later second and third step.

The majority of training on CAD, CAO, CAE, CIM, etc. and other new methods of management and engineering are taught this way, leaving universities with their summer courses and two or more-week courses in the third row - if at all.

University professors who are able and successful in transferring their knowledge to adult engineers form an exception. Mostly engaged in training and education of young students they often miss intensive industrial experience so it may be difficult for them to test their ideas and present practical experience and workable solutions. Experience collected at German Technical Universities over a number of years shows that only 10-15% of the professors and assistants accepted and reinvited to teach such adult engineering training courses were interested in trying these tests. Language, use of expressions, construction of sentences and other linguistic details which are common and necessary in scientific work too often constitute barriers between scientific knowledge and success in the training of adult engineers.

5 Organisational needs and tools

In practice adult engineers are active and are used to continuous personal performance. Practising engineers are not trained any more to sit silently for hours listening to theoretical and abstract presentations of theories and their mathematical foundation but are rather exposed to all kinds of modern presentation, public relation and advertisement. This demands meaningful presentation of figures, tables and examples using colours, 3D models and other modern tools from computer technology.

6 Conclusions

The presentation of new scientific and engineering material and knowledge and their results should be guided by the principles, methods and means of modern advertisement and public relation. The concentration of a few decisive - not too many - aspects is necessary together with a verbal explanation of usefulness and effectiveness accompanied by specific figures but not very academic general explanations.

Any method of presentation has to be concentrated on the most interesting aspects, leaving all unnecessary details to a sideline or a footnote. Concentration on the most important factors is necessary.

Long tables of data and figures difficult to grasp and understand do not assist a positive method of transfer of knowledge. The teaching material must be reduced to the essential assisted by a few easily understandable consequences which should be presented in combination with practical findings taken from industrial life.

Rules for presentations of slides and overhead sheets have been previously formulated in national and international standards of DIN and ISO (DIN 108 Teil 1 und 2) evaluating the results from cognitive research and to achieve the best possible attentiveness. They constitute the minimum base for all kinds of visual presentation together with the extraction of the comprehensive formulation of findings for best memorisation. A combination of different ways how to present the material combines reading, listening, demonstrating practical tests with samples and the oral presentation of the findings.

The rate of memorisation with regard to these methods varies between 10% and 90% with the upper limit to be reached if the right tools are used in the most suitable order.

Group work and group discussions therefore constitute a necessary element in the training of engineers.

This led to the foundation of an academic institution at the University of Bochum dedicated to the investigation of the various ways, means and procedures to best knowledge transfer.

However, the results worked out at this pedagogical institute are little known and even less accepted or used in academic life. In general, many university professors do not have sufficient knowledge in pedagogy and therefore cannot appreciate the benefits of research in pedagogy and do not put to work these findings. This happens to the disadvantage of their lectures and their prestige among the students.

11 EDUCATIONAL PROGRAM IN FRACTURE MECHANICS OF COMPOSITES FOR DESIGNERS

N.A. MACHUTOV
Institute of Machinery of the Russian Academy of Sciences, Moscow
I.I. KOKSHAROV
Computer Center of the Russian Academy of Sciences, Krasnoyarsk, Russia

Abstract
The paper outlines in brief key aspects issues related to the teaching of fracture and fatigue of composites are briefly outlined in the paper. Much attention is given to: who are the students, objectives, strategy and program of the course and the order of principal steps in analysis of failure problems.

Keywords: composites, driving force, fracture mechanisms, structural integrity.

1 Introduction

A special one-year course 'Fracture mechanics of composites' for senior students of the Siberian Aerospace Academy studying aerospace vehicles' design is based on the main ideas of 'Mechanics of deforming and fracture' course delivered by professor N.A.Machutov at the Moscow Aviation Technological Institute now for more than 20 years.

Textbooks [1] - [18] and complete scientific scientific surveys [19], [20] present a variety of descriptions for fracture mechanics methods. Practical engineering work implies complete understanding of failure cases [21] - [23] and use of reference books [24] - [26]. Fracture mechanics of composites is forming [27] - [31] and invites further development.

Teaching and Education in Fracture and Fatigue. Edited by H.P. Rossmanith. Published in 1996 by Taylor & Francis, 2 Park Square, Milton Park, Abingdon, Oxon, OX14 4RN. ISBN 0 419 20700 7.

2 Objectives of the course

Fracture mechanics is considered as **mechanics of fracture mechanisms** and not as a theory of macrocracks only. Principal objectives of the course are to teach the students to understand features and mechanisms of deforming and fracture in composites; to give practical skills for rational use of the carrying ability of the materials. Therefore, the thesis of R.W. Hamming '**Purpose of computing is insight, not numbers**' [32] is highly important for the approach.

3 Strategy of teaching

Since the students are trained to be practical engineers the course omits the theoretical aspects of fracture mechanics (for example, analytical methods of the theory of elasticity for crack problems) to pay more attention to **conclusions useful for practical engineering work**. Perceiving information a designer in his way of thinking is apt to prefer **drawings and graphs** to formulas, text and long explanations. The lecturer is striving to make this.

In addition, such obligatory principal notions as fracture mechanisms, driving force, critical state equation are necessarily involved in teaching each section of fracture mechanics. The well-structured presentation of every section (Fig.1) is efficient for examination of arising problems and is useful for future work of our students.

4 Program of the course

The course consists of two parts: fracture mechanics of isotropic materials (steels, aluminum alloys) and fracture mechanics of composites. In the first part the principal topics are: 1) brittle fracture; 2) mixed mode fracture; 3) elasto-plastic fracture; 4) fatigue. The key notions are presented in the following order (Fig.1): 1) macroscopic **fracture mechanism**; 2) definition of **driving force** appropriate to the mechanism; 3) **engineering approaches** for practical use of fracture criteria.

Thus, in the classic problem of a crack under tension the breaking takes place when the stress intensity factor (**SIF**) reaches its critical value. The factor determines stress state in the crack tip vicinity. Diagrams 'critical stress - defect size' make possible to solve several kinds of engineering problems.

For the mixed mode loading there are two principal macroscopic fracture mechanisms initiating new cracks in two different directions: breaking (cleavage) and shear. It is necessary to distinguish **the loading mode** (mode I, II, III) and **fracture mechanisms** (breaking (A), shear (B),). Direction of propagation and the moment of fracture initiation depend on stress distribution in the crack tip (radial direction). The distribution can be expressed by equivalent stress intensity factors $\mathbf{K_A}$ and $\mathbf{K_B}$ [33]. The engineering approach considers several possible fracture mechanisms.

Shear is the basic macromechanism of plastic deformation and fracture. Forms of the plastic zones may be sufficiently different. Strain energy release rate or **J**-integral characterize the stress-strain state. The engineering approach considers two cases: full-scale plasticity and breaking from the crack tip.

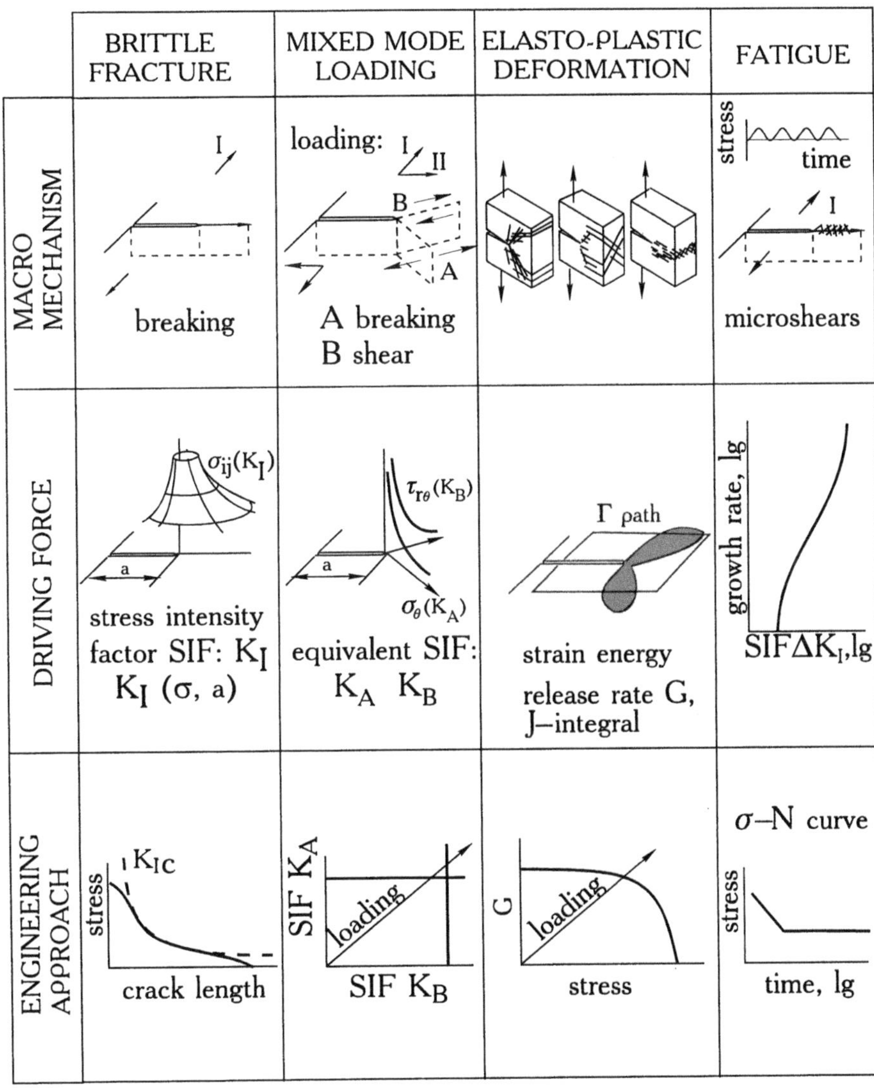

Fig. 1 Fracture mechanics models for isotropic materials

Fatigue crack growth rate depends on the range of the (equivalent) stress intensity factor, the propagation has different stages and the diagram has different characteristic parts. The engineering approach is to use **σ-N** curves.

The next stage is to consider the influence of different factors on static and cyclic crack resistance characteristics (Fig. 2). Varying thickness or sizes of specimens, temperature, loading rate we show qualitative modifications (different mechanisms) and quantitative changes of crack resistance characteristics.

The second part of the course deals with composite materials. Multiplicity and complexity of fracture of these materials greatly complicate application of the fracture mechanics methods. Nevertheless, some useful generalizations can be made (table of Fig. 3).

The order of sections is the following: description of **macro mechanism**, definition of **damage parameters**, **driving force** and analysis of **engineering approaches**.

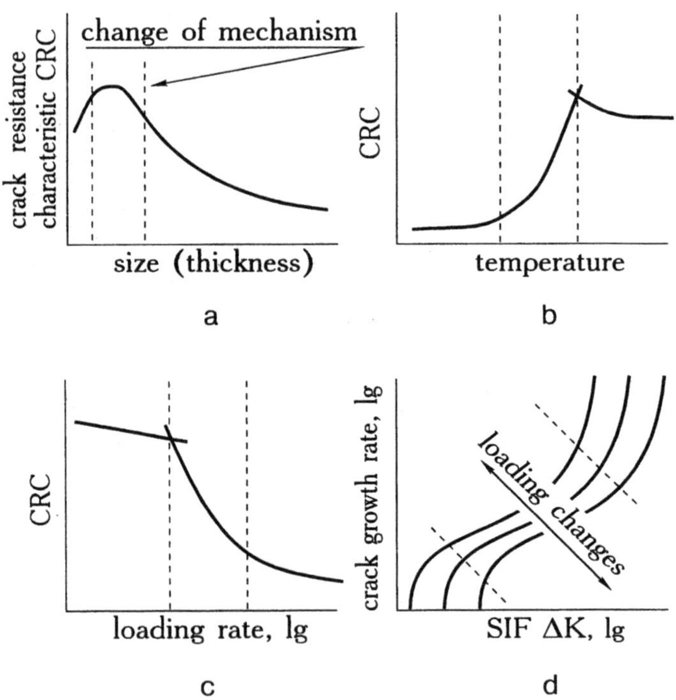

Fig.2 Crack resistance characteristics

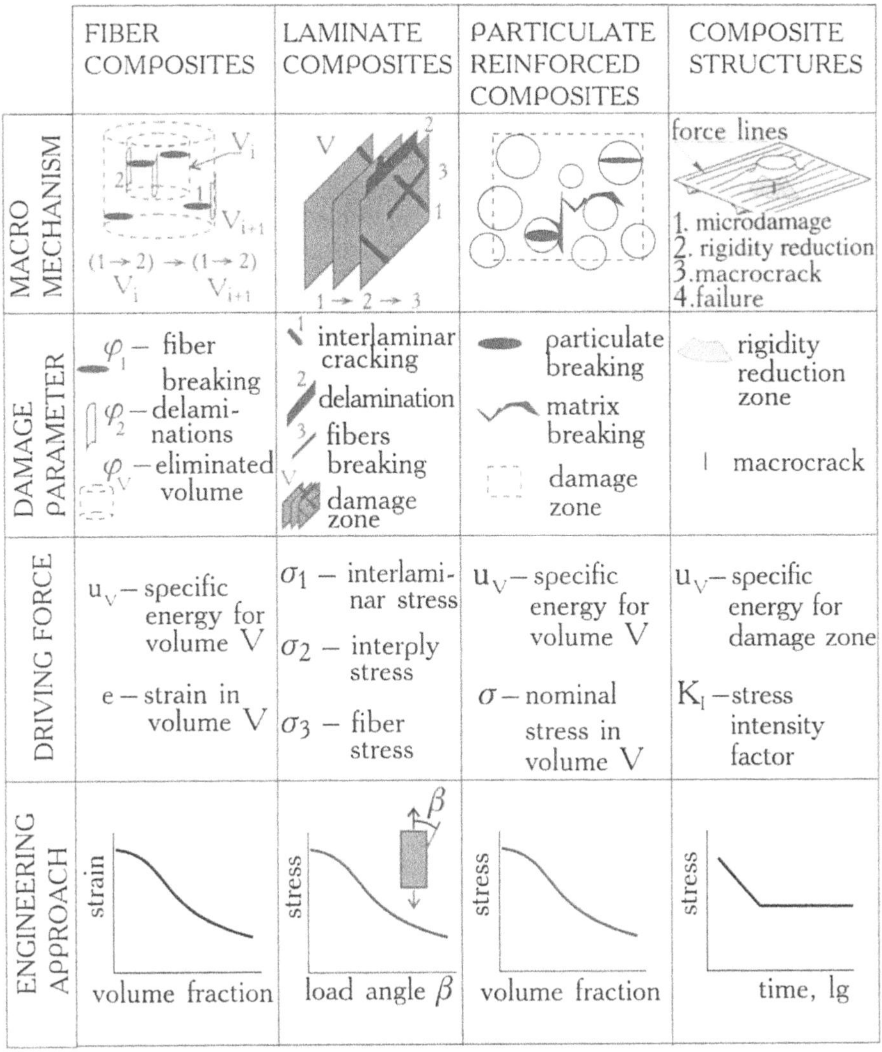

Fig.3 Fracture mechanics models for composite materials

For fiber composites the damage parameters are fractions of fiber breaking, delamination and volume eliminated from the carrying system. The mechanism of fracture is step-by-step enlargement of the damaged zone. The governing parameter of the stress-strain-damage state of the composite is one of energy characteristics and/or nominal strain. The engineering approach can yield the critical strain value and, consequently, the critical stress value.

Damage parameters for laminate composites are defined like for the foregoing. The governing parameters are nominal stresses (interlaminar, interply, fiber). The engineering approach answers the question how optimal fiber placement produces desired properties of composite structures.

Damages of components in particulate reinforced composites are sufficiently different. Brittleness and plasticity of the components affect the fracture mechanisms and strength parameters of the composites. The engineering approach can yield the critical stress value.

In large-scale composite structures microdamages make reduce rigidity over certain zone, enlarge further the area and a macrocrack grow. Energy parameters for the damaged zone define the moment of full-scale fracture initiation. The engineering approaches use **σ-N** curves.

After this we focus our attention on principal features peculiar to many composites (Fig. 4). Analysis of the critical strain allows to point out three classes of materials: 'reinforced matrix', 'composite', 'bundle of fibers'. The scale effect gives the following: there are qualitative distinctions between fiber and composite because of strength reservation. Consideration of probability makes this difference even more pronounced. For fatigue the stress-strain state parameters determine the rates of damage growth only for

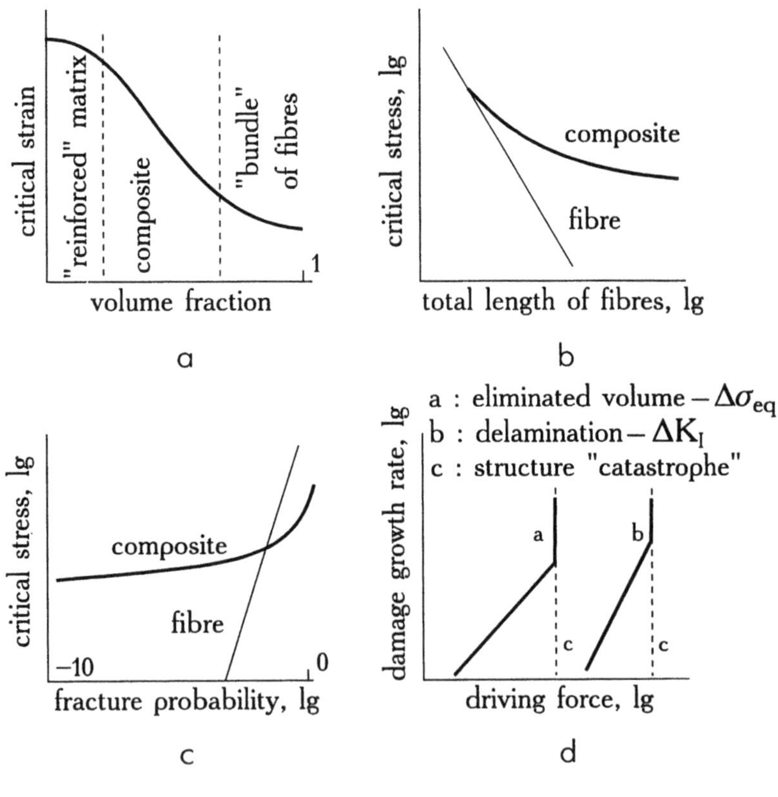

Fig. 4 Mechanical properties of composite materials

some characteristics of the damaged state.

Since we work with complex unhomogenous microsystems - the composite materials it is efficient to demonstrate the structure of the materials, especially during the fracture process, the pictures of non-destructive control methods, computer films with results of structure imitation modeling. Analysis of real structures and failure causes takes a lot of teaching time.

5 Conclusion

Work to understand the fracture mechanisms, the well-defined structure for analysis of failure problems, illustrative explanations provide for efficient teaching of safety engineering of composite structures.

REFERENCES

1. Atkins A.G., Mai Y.W. (1985) *Elastic and Plastic Fracture*, Ellis Horwood Ltd.
2. Barsom J.M., Stanley T.R. (1987) *Fracture and Fatigue Control in Structures*, Prentice-Hall, Inc.
3. Broek D. (1987) *Elementary Engineering Fracture Mechanics*, Kluwer.
4. Broek D. (1988) *Practical Use of Fracture Mechanics*, Kluwer.
5. Cherepanov G.P. (1979) *Mechanics of Brittle Fracture*, McGraw-Hill.
6. Collins J.A. (1981) *Failure of Materials in Mechanical Design. Analysis. Prediction. Prevention*, John Wiley & Sons.
7. Freund .B. (1990) *Dynamic Fracture Mechanics*, Cambridge Univ. Press,Cambridge.
8. Hellan K. (1984) *Introduction to Fracture Mechanics,* McGraw-Hill, New York.
9. Hutchinson J.W. (1987) *Micro-mechanics of Damage in Deformation and Fracture,* Lyngby: Technical University of Denmark.
10. Kanninen M.F., Popelar C.H. (1985) *Advanced Fracture Mechanics,* Oxford University Press.
11. Knott J.F. (1973) *Fundamentals of Fracture Mechanics*, Butterworth, London.
12. Machutov N.A. (1981) *Deformation Criteria of Fracture and Strength Calculations of Strength for Structures,* Machinostroenie, Moscow (in russian).
13. Madsen H.O., Krenk S., Lind N.C. (1986) *Methods of Structural Safety*, Prentice-Hall, Inc.
14. Miller K.J. (1982) *Creep and Fracture*, North Holland Publishing Company.
15. Parker A.P. (1981) *Mechanics of Fracture and Fatigue, An Introduction,* E. and F.N. Spon, London.

16. Parton B.Z., Morozov E.M. (1985) *Mechanics of Elasto-plastic Fracture*, Nauka, Moscow (in russian).
17. Sih G.C. (1991) *Mechanics of Fracture Initiation and Propagation*, Kluwer.
18. Yokobori T. (1978) *Scientific Fundamentals of Material Strength and Fracture*, Naykova Dymka, Kiev (in russian).
19. *Fracture*. (1968-1972) An Advanced Treasure in 7 volumes / Ed. by H.Liebowitz, Academic Press, New York and London.
20. *Composite Materials* (1973-1976) Volumes 1-8 / Ed. by L.J. Broutman and R.H. Krock, Academic Press, New York and London.
21. Baker R.D. Peter B.F. (1971) *Why Metals Fail*, Gordon and Breach Science Publishers.
22. Fisher J.W. (1984) *Fatigue and Fracture in Steel Bridges - case studies*, John Willey & Sons.
23. Nishida, Shin-ichi (1992) *Failure Analysis in Engineering Applications*, Butterworth-Heineman, Boston.
24. *Fatigue Design Handbook* (1988), Ed. by R.C.Rice, Society of Automative Engineers, Inc., 1988.
25. Sih G.C. (1973) *Handbook of Stress Intensity Factors for Researchers and Engineers*, Bethlehem, Pa: Lehigh University, (vol.1-1973, vol.2 - 1974).
26. *Stress Intensity Factors Handbook* (1987) Ed. by Y.Murakami, in 2 volumes, Pergamon Press.
27. Chou, Tsu-Wei (1992) *Microstructural Design of Fiber Composites*, Cambridge U.Pr.
28. *Damage in Composite Materials* (1993) Ed. by G.Z. Voyaljis., Elsevier, New York.
29. *Design With Advanced Composite Materials* (1989) Ed. by L.N.Phillips, Design Council, Springer.
30. Fujii T., Zako M. (1982) *Fracture Mechanics of Composite Materials*, Mir, Moscow (in russian).
31. Kelly A., Macmillan N.H. (1986) *Strong Solids*, Oxford Science Publications.
32. Hamming R.W. (1962) *Numerical methods for scientists and engineers*, McGraw-Hill.
33. Machutov N.A., Koksharov I.I. (1990) Fracture mechanics criteria for mixed mode loading, *Zavodskaja Laboratorja* (*Factory Laboratory*), N 4, p.77-81 (in russian).

12 CHANGES BESIDE CONTINUATION APPROACH TO FRACTURE/FATIGUE EDUCATIONAL PROGRAMS

Y. KATZ
Nuclear Research Center Negev, Beer-Sheva, Israel

Abstract
It becomes apparent that beside traditional programs as related to fracture and fatigue education, new avenues require intensive development and modifications. However it seems also quite accurate to describe the current situation in schools as being attached more to constrains, rather than motivating effective activities for future needs.

The current study intends to emphasize the desire for supplements in order to match future requirements in more of integrated fashion, associated with advanced material technologies. In fact, the ultimate objective of this study remains in introducing the group system approach, particularly in graduate school, in order to cope with complexities involved in fracture and fatigue processes. This in the framework of Academic/ Industrial interactions along Fundamental/Engineering interfaces .
Keyword: Education, fracture mechanics, fatigue localized approach, group system.

1 Introduction

Following educational programs in few Universities in Israel the traditional programs in fracture mechanics and fatigue prevail, mainly included in undergraduate courses or in one year of upper class course. Due to confusion or misconceptions, it appear frequently that fracture is such a specialized area to be developed only in the future, even outside the category of formal education. Often the basic background on the conceptional levels are missing which mask the role of encouragement for creative skill development, systematic chain of thinking or imagination. As such, this area is lacking many of the qualities required and could be cultivated in the early stage of the formal, higher education. For example Broek [1] addressed some of the misconception about fracture mechanics, damage tolerance and fatigue reflecting on the real situation which exist even nowadays. Recall, that fracture mechanics transformed at least along three stages.

The first stage consisted on more of the classical failure analysis in continua stress strain criteria. Second, the intensive development of the linear fracture mechanics in which the stress singularity characterization induced by the crack geometry dominate the fracture problem description. This followed an extension to the elastic-plastic fracture mechanics. Thus, high stress-strain gradients provided by the crack are coupled with material based failure criteria. The third stage emphasize the local, material intrinsic parameters at the crack-tip, realizing the nonlinearity of a significant amount which develops at the process zone. In fatigue, at least two levels of activities have to be recognized. First, along the ongoing developments as centered on the nano/micro scale strongly assisted by novel techniques. Second, activities which follow more of a macroscopic approach in either local or global interpretation efforts. Such a wide scope requires a general background, which introduce by nature, a difficult challenge for educational programs. Here, the general view is based on the idea that most of the background knowledge is confined to existing courses, while the diversity will be achieved by an appropriate selection of students. This might form in graduate school enough members in the group with different background. This framework really promise crystallized the positive interactions, not only on the information level. The school opinion might emerge from analysis, processes, discussions with possible or even recommended approaches, always after a critical way of thinking, clarification and confirmation.

2 Some remarks prior to the proposed program

Insights into the subcritical microcrack domain (presented later by one example) not only in fatigue, is the most complex issue in mechanical behavior. Frequently this leads to misconceptions instead of initiating programs to establish sound concepts and consistent approaches. Such complexities are on different levels, thus it seems unrealistic to expect a unique approach for appropriate solutions. Nevertheless, the developments of assessment capabilities become essential. As known, fracture and fatigue are an interdisciplinary and interactive fields. As such, methodologies rely basically upon scientific and technological fields, which include,

- Continuum mechanics.
- Fracture and damage mechanics.
- Material science.
- Deformation behavior and defects.
- Numerical methods.
- Environmental aspects and interactions.

Needless to state that in many respects this field become a junction for scientists and engineers in several areas;

Materials.
Mechanics.
Applied mathematics.
Applied physics.
Chemistry or chemical engineering.
Computer science.

However, the main incentive remains in recognizing that cooperative work is required if not essential. Enormous efforts in coordination activities based on various fields have to invested if programs are conducted constructively. In particular it appears important that under such circumstances the ultimate goals need to be established, with some time factor for rethinking and modification junctions. Many ideas come in mind but only (just indicating the sequence of thinking) few are mentioned. (i) Adopting long term activities that evolve systematic work by following scientific/engineering methods. (ii) To conduct theoretical/experimental studies with deep evaluation capacity. This by material approach, model and simulation approaches beside the important drive for critical experiments and experimental confirmation. (iii) The study as such should be focused on better understanding which might propose (as the end result) methodologies based on physical views. The latter is very important, realizing the complexities and the difficulties in achieving ideal situations. However the sound intention and consistency here, seems the key for educational activity, with attention to the strength and the limitation which are in the nature of complex problems.

Finally, one additional remark seems in order. According to the author opinion the recommended progress can evolve from continuation beside supplements. This by realizing that the role of existing training methods should not be de-emphasize. Substantial volume of these activities turned out to be effective and important. In fact intensive efforts have been devoted already to fracture mechanics on the theoretical and the experimental levels. The literature which has been established should be highly respected and should be transferred to students effectively. Following more of global educational aspects which seems very promising, should mainly be based on evolutional changes due to an additional reason. It is well recognized that changes need to be considered in a realistic way. The role of school interia is very powerful, frequently connected to cultural and social norms. Sometimes the discussion should not be on new courses but more on new approaches and emphasis. However, even for undergraduates, under some circumstances, new courses might be offered with emphasis on materials, dislocation theory, relevant numerical methods for damage and fracture. Additionally, overviews centered on the state of the art (mainly along the dominant concepts) are extremely beneficial. This with academic openness even along critical approach which provide means to introduce the field in relatively short time and appropriate depth. In contrast, for graduates guided by a proper advisory function, the requirement for minors become important. Namely, not only advanced undergraduate courses but new courses for the major and the minors.

Following such scheme students planning to participate in fracture and fatigue topics should become aware at early stages to the interdisciplinary nature of the field. This in order to establish the attitude of broadness and complexities involved, which are very real. For the long run, such approach might contribute to the reduction of possible confusion or substantiate the field integrity properly.

3 Group system and related interfaces

Even assuming only minor changes, the existing undergraduate studies based on the traditional engineering courses, enable an excellent start. However for students intended to continue their studies in mechanical properties (including fracture and fatigue) supplemented programs are required. For example, more emphasis on material science for mechanics or computer science student, while more in numerical methods and mechanics for material science students. It should be on the level of supplements for better communication and attraction. However, The principal activities still remain in the graduate school guided and coordinated in the framework of a group system. This seems essential, allowing to expose the group member during the formal studies to many problems as possible. Most of the problems are conducted by others but the assessments, discussions, the critical remarks, the complex considerations and findings evaluation as also the strength of techniques, play an important role.

This can be achieved by establishing (similar to several universities in the U.S.A or other places) centers or institutes in the framework of Universities or Research Institutes. At this stage it appears natural to consider such centers in engineering departments (mechanical and chemical or in material science departments). The meaning of it relates to the role of leadership that is essential, as always. Notice that fracture and fatigue are not isolated problems but deal with the critical issues of structural integrity. This, regardless of large scale construction materials or micro-scale structural materials, activities so demanded now in the electronic industry. It is a broad field and here formal education requires to cover complex and fundamental aspects in order to match with such technological challenges. Again, the only way to broaden the view for all scales will be by adopting education programs based on group activities. In Table 1 proposed description and challanges for a fracture group are given.

Table 1 Group system and challanges.

Group members	Research and Technological Challenges
Prof. (Head)	- The key challenges need to be defined and modified periodically.
Profs. (Associates from the University)	- Individual and collaborated planes.
Visiting Profs.	- Regular meetings and seminars.
Postdoctoral associates	- Theory and models.
Industrial fellows	- Instrumentation → Current / → Future
Graduate students	- Computation mechanics dedicated software.
Undergraduate student (candidates to join the group more actively later)	- Evaluations.
Technical staff	- Applications - active attitude and involvement. - Academic/Industrial interface. - Basic/Eng. interface is achieved by the group - nature and strength.

Finally a possible scheme for the educational program and application in fracture and fatigue center is given in Fig. 1.

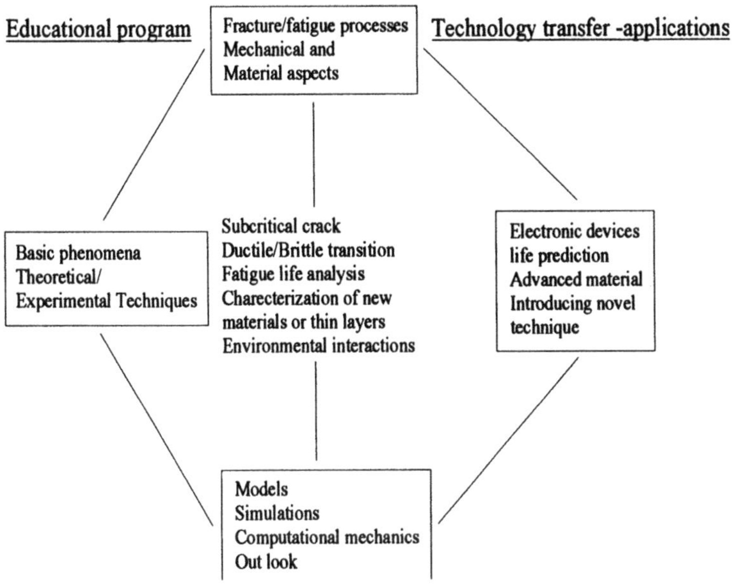

Fig. 1. Schematic - Educational program beside technological transfer and application.

4 Examples for group efforts in fracture and fatigue

4.1 Crack-tip field perturbation

Cumulative damage activated by irregular or random spectra situation, is more realistic in service. If the ultimate objective is fatigue life prediction, the issues of crack initiation and propagation require further exploration. In fact, these two stages are essential in fatigue life analysis recognized here as the key element for safety and reliability. The general approach does not distinguish between structural or electronic material. It is more the methodologies that require emphasis. In terms of structural integrity, explorations always require local/global approaches, including the higher order problems of local interactions and environmental effects. Only in the framework of a group, a material approach might effectively be conducted providing the data base for models analysis and tendencies. As such, a group program which include various materials beside theoretical and experimental activities can be performed with significant progress.

Table 2, as related to the problem of crack-tip field perturbation during cyclic crack extension, describe a summary of various materials tests, selected variables and suspected transient origins and mechanisms.

Table 2. A summary of various experimental programs.

Materials	Tests/Variables	Specimen geometry	Suspected transient origins and mechanisms
Single crystals			
Fe-3%Si [6]	cyclic tension-tension with overloads. Overload intensification factor.	mini compact-disc specimens.	Residual-stress in a subcritical growth confined to the cleavage plane.
Polycrystal systems			
Al, Iron base alloys, U-Ti alloys [4,7]	Cyclic tension-tension or compression-compression with overloads. Various crystal structures, overload intensification factor. Under subsequential fatigue, constant or variable amplitude range.	Compact tension (CT), three point bending specimens.	Blunting, closure, residual-stresses.
Metastable austenitic stainless steels [8]	Cyclic tension-tension with overloads. Thermal effects. Overload intensification factor.	CT, Single edge notched (SEN), three point bending.	Dilational stresses due to martensitic transformations.
U-Ti alloy AISI 304 metastable stainless steel	Transients activated by overloads, with environmental interactions.	SEN, three point bending, tapered specimens.	Residual stresses. Deformation/ corrosion effects.
Superplastic alloy Zn-22Al [9]	Transients by overloads in visco elastic-plastic model material. Thermal effects on fracture modes.	CT specimens.	Residual stresses. Time dependent crack-tip shielding. Process zone development and effects.
Al-Li planar slip model material [10]	Monotonic warm prestressing (WPS) or cold prestressing (CPS) as also warm precracking (WPC).	Three point bending specimens.	To establish consistency regarding the role of crack tip field perturbations on the toughness values.

4.2 Subcritical cracks

Subcritical microcracks is a difficult domain in terms of acceptable margins against failure. Same problems are introduced here in fatigue crack extension or slow crack growth with and with no environmental interaction. Program of such volume may be tackled in a group framework, with versatile approaches in terms of experimental and theoretical efforts. Here few objectives are mentioned which open opportunities for many more.

For example;

(I) Cyclic crack growth

(i) From single crystals to polycrystalline systems.
(ii) Static vs. dynamic approaches.
(iii) Temperature and crack-tip deformation rate effects.
(iv) Non-isothermal conditions (high frequency).
(v) Models for crack-tip processes.

Questions while considering deformation/environmental interaction (sustained load-slow crack growth);

(i) How does the crack tip advance and induced by aggressive agent interactions and how does the role of plasticity manifested itself?
Note, the intimate connection between this problem to the exploration of crack-tip dislocation interactions in general.
(ii) If slow crack growth is discontinuous, how does the process of crack initiation/arrest occur?
(iii) What shields the crack tip from instability? Namely, by introduction here the main concern of intrinsic dislocation shielding. It is no intention to de-emphasize other important sources of crack shielding such as blunting, branching, ligaments, steps and ledges between planes, shielding origins which are well recognized.

4.3 Ductile/brittle transition

Additional example for a group activity is along the semibrittle materials or the Ductile/Brittle Transition (DBT). Systematic attempts to link dislocation shielding models to polycrystalline fracture-toughness initiate the following programs with some progress [5]. Briefly few elements are menttioned here.

(I) (i) From TEM observations in single and polycrystalline Fe-Si (model materials) fracture toughness prediction were attempted at least in the lower temperature regime.
(ii) First order models were established and experimentally confirmed.
(iii) Implication on DBT temperature even in low symmetry crystal structures. In addition, more insights into dynamic loading including adiabatic shear.
(iv) Modified concepts in terms of micromechanical processes.

(II) Indication to;

(i) Simulation-scale wise. Require the important advantage of localization with a description capacity of the submicron near tip stress yield.
(ii) Incorporation of microstructural aspects, providing a better physical meaning as related to the micromechanisms of fracture .
Thus, simulation by dislocation models which are in between a continuum finite elements crack-tip calculation and an atomistic model turned out to be beneficial.

The two programs 4.2 and 4.3 allowed to shed more light on the controversial issue of hydrogen-plasticity interaction. For the sake of illustration, only two striking findings are here discussed emerging from extensive programs as in the fracture group. In single crystals (Fe-3%Si), with external and internal hydrogen interactions, the discontinuous cracking on cleavage planes occurred at slower velocities for crack growing in the macroscopic <110> direction compared to <100> macroscopic growth. However, in both cases the microscopic crack path was {001} cleavage [2]. An additional feature was that selected area channeling patterns at and near the crack plane showed that plasticity was more severe in the more resistant orientation [3].

Thus;

$$da/dt_{(110)} < da/dt_{(100)} \tag{1}$$

but;

$$\varepsilon^{local}_{(110)} > \varepsilon^{local}_{(100)} \tag{2}$$

Here, enhanced localized plasticity did not assist the cracking process but retarded it. Additional interesting result was that in highly supersaturated trap site (grain boundaries) situation, hydrogen enhanced decohesion may be preferred along adjacent cleavage planes rather than at the grain boundary [4].

Aspects of this kind motivate further studies but are significant to major issues concerning micromechanical mechanisms, evaluations in cases of environmental interaction, texture effects or directional dependency as related to slow crack growth.

5 Outlook and final remarks

On the informative level the issue of how to narrow gaps today in fracture and fatigue educational program, is only a partial goal. Note that students today will lead technology in the future. Thus any outlook requires at least 15-20 years vision in theory, experiments and computational mechanics.

Only some proposed activities are here considered which require emphasis and considerations. For example;

(i) Emphasis to local approach - a significant lever to nanomechanics in general.
(ii) Attention and application to novel techniques including in-situ capabilities.
(iii) Anisotropic discretized dislocation models.
(iv) Fracture and damage mechanics to microscale electronic assemblies.
(v) Interface Engineering.
(vi) Models for advanced materials.
(vii) New concepts in fracture and damage mechanics.

6 Summary and conclusions

Here, some of the complexities of fracture and fatigue processes were described. As such, knowledge transfer become difficult reflecting on educational constraint. Over the years, fracture and damage mechanics have been achieved already significant progress. Such advances have been manifested in design, testing, standards and codes, marginal evaluation and confirmations. Clearly fracture and fatigue disciplines are far from being mature [6] but they follow scientific and engineering methods and physical rationale. The integration of fracture mechanics and proposed methodologies into the engineering design and analysis demands cooperation and interdisciplinary efforts. Contribution of artificial intelligence and expect systems create hopes and may assist. However, these still have to be explored with careful assessment.

Nevertheless, all of this need to be cultivated in the formal education framework, proposed currently by a group system in centers which are based on the graduate school level.

Thus the following is concluded;

(1) Group system in the framework of centers might offer an appropriate environment for complex interactive problem studies.

(2) Such approach in Universities is evolutional where education, research, technology can be developed with no major changes.

(3) Basic/Engineering and Academic/Industrial interfaces might be established and advanced. Moreover, technological transfer and application are relatively experienced in the early stage of education.

(4) Future programs are essential and ultimate long term goals may be defined on a broad viewpoint with combined efforts.

7 Acknowledgments

The author would like to acknowledge the contribution, advise and collaboration of Prof. W.W. Gerberich and his group members from the University of Minnesota as also the mechanical and fracture group members in NRCN - Beer-Sheva, for their dedication, support and serious efforts, always.

8 References

1. Broek, D. (1988) The practical use of fracture mechanics, Kulver Academic Press, London.
2. Chen, S.H., Katz, Y. and Gerberich, W.W. (1990) Scr. Metal., Vol 29, p.1125.
3. Chen, X. and Gerberich, W.W. (1991) Metall. Trans., Vol. 22A, p. 59.
4. Wang, Y.B., Chu, W.Y., Katz, Y. and Gerberich, W.W. (1990) Scr. Metal., Vol. 19, No. 10, p. 1165.
5. Lii, M.J., Chen, X.F (1990) Acta Metall., Vol. 38, p. 2435.
6. Liebowitz, L. (1989) in ICF7, Vol. 3, (eds. K. Salama, K. Ravi-Chandar, D.M.R. Taplin and R. Rama Rao), Pergamon Press, Oxford, p.1887.

13 THE USE OF CASE STUDIES IN MULTI-DISCIPLINARY TEAMWORK FOR LOSS MANAGEMENT OPTIMIZATION

B.M. PATCHETT and L. WILSON
Mining, Metallurgical and Petroleum Engineering Department,
University of Alberta, Edmonton, Canada

Abstract

Case studies are conventionally used in specific disciplines to illustrate real industrial problems to students of business and engineering. This approach is appropriate for the individuals in a specific discipline, but does not address the necessary teamwork required by industry to solve the problems in an optimum and cost-effective manner. A recent report from the insurance industry in North America looking at worldwide engineering disasters covering the last thirty years concluded that only 4% were caused by design errors. Therefore, the Faculty of Engineering at the University of Alberta has taken a somewhat different approach to the use of case studies.

Industrial failures involving significant damage to people, the environment, material assets and production are taught in a multidisciplinary setting for both students and faculty members involved in the course. Analysis of a given industrial problem is presented by several lecturers from differing engineering and business backgrounds to a student body comprising approximately 40 students per section from all disciplines in the engineering and business faculties. In the near future, occupational medicine students will be introduced to the program.

Keywords: Case studies, failure analysis, teamwork, teaching.

1 Introduction

The use of case studies to illustrate lessons from real incidents is not new. They were initially introduced in business schools and have been used in a variety of disciplines since, including engineering. The authors used case studies themselves for loss management courses (Wilson) and failure analysis (Patchett), the former from a chemical engineering approach and the latter from a metallurgical/welding engineering approach. Comparison showed that similar information from particular case studies was used to emphasize important features in our own disciplines, and that integrating the two analyses could be beneficial to student learning to work in teams. Wilson's course in loss management, which involves students in several engineering disciplines as well as business students was chosen as the vehicle to try out the approach. The results have been gratifying to both the students and the instructors, showing both that there is more than one type of lesson to be learned from multidisciplinary analysis. One of the significant outcomes of the approach is the use of knowledge judged to be secondary in one discipline as a primary indicator of improvements for another discipline.

Teaching and Education in Fracture and Fatigue. Edited by H.P. Rossmanith. Published in 1996 by Taylor & Francis, 2 Park Square, Milton Park, Abingdon, Oxon, OX14 4RN. ISBN 0 419 20700 7.

2 Case studies

The Flixborough disaster has been well documented by investigations ranging from a Court of Inquiry to several scientific papers from differing disciplines and television documentaries. It is therefore an excellent candidate for a multidisciplinary approach and will be used as an illustrative example. Several other case studies are used in a similar fashion in the course, and we are always looking for additional material to use in case study situations.

2.1 Flixborough

The events and some background information on the results of the Court of Inquiry are given in Appendix A for those who are unfamiliar with the disaster. Further details can be found in references [1-4]. The students are presented with some of the detailed aspects of the failure in formal lectures by at least two engineers from differing disciplines - in this case, chemical and metallurgical engineering. The mechanical engineering features are well documented and reasonably clear after the event and can be presented by engineers from other disciplines. The discussion periods are best when industrial people with piping design expertise are present. The following information is presented to the student groups:

2.1.1 Mechanical Engineering Aspects

The main mechanical engineering problems at Flixborough were generally agreed within the time period for publishing the Official Report [1]. The primary cause of the disaster was the collapse of a temporary bypass between two reactors. The design of the bypass was done in chalk on a shop floor with no mechanical engineer present or available. The bypass configuration involved a dog-leg in the middle, had bellows at each end attachment, and was supported only by scaffolding. The flexibility of the bellows, the lack of support and the moment induced by the dog-leg combined to cause squirming, a leak and ultimately, collapse. The potential for this problem to occur was not apparent to most of the repair and maintenance personnel nor to the other engineers (electrical, chemical and fuels) on the site. One technician expressed concern and was told that the by-pass was a routine "plumbing" operation.

2.1.1 Chemical Engineering Aspects

While there were more chemical engineers at the Flixborough site than any other group, they did not have the mechanical design expertise to avoid the bypass failure. This exposes a question of importance for educators: given the time constraints in an undergraduate program, how is an appreciation of general engineering principles to be established? One would hope that any engineer could see the potential problems in an unsupported dog-leg bypass design. If all engineers cannot gain this insight in conventional courses outside of mechanical or civil engineering, then multidisciplinary case studies in loss management may be the only way.

There is also a chemical engineering problem regarding the dangers of a large inventory of dangerous substances. As Kletz [5] points out, the Flixborough plant contained about 400 tonnes of cyclohexane of which about 40 escaped in the minute or so which elapsed before ignition. The inventory was so large because the conversion was low, about 6% per pass, and most of the raw material had to be recovered and recycled. Recycling large quantities of dangerous material is a problem in itself concerning leaks. If there had been less material there, probably less would have escaped, and recycling would not be as necessary. One of the most important lessons of Flixborough not mentioned in the official report, is the need to reduce inventories of hazardous materials, whenever possible. The best way of preventing a fire, explosion or toxic release is to use so little hazardous material that it does not matter if it all leaks out or to use a safer material instead - the inherently safer approach: "What you don't have, can't leak."

It is not easy to reduce the inventory in the Flixborough process, or find a better process. At Bhopal, reducing the stock of methyl isocyanate, the material which leaked in 1984 and killed over 2000 people, would have been easy as it was an intermediate. It was convenient to store it, but not essential to do so. Ten years after Flixborough the need to reduce inventories had still not been recognized by many in the chemical processing industry. The fact that it is possible to reduce or avoid large inventories of hazardous materials is now known to most chemical engineers but is still not recognized in most boardrooms where they still seem to believe that safety is best measured by the lost-time accident rate. They would do better to monitor the progress made in reducing inventories. Students need to be aware of the inventory problem before they enter industrial design exercises.

2.1.3 Metallurgical Engineering Aspects

The "second theory" of the cause of the disaster was one of the main issues debated in several publications. The failure of the 200 mm line could have caused the failure, but only after a series of low probability events. This series of low probability caused its rejection as the primary cause of failure. However, the litany of metallurgical problems indicate that material choices and maintenance of the plant were both deficient.

Why were there so many problems with the 200 mm line? Why were they not acted upon before the disaster? Creep damage requires high temperatures and an extended time period under load - why was the pipe hot for so long? Why were galvanized wires used near the stainless pipe, when zinc embrittlement is possible? The whole situation surrounding the 200 mm line suggests poor maintenance procedures in the whole plant and a lack of knowledge of appropriate material choice at the design stage.

The use of untreated cooling water to condense cyclohexane vapour at the reactor site also raises questions about the lack of metallurgical engineers in the operations at Flixborough. Cooling water evapourates on hot vessels, concentrating solutes and impurities. Knowing of the nitrate corrosion problem would suggest water analysis before use. Most metallurgical engineers would then ask "why is there a leak?", so that use of water for extrinsic cooling can be minimized or avoided. Pouring water over a hot reactor in a relatively uncontrolled manner is a poor engineering substitute for good design. Better design or installation of items such as glands on the stirring mechanism on the reactors could minimize leaks. Indeed, a more efficient reactor design might avoid stirrers altogether, which would eliminate that particular kind of leak. A better reactor design would require cooperative teamwork among chemical, mechanical and materials engineers as well as a management dedicated to the optimum use of engineering talents.

2.1.4 Management Aspects

Many of the events that occurred at Flixborough before the disaster indicated that there were many small things occurring that should have been addressed by management before a major problem arose. Lack of teamwork and a reluctance to involve "experts" or to ask questions were common. For example, the staffing of the plant revealed many problems - a mechanical engineer had not been replaced immediately to allow a continuous presence at the plant, and no consulting arrangements had been made for the interim. This can only be due to an indifference to the importance of mechanical engineering expertise in plant design. In today's climate, it is even more important that the principle of using appropriate expertise is acknowledged. "Re-engineering" and "downsizing" have limited the number of employees in all areas of plant management and operation. Thus studies on potential hazards done before the design is finalized become even more vital. There are two important types in current use [6]. The first is Hazard Operability or HAZOP, which is used to identify qualitative hazards (on a "what if?" basis} for each and every project by a team of engineers and managers. As author Kletz quotes from Coleridge, "History is a lantern on the stern" to illuminate our past, while HAZOP is a "lantern on the bow". The second is Hazard Analysis, or HAZAN, which is quantitative and based on risk analysis. This is used more selectively, done by a small team or an experienced and expert individual. It asks and answers such questions as "How often? How big? What are the consequences?". One cannot expect 100% reliability of plant, nor top performance from staff all of the time. Therefore failures must be expected from time to time and the resulting damage minimized [7] by a systematic approach to plant design, engineering, maintenance and emergency preparations. Another important management function, usually overlooked, is to have one member of a plant design team charged with the responsibility of looking out for the next plant [8]. There is always insufficient time to use all of the possible beneficial modifications discovered in the middle and late stages of a project. Management must always be proactive in seeking out solutions to risks and also taking ultimate responsibility controlling risk.

3 Discussion

After a formal presentation by the faculty outlining the points above from the perspective of the different engineering disciplines involved, the case studies are analyzed by six member teams of students representing all groups in the class. Each team must contain at least one business student and in the future will also contain one occupational medical student. This approach ensures that all points of view are represented at the initial stages when it is vital to have all the information analyzed from all angles. The assignment is to devise a plan to minimize future risk to acceptable levels and optimize future operations from both financial and technical aspects. The students are expected to research and utilize "best industrial practice" for all aspects of the analysis.

The team approach to loss management is trying to instill involves the following principles:

- picking a team leader from within the team.
- dividing the assignment into areas of specialty i.e., each student is responsible for investigation of the "facts" in their area of knowledge.
- teams come together to share their data and knowledge and explain them clearly to the other team members. This sharing is critical to the broad understanding of the problem and often initiates a synergistic process for further insight.
- the leader does not strongly interfere in the process of analysis, but encourages interaction and integration.
- recommendations are made only as an undivided team, with a focus on the deepest fundamental cause(s) of the failure, especially the ones that have come out of the mixing of disciplines.

The assignment produces a sophisticated, fairly structured "brainstorming" session leading to the best long-term solutions to the most basic causes of failure uncovered during the work. A report is presented in written form to the course supervisors and in oral form to at least three industrial managers from appropriate industries, the rest of the class and the professors. The students are given the role of junior managers making a case to a senior management team for alterations to the plant (materials and design) and operations to avoid future problems of a similar nature. Marks are assigned by academic staff on both written and oral presentations. After this, short critiques are made by the industrial managers, and the students are given an opportunity to ask questions to a panel of the industrial and academic assessors.

Flixborough started a debate on the extent to which governments should control hazardous industry. As always after a major incident, there were those who wanted lots of detailed regulations. While regulation does have a place in accident prevention, we feel that the best method is to educate engineers and management personnel to use teamwork and consulting with experts when deciding how to design, modify or operate a plant. An appropriate case study, using a varied approach to the analysis of the incident, is best way we have found to accomplish this important task.

It is important for operational personnel to recognize the necessity to use appropriate expertise in plant modification and replacement procedures. For management, this means that the resources must be available for consultants to be used and that all employees are made aware of this, so that half-measures are avoided. This will probably mean that the fiscal benefits of re-engineering are somewhat reduced in the short term, but the long term results will be better - how much positive effect does a disaster have on the bottom line?

Experience has shown that unstructured responses to emergencies without expert input and

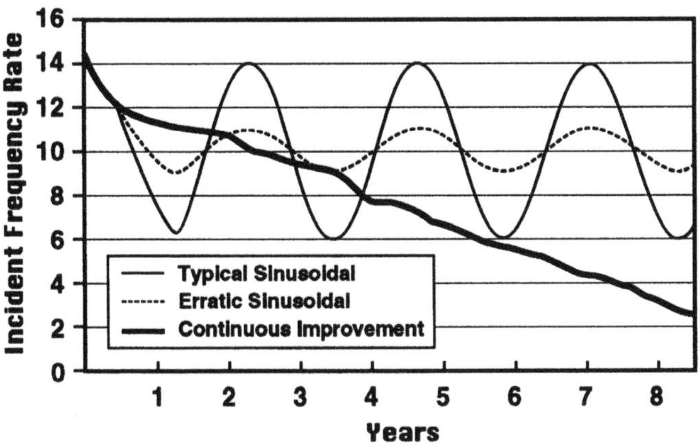

Fig. 1. Variations in Incident Frequency with Program

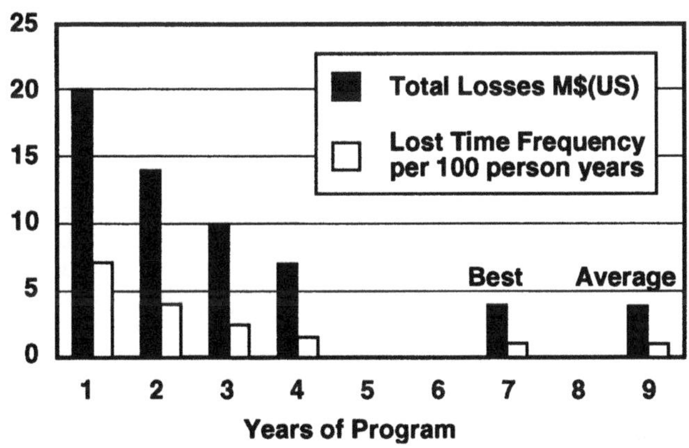

Fig. 2. Benefits over Time with Integrated Loss Management Programs

a plan to ensure that design and modifications of plant are conducted systematically tend to cause sinusoidal effects as a function of time. As shown in Figure 1, the cycles may be damped, indicating a reactive rather than proactive policy, or erratic, which indicates an ineffective policy. The more desirable gradual continuous improvement can only be achieved through management awareness of the necessity to plan with foreknowledge and the encouragement of the team approach to problem solving without egos getting in the way of mitigating ignorance. The benefits of an integrated teamwork approach to loss management are evident in both cash flow and safety, Figure 2.

4 Conclusion

Teaching loss management using case studies to students who will be involved in plant design, maintenance and operations is promoted by a multidisciplinary approach. The approach is valid for class composition (in this case engineering and business students), the teachers (various engineering disciplines) and for assessment (teachers and industrial managers). The multidisciplinary approach often finds that secondary evidence or knowledge derived from one discipline causes new ideas to emerge in other disciplines, or that previously derived relationships between cause and effect are confirmed or reinforced. The particular study chosen, the well-known Flixborough disaster, truly demonstrates these synergistic effects. Other case studies, from particular areas of the world or from particular industrial operations should be equally suitable, provided that the approach is consistent.

5 References

1. Parker, R.J., Pope, J.A., Davidson, J.F. & Simpson, W.J. (1974) *The Flixborough Disaster - Report of the Court of Inquiry* HMSO, London.
2. Sadee, C., Samuels, D.E. & O'Brien, T.P. (1977) Characteristics of the Explosion of Cyclohexane at the Nypro (UK) Flixborough Plant on 1 June, 1974 *Journal of Occupational Accidents* (3) p203 July.
3. Warner, Sir Frederick & Newland, D.E. (1975) Flixborough Explosion - Mechanical Engineering Provides the Key *Chartered Mechanical Engineer* Vol. 22, No. 6, pp76-81.
4. Cottrell , A.H. & Swann , P.R. (1976) A Metallurgical Examination of the Eight-Inch Line *The Chemical Engineer* pp266-274 April.
5. Kletz, T. (1994) *Learning from Accidents* Butterworh-Heinemann, Oxford.
6. Kletz, T. (1992) *Hazan & Hazop* Hemisphere Publishing Co., London.
7. Kletz, T. (1991) *Plant Design for Safety* Hemisphere Publishing Co., London.
8. Malpas, R. (1986) *Research and Innovation for the 1990's* (editor B. Atkinson) Institution of Chemical Engineers, Rugby, UK.

Appendix A

1 The Flixborough Disaster

At 1653 hours on Saturday, June 1, 1974, the chemical works of Nypro Ltd. at Flixborough, Yorkshire (about 400 km north of London) was completely destroyed by an explosion which broke windows and structurally damaged 1,988 buildings up to 15 km away. Of the 72 workers present at the site, 28 were killed and 36 injured. The fires started at the site by the explosion raged for 2-3 days and the damage on the 25 hectare site was comparable to the explosion of about 16 tonnes of TNT detonated about 50 metres above the ground. The explosion was caused by the ignition and deflagration of a massive cloud of cyclohexane which had escaped from containment vessels operating at a pressure in excess of 800 kPa and a temperature of 155°C.

The site was originally developed in 1938 by Nitrogen Fertilizers Ltd. for the production of ammonium sulphate. It was acquired in 1964 by Nypro (owned 55% by Dutch State Mines and 45% by the UK National Coal Board) and converted by 1967 to produce caprolactam, a basic raw material for the production of Nylon 6. It was the only such plant in Britain.

1.1 Plant Layout

The six oxidation reactors were located sequentially in one building, in a central position on the plant site. Each reactor was aligned 350 mm below its predecessor, so that flow was gravity controlled. In subsequent stages, the process stream was distilled into cyclohexane, which was recycled, while the cyclohexanone and cyclohexanol were converted to caprolactam. At the time of the disaster, #5 reactor had been removed for maintenance and replaced with a 500 mm diameter "dog leg", Figure 1. Directly behind the row of reactors was a 200 mm diameter pipe connecting two of the separation vessels. The main office block, laboratories and control room were close to the reactor site. All 18 people in the control room died and the process instrumentation and records destroyed.

1.2 The Caprolactam Process

The process produces high purity caprolactam from cyclohexane, ammonia, sulphuric acid and hydrogen. The cyclohexane is oxidized to cyclohexanone at 155°C and a pressure of 880

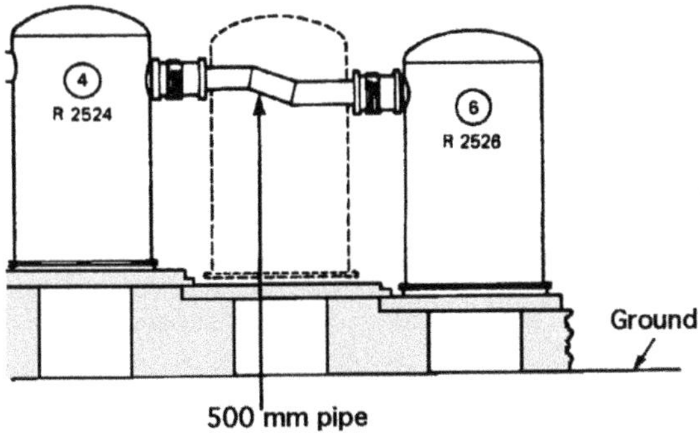

Fig. 1. Arrangement of Reactors and Temporary Pipe at Flixborough. (HMSO)

kPa, which is then reacted with hydroxylamine in the presence of an organic solvent at low pH, converting it to cyclohexanone oxime. The unreacted cyclohexanone is converted in a subsequent step at a higher pH. The overall reaction rate was slow and conversion was kept at a low 6% to minimize unwanted by-products. The slow step is the decomposition of the cyclohexanone (peroxide), which is dependent on intimate mixing of air with the hydrocarbon. Unreacted cyclohexane was recycled via evapouration and condensation followed by return to a reactor. About 7 tonnes is evapourated for each tonne reacted, so large amounts of heat were involved.

The oxime is separated from the solvent via distillation and is pumped to the rearrangement unit, where the caprolactam configuration is formed with the side production of ammonium sulphate. The crude caprolactam is then refined and purified, while the ammonium sulphate is made into fertilizer at a yield of 1800 kg of fertilizer per 1000 kg of caprolactam produced.

1.3 Events Prior to June 1, 1974

On March 27, a cyclohexane leak was discovered in Reactor #5. All of the reactors were fabricated from 13 mm mild steel plate internally clad with 3.2 mm of stainless steel. A vertical crack was found in the outer mild steel shell of the vessel extending for about 1.7 m. Nypro staff were aware that the stainless steel liners could suffer from stress corrosion cracking (SCC), and took stringent precautions to minimize the buildup of chlorides in the process stream. The crack in the mild steel outer shell of the reactor was later determined to be caused by stress corrosion cracking following cooling the reactors during shutdowns (to condense cyclohexane vapour leaks) with process cooling water, which was contaminated with nitrates. Although a documented phenomenon, the Nypro staff was not aware of this corrosion effect. On March 28, it was decided that the vessel would be removed and replaced with a dog-leg bypass in order to maintain production. No design alternatives were discussed. Although the openings to be connected were just over 700 mm diameter, 500 mm diameter pipe was used, since it was the largest size available on-site. Calculations were done to see if the pipe was large enough to take the required flow of liquid to maintain production, and if it could (considered as a straight length of pipe) withstand the process pressure. The only drawing for construction was in chalk on the shop floor. It was completed late in the evening on March 29. Between April 1 and May 29, the system operated without incident. On May 29, a leak was discovered on the bottom isolating valve in a sight-glass on one of the reactors, and the plant was shut down. The leak was repaired, and some minor maintenance done, on May 30 and 31.

1.4 Events On June 1

Early on June 1, leak testing at 400 kPa was carried out without incident and the process stream introduced to the reactors. When the pressure reached 850 kPa, a leak occurred and the system shut down again. At 0500 circulation began again with steam additions. Pressure rose very rapidly to 850 kPa at only 110°C in reactor #1, with lower temperatures in the following ones, down to 50°C in #6. This abnormal rise was never explained. There was no accompanying rise in temperature, nor a decrease in circulation rate, suggesting that a high pressure nitrogen leak might have been responsible. Venting did not occur, because another product leak produced another shutdown.

The leak was not repaired by the night shift, due to the required spark-resistant tooling being locked in a shed. After the new shift had repaired the leaks, start-up recommenced. No further venting took place, and at the end of the day shift, the temperature and pressure were still rising toward normal operation conditions. After the afternoon shift took over at 1500 hours, it is not known if further venting took place. Steam valves were found in a closed condition after the disaster, however, suggesting that at some time process adjustments took place.

1.5 The Disaster

Two explosions were heard by local witnesses - a small one followed by the very large one approximately 30 seconds later. At a very early stage in the government-sanctioned Inquiry following the disaster, two important things were found:

- The 500 mm bypass assembly, with the bellows at both ends torn asunder, was found jack-knifed on the plinth beneath the reactors.
- An 1100 mm split was found in the 200 mm stainless steel piping joining separators S2538 and S2539 behind the line of reactors. Figure 2 shows the piping. The location of the

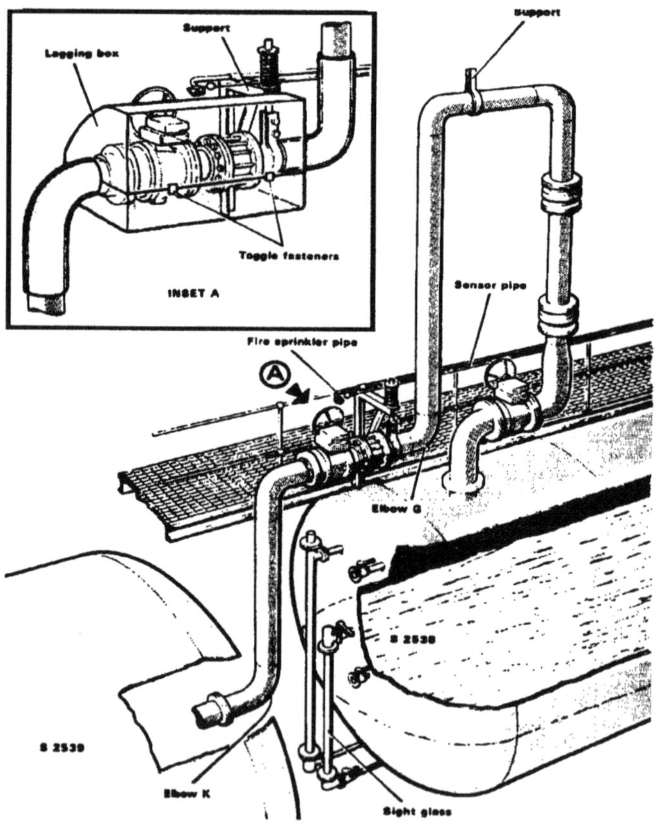

Fig 2. Configuration of the 200 mm Separator Plping. (HMSO)

damage after the explosion had taken place was the 1100 mm crack upward from elbow G and the 75 mm crack between the long crack and the top support..

It was soon established that nothing unusual had been heard at the plant by eye-witnesses until approximately two minutes before the main explosion. Within that time frame, it was possible for the ruptured 500 mm dog-leg to form a vapour cloud of sufficient size to cause the final explosion. The split in the 200 mm line could have caused a small explosion in the same time period, but not the large one. However, the 500 mm bypass had survived two months of normal operation without apparent trouble. In addition, there were other cracks in the 200 mm pipe in addition to the large 1100 mm split and other cracked lines were discovered in other places in the wrecked plant.

The inquiry soon decided that it had to establish why the 500 mm bypass had ruptured, since it had to be the source of the cyclohexane for the main explosion. Three hypotheses were advanced for the cause of the rupture in the 500 mm line:

- Rupture of the bypass via excessive internal pressure.
- Rupture from a tear in the bellows due to excessive pressure followed by a minor explosion causing final rupture.
- Rupture of the 200 mm line causing the long split, again leading to a minor explosion bringing down the bypass and releasing tonnes of cyclohexane.

Hypothesis 2 was rejected by the Inquiry, since tests at Nottingham University showed that a pressure of 1450 kPa would be necessary to tear the bellows. There was no evidence to support the existence of such a pressure at any time.

Hypothesis 3 caused much discussion and debate which lasted several years [1]. If Hypothesis 3 had happened, the Inquiry held that a series of events had to occur, including a fire directed at elbow G, which destroys the aluminium cladding and lagging in the area.

Both the 75 mm and the 1100 mm cracks were formed when the sections of elbow and pipe were red hot. At elbow G, the maximum temperature was estimated as 850-900°C. The long split started with w - type creep cavitation (triple point wedge cracks) and was completed by plastic shear. Laboratory tests caused similar failure in 4 minutes at 950°C under normal operating pressure for the Nypro plant. The 75 mm crack was the result of Zn liquid metal embrittlement, which would cause failure in seconds above 800°C at operating pressure. Since this only happens rapidly on clean steel, the pipe must have been splashed with molten zinc before it reached 800°C. The bulging of the 75 mm crack indicated that it formed while the 200 mm pipe was still under process pressure.

The Inquiry accepted from photographic and other evidence that the flange bolts were loose before June 1. Since the hypothesis requires so many consecutive events after a leak, the Inquiry thought that, although possible, the 200 mm pipe failure was highly unlikely to be the initiator of the disaster.

Hypothesis 1 was adopted by the Inquiry as the most likely. No malpractice on the part of the control room staff by the final shift was assumed. Thus normal pressure/temperature relationships, below the 1100 kPa setting on the relief valves, were assumed. The primary forces acting on a dog-leg containing two off-set bellows are shown in Figure 3. The offset '2e' produces a bending moment balanced by shear forces. The bending moment can cause failure in two ways in an unanchored bellows:

- The shear force may cause 'squirm' in the bellows
- The bending moment may cause buckling at one of the pipe mitre joints.

The scaffolding provided partial support for the bypass, but did not anchor it. Squirm in a bellows is a phenomenon in which part of the bellows convolutions are pulled or pushed out and never return to the original configuration. Under steady near-normal operating temperature and pressure, tests showed that the dog-leg should not fail. However, a dynamic

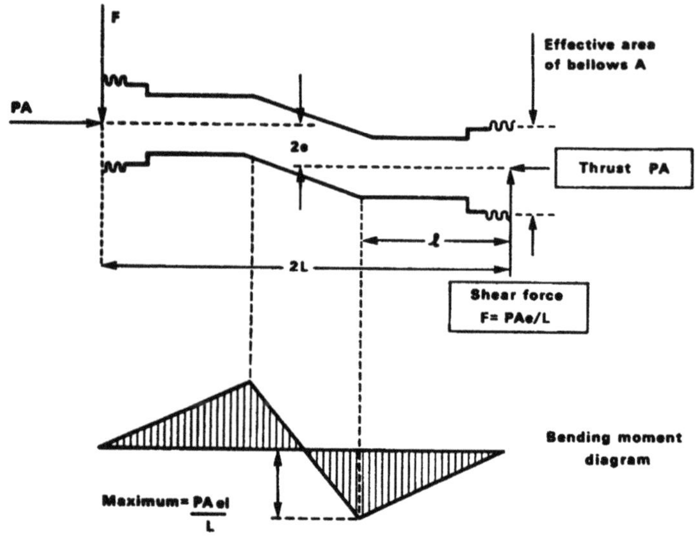

Fig. 3. Forces on the Dog-Leg Bypass containing the Bellows. (HMSO)

movement, where the bellows at each end extend, would increase the contained volume, with vapour from the reactors increasing mechanical work into the assembly, thus increasing the kinetic energy in the flanges.

2 Conclusion

The failure of the bypass by dynamic squirming followed by jack-knifing and rupture was a single event of low probability. It is, however, the most likely primary cause of the failure. The failure of the 200 mm pipe involved a succession of events, many of which required that some or all of the necessary preceding events occurred. Although the scenario points to several problems in the operation of the plant, is too improbable as a cause of the final disaster.

3 References

1. Anon (1976) Flixborough - The Metallurgical Implications *Metallurgist & Materials Technologist* Vol. 8 No. 10 pp536-538.

14 THE TRAINING OF ENGINEERS ON TOTAL QUALITY

L. FARIA
Mechanical Engineering Department, Instituto Superior Tecnico, Universidade Tecnica de Lisboa, Lisbon, Portugal

Abstract
The education of engineers at university level courses (4 or 5 years) is based in mathematics, physics, chemistry, and matters related to the specific area of each course: civil, mechanical, electrical, etc. In general, topics concerning certain practical industrial applications are not lectured as the university should form mainly the mind of the future engineers and has no possibility of assuring their training in all matters, due to lack of time and to the impossibility of present real industrial scenarios.

So the training of engineers in industrial applications should be assured either in the companies – training during the job –, or in post-graduated courses. On the other hand, the rapid evolution of science and technology requires a periodic knowledge updating.

One of the problems the industry is facing at present is the quality of their products. The companies should introduce quality systems in their organizations in order to assure the requited quality level, to reduce, even eliminate, errors and faults, increasing their productivity.

A quality system assures quality assurance in design, production, installation, servicing of manufactured products, and services. The model to be adopted should foresee a "Total quality system" for the company, from the management to the work shop.

The introduction of these concepts in industrial companies is one of the main tasks of engineers; this requires training in the principles and in application of quality systems, a good knowledge of the guidelines allowing a correct selection, and use of "quality management and quality assurance standards" in order to get a better reliability of the products.

A general orientation for the training of engineers in this area is presented.

Teaching and Education in Fracture and Fatigue. Edited by H.P. Rossmanith. Published in 1996 by Taylor & Francis, 2 Park Square, Milton Park, Abingdon, Oxon, OX14 4RN. ISBN 0 419 20700 7.

1 Introduction

The engineering courses at university level have, as a rule, a duration of 4 or 5 years; during the first year or the first two years are lectured the bases of these courses, mainly mathematics, physics, chemistry, and certain applications, for instance, informatics, metallurgy, mechanics, statistics.

During the last 3 or 4 years the subjects that are taught are specifically related to each course (civil, chemical, mechanical, electrical, etc.) with the target of training and educating the mind of the future engineers and giving them the basis for their professional activities.

So the training in practical industrial applications is very limited due not only to the general structure of university courses and to the lack of time, but also to the impossibility of creating in the universities an environment corresponding to the real industrial conditions.

2 Technical evolution

Facing the quick technical evolution, world-wide competition, and the globalization of markets, new demands for the companies are arising, such as:

- The reduction of production cost
- The reduction of lead times, in product development and in manufacturing
- The required right definition of quality and reliability levels for each product
- The reduction of the total life cycle of each product.

The introduction of the informatics has led to the use in design, manufacturing, assembly, purchasing, production planning, management, of tools such as CAD, CAM, MRP, CAPP, CAT, etc. To take advantage of the use of these tools the companies tried to integrate them and new concepts arose for this purpose, called CIM – Computer Integrated Manufacturing: now integrates also the engineering aspects – CIME.

Following these concepts new types of production organization are replacing the traditional technology, consisting of oriented distribution of manufacturing tasks, by production islands, which are responsible for manufacturing, planning, control, material supply, tools and jigs, handling and transportation, machines, maintenance, warehousing, and for the process integrated quality control tasks.

The cooperation between the different departments of a company has become a must; and this cooperation should exist from the design phase of a new product. Working groups, including people from the different departments, should collaborate with the design department from the start of a new product.

To cope also with the above mentioned goals, a similar approach for manufacturing has been developed, consisting of the replacement of sequential operations by largely overlapping processing during product development and design. This approach is generally named concurrent/simultaneous engineering (CSE) and requires appropriated project management systems allowing better linkage between product and process, time optimized process plan, control of the critical parameters of the project, analysis of the risks involved, and integration in CIME.

For this purpose, during product design one should consider the problems of manufacturing, of process planning, of assembly, of maintenance, of quality control and assurance, of other existing constraints, and should assure that the pre-defined costs objectives are not exceeded.

3 Total quality

The problem of the quality management and of quality assurance is of paramount importance all over the world for all organizations – industrial, commercial, services or governmental. This justified the effort made by the ISO leading to study and to approve international standards on this matter.

Standard ISO 9000:1987 establishes the principal concepts, definitions and the use of the standards for quality systems management and contractual purposes. The other standards (9001, 9002, 9003, and 9004) present the different models for quality assurance in design/development, production, installation and servicing, and guidelines for application.

It is important to refer the quality system elements mentioned in these standards:

- Management responsibility
- Quality system principles
- Auditing the quality system (Internal)
- Economics – Quality-related cost considerations
- Quality in marketing (Contract review)
- Quality in specification and design (Design control)
- Quality in procurement (Purchasing)
- Quality in production (Process control)
- Control of production
- Material control and traceability (Product identification and traceability)
- Control of verification status (Inspection and test status)
- Product verification (Inspection and test status)
- Control of measuring and test equipment (Inspection, measuring, and test equipment)
- Nonconformity (Control of nonconforming product)
- Corrective action
- Handling and post-production functions (Handling, storage packaging, and delivery)
- After-sales servicing
- Quality documentation and records (Document control)

- Quality records
- Personnel (Training)
- Product safety and liability
- Use of statistical methods (Statistical techniques)
- Purchaser supplied product.

The application of the quality system principles recommended by these standards leads to the new concept of Total Quality which is a philosophy integrated in the quality management and concerns all aspects related to the quality in a unifying perspective. The Total Quality System (TQS) embraces everything that affects the quality of the product or the service supplied by each company or organization, including management, design, manufacturing, process planning, maintenance, inspection, marketing, etc.; all sectors or departments of a company should act in such a way that the quality concept influences all actions or decisions to be taken.

When introducing the quality system principles in a new product or service "all phases from initial identification to final satisfaction of requirements and customer expectations" should be considered. These phases include mainly the following aspects:

a) marketing and market research: the knowledge of the market and of the possible customers is essential when a new product is introduced;
b) design of the new product, the corresponding specification engineering, the analysis and synthesis adopted and their validation, the fault tree analysis, the fitness for purpose, reliability, safety, the possible future developments, tolerances according to the performance requirements and to the existing manufacturing equipment, planning and control (verification) of design and drawings;
c) procurement and purchasing, detailed specification of raw materials and components, correct adoption of standards, choice of subcontractors (acceptance, audits), inspection and testing of materials, components and parts supplied by the subcontractors;
d) process planning, MRP, JIT, and all possible applicable facilities;
e) manufacturing/production, processes and equipment and their influence on the quality of the products, in-process inspection;
f) packaging, storage, installation, operation, and maintenance;
g) sales, distribution, and technical assistance after sales.

such quality system requires a structure established by the management who is ultimately responsible for the quality policy and for the resources essential to the achievement of quality objectives, including human resources and appropriate equipment and techniques.

Statistical process control is one of the powerful tools for improving quality, as it reduces the process variation and gives satisfaction to the customers, but with a large support of the management, statistics mean nothing.

4 Training and education

The need of adequate training of the personnel involved in production is underlined by the ISO standards mentioned above; these standards refer specifically to the executive and management personnel, the technical personnel and the production supervisors and workers. The need of formal qualifications for certain specialized operations, processes, tests, and inspections is also required. It is of paramount importance to motivate personnel.

The training should provide a clear understanding of the quality system and of the tools and techniques for its operation; the training should include the correct knowledge and proper operation of the equipment, manufacturing processes and product, and assignments on all the tasks contributing to the production.

The list presented in the Section 3 gives a comprehensive view of the matters to be taught to all personnel performing their tasks. As all activities are embraced by the TQS concept all personnel should be trained according to their tasks and responsibilities. The introduction of the new concepts as referred above – CIME, CSE, TQ, etc. –, the technical evolution, and the new manufacturing processes require a continuous updating of all personnel, mainly the engineers who have special responsibilities, as a rule, for production and management of industrial organizations.

The Appendix shows a typical course for training engineers in Total Quality Systems and related matters.

References

[1] ISO 9000:1987 (E)
[2] ISO 9000-3:1991 (E)
[3] ISO 9001:1987 (E)
[4] ISO 9002:1987 (E)
[5] ISO 9003:1987 (E)
[6] ISO 9004:1987 (E)
[7] ISO 9004-2:1987 (E)
[8] E.S.Carnevale, Total Quality Through SPC, Technical of Skills Training, July 1991

Appendix

Typical course for training engineers in Total Quality Systems

1. Integrated management, including:
    - Management and strategy, forecasts
    - Manufacturing (general sense) planning and organization.
2. Quality systems, including:
    - Basic concepts, codes, and standards
    - Quality as a competitivity factor
    - Technical barriers
    - Evolution along space/time

- Non-quality costs
- Tools for introducing TQ
- Audits.

3. Design for quality, manufacturing, assembly, and cost; design of production facilities; production/manufacturing process planning.
4. Purchasing, sub-contracting: audits, inspection of purchased products and components; specifications and standards.
5. Quality inspection techniques:
   - Physical tests, metallurgical test, metrological dimensional control applied to quality inspection
   - Quality inspection procedures
   - Informatic (software) tools for inspection
   - Implementation and management of laboratories for quality inspection.
6. Manufacturing processes and their influence on the quality and cost of products; adequate choice of tolerances.
7. Maintenance and reliability related to the manufacturing equipment and to products:
   - Deterministic and probabilistic design methods
   - Fitness for purpose
   - Maintenance organization; monitoring control
   - Lubrication and its importance; practical applications.

NOTE: The course will include theoretical and practical lectures, laboratory and informatic practices and exercises.

15 TEACHING THE IMPACT OF FRACTURE MECHANICS ON MATERIALS CHARACTERIZATION AND QUALITY CONTROL

H. BLUMENAUER
Otto-von-Guericke University Magdeburg, Germany

Abstract

This paper presents a survey on the teaching and education program in fracture mechanics at the Institute of Materials Engineering and Materials Testing at the Otto-von-Guericke University in Magdeburg in Germany. Special attention has been paid to the current knowledge in the field of micro-structural characterisation of new materials by improved fracture and damage mechanics concepts and the combination with advanced NDE methodologies within the framework of quality assurance.

Key words: Fracture, damage, micro-structure, quality assurance.

1 Materials Science Education Network

The teaching of fracture mechanics is based on various disciplinary backgrounds: physics, mechanics, structural design and last but not least materials science and engineering. This contribution looks at this complex subject from the viewpoint of a materials science engineer.

Figure 1 shows the broad spectrum of materials-related education programs at German universities in their different relation to the natural sciences and engineering sciences.

In order to provide a better information transfer across the European educational system in materials science and related studies, the European Union (EU) has sponsored a project which is called "Establishing a Materials Science Education Network". A

Teaching and Education in Fracture and Fatigue. Edited by H.P. Rossmanith. Published in 1996 by Taylor & Francis, 2 Park Square, Milton Park, Abingdon, Oxon, OX14 4RN. ISBN 0 419 20700 7.

database of the present structures and conditions in education throughout Western and Eastern Europe will be evaluated and made available in the form of a guide. A questionnaire has been developed by an International Working Group and recently distributed to more than 500 departments. The first results have been presented and discussed in a public workshop in the frame of the Euromat Conference at Padua in September 1995. This network will hopefully provide the framework for a new kind of an integrated education program in the field of materials science and engineering.

2 Fracture Mechanics at the University of Magdeburg

How is fracture mechanics included into the curriculum development at the University of Magdeburg?

At the undergraduate level students are integrated in the curriculum of mechanical engineering for the first and second year. An introduction to fracture mechanics is part of the course on materials engineering. In this course the basic concepts of linear-elastic and elastic-plastic fracture mechanics are presented with regard to the mechanical properties and testing methods. An essential issue is to provide students with a good understanding of the interaction between the microscopic features of fracture and such parameters as K_{Ic}, CTOD, crack resistance curve and fatigue crack growth rate. In this introductory course attention is being paid to an integrated treatment of all types of materials [1-2]. During the lectures slides and TV clips showing illustrative examples of structural failures by fracture and fatigue will be presented. It is very beneficial for the students, that, for a better under-standing, they can directly observe crack initiation and extension by SEM in-site investigations.

The materials engineering curriculum in the main study (3rd - 5th year) differs from the other curricula in the Faculty of Mechanical Engineering. The following teaching elements are related to fracture mechanics:

- materials modification and microstructural characterisation,
- mechanical and fracture mechanics testing,
- non-destructive testing (NDT) and evaluation (NDE), and the
- application of fracture mechanics concepts to the improvement of materials and the integrity assessment of structures.

Since materials characterisation and fracture mechanics are our main research activities, students are able to do interdisciplinary work by combining experimental tools (metallography, fractography, mechanical and non-destructive testing) and numerical tools.

3 Quality Assessment and Development

Two points of this curriculum should be emphasized in more detail: the micro-structural characterisation of the damage process and the role of fracture mechanics in quality assessment and development. The current trend in the "micro-structural design" of

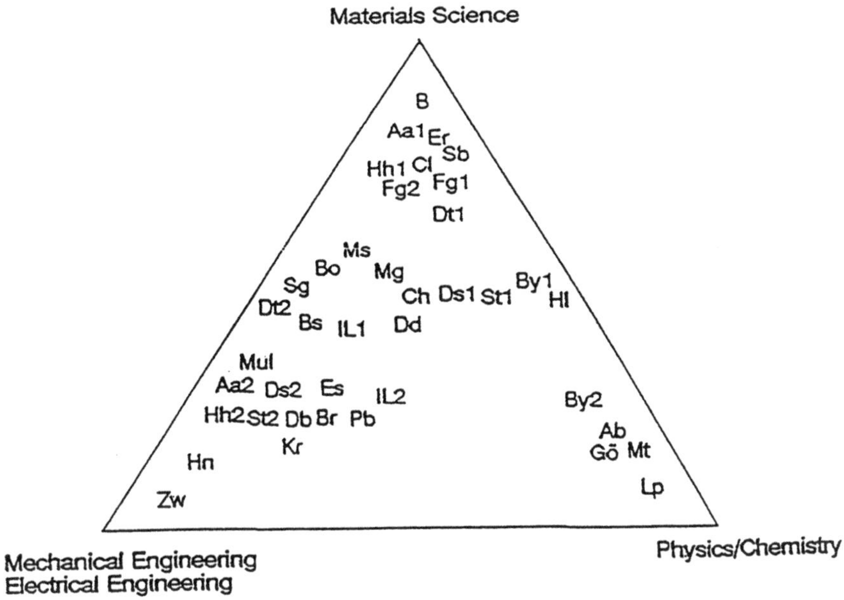

Fig. 1. The character of the teaching programmes in materials science in German universities. Aa: Aachen. Ab: Augsburg. B: Berlin. Bo: Bochum. Br: Bremen. Bs: Braunschweig. By: Bayreuth. Ch: Chemnitz. Cl: Clausthal. Db: Duisburg. Dd: Dresden. Ds: Darmstadt. Dt: Dortmund. Er: Erlangen-Nürnberg. Es: Essen. Fg: Freiberg. Gö: Göttingen. Ha: Hannover. Hh: Hamburg. Hl: Halle. IL: Ilmenau. Kr: Karlsruhe. Lp: Leipzig. Mg: Magdeburg. Ms: Merseburg. Mt: Münster. Mu: Munich. Pb: Paderborn. Sb: Saarbrücken. Sg: Siegen. St: Stuttgart. Zw: Zwickau

(Courtesy of P.P.Schepp, FEMS News 1/94)

damage-tolerant materials need a deeper understanding of the damage under service conditions; therefore, damage mechanics which attempts at simulating the failure process numerically by using constitutive relationships incorporating micro-structural damage parameters, should be more fully integrated into the teaching program. Other challenges come from "smart" or "intelligent" materials with integrated sensors and actuators made from piezo-electric ceramics or shape memory alloys. In this case the combined mechanical, thermal, and electric loads during adaptive operation in adaptive systems must lead to new fracture and damage mechanics concepts representing this interesting behavior.

4 Quality Management and Quality Assurance

A second way of fracture mechanics teaching can be described under the general heading of quality management and quality assurance. The latest phrase "total quality management (TQM)" encompasses the two major interrelated aspects: humanity and technology. TQM becomes increasingly important for quality accreditation and will be a predominant factor for survival on the market. The International Standards in the ISO

9000 family contain principal concepts on quality management and quality control. For the entire life time of a product quality plans must define the specified requirements for testing and inspection. In this context there is a need for the application of advanced testing methods. The development of a more quantitative non-destructive analysis has led to the term "non-destructive evaluation (NDE)" with three categories:

- NDE for defects ("effects of defects" analysis),
- NDE for materials properties (including non-destructive estimation of fracture mechanics parameters), and
- NDE for strains or residual stresses.

The curriculum at the University of Magdeburg provides a broad education in these NDE categories, and a course on "Materials-related Quality Assessment" combines NDE with fracture mechanics approaches to the fitness-for-purpose philosophy. Fitness-for-purpose means that we can tolerate certain imperfections provided that they prove to be harmless in service. The new NDE methodologies demands highly trained personnel and for this reason the University of Magdeburg offers a post-graduate training course in non-destructive testing and evaluation leading to a qualification in a specified field of engineering within the requirements of the general principles of EN 473 "Qualification and certification of NDT personnel".

5 Fracture Mechanics Teaching System

Figure 2 shows a survey on the inputs and outputs of the teaching system in fracture mechanics. This system is assisted by the following two textbooks: Werkstoffprüfung [3] and Technische Bruchmechanik [4]. Reference [3] provides a comprehensive treatment of materials testing and characterisation including fracture mechanics and non-destructive testing for a wider readership. The textbook, Ref.[4], serves as an introduction into fracture mechanics for under-graduate and graduate students as well as a reference book for practical applications in the field.

6 Conclusions

Several conclusions can be drawn:

There is an urgent need for integrated education in materials and mechanical engineering because modern disciplines in engineering require the fall of the traditional borders between classical subjects such as mechanics, materials science, informatics etc. An interdisciplinary view of fracture mechanics and damage mechanics contributes to the teaching and education of students who then should be able to combine physically related models with sophisticated experiments.

- The subject "Reliability of materials" which contains both the microscopic and macroscopic aspects of damage of all types of materials as well as their reliability and life time under service conditions is of great importance.

- Increased attention should be given to the application of fracture mechanics in connection with non-destructive evaluation and quality assurance. Here, the design of an additional post-graduate training course is recommended.

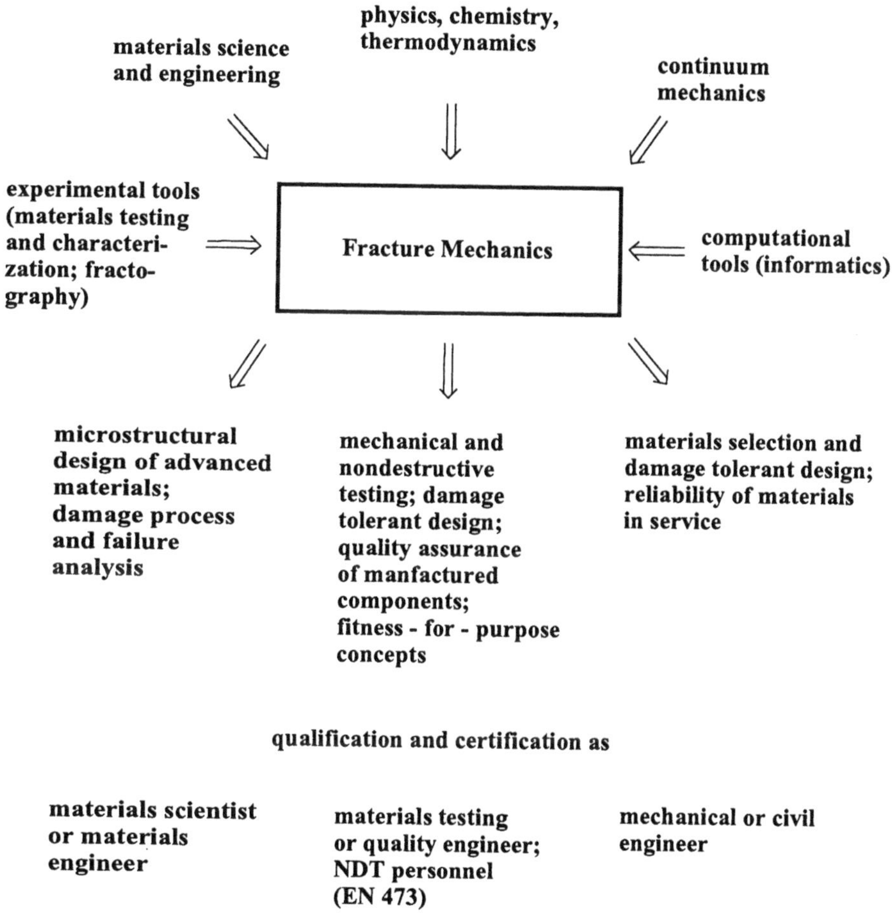

Fig. 2. Fracture mechanics teaching system

References

[1] Schatt, W. (Ed): Einführung in die Werkstoffwissenschaft. Deutscher Verlag für Grundstoffindustrie, Leipzig, 1991.
[2] Schatt, W. (Ed): Werkstoffe des Maschinen-, Anlagen-und Apparatebaues. Deutscher Verlage für Grundstoffindustrie,Leipzig, 1991.
[3] Blumenauer, H. (Ed): Werkstoffprüfung. Deutscher Verlag für Grundstoffindustrie, Leipzig/Stuttgart, 1994.
[4] Blumenauer, H. and G. Pusch: Technische Bruchmechanik. Deutscher Verlag für Grundstoffindustrie, Leipzig/Stuttgart, 1993.

16 THE ROLE OF PRACTICES IN TEACHING AND EDUCATION IN FRACTURE MECHANICS

G. PLUVINAGE
Laboratory of Mechanical Reliability, University of Metz, France

Abstract

Five types of engineering practices can be distinguished in fracture mechanics: Application of a formula, treatment of data, application of a procedure, search for an optimum, and modelization. The prerequisites for these practices will be discussed.

1 Introduction

In teaching it is of major importance to include into the lectures practical problems as well as to integrate engineering practice.

The main purposes of including engineering practice and example problems are:

- to anchor the knowledge,
- to clarify the knowledge and procedures,
- to understand the problems,
- to make a comparison between academic and actual situations,
- to mobilize the knowledge gained.

Before building in practices, the following operations have to be performed:

- selection of the associated contents,
- identification of the knowledge,
- checking of the prerequisites,
- consideration about the formalism of the knowledge,
- selection of the strategy to solve problems.

Teaching and Education in Fracture and Fatigue. Edited by H.P. Rossmanith. Published in 1996 by Taylor & Francis, 2 Park Square, Milton Park, Abingdon, Oxon, OX14 4RN. ISBN 0 419 20700 7.

There are general rules used by teachers to prepare practices and these general rules can also be used in teaching fracture mechanics. However, in this particular field few books featuring practices as well as practical examples are available. This paper informs about the author's experience in this field [1].

The following five types of fracture mechanics practices should be included in a FM course offered to students:

- Application of a formula,
- Treatment of data,
- Application of a procedure,
- Search for an optimum, and
- Modelization.

Examples highlighting these five types of practices will be presented in this paper.

Linear Elastic Fracture Mechanics (LEFM) and Elasto-plastic Fracture Mechanics (EPFM) form the framework of these investigations.

In LEFM, three basic formulas will be used:

- the distribution of the stress σ_{yy} around the crack tip

$$\sigma_{yy} = \frac{K}{\sqrt{2\pi r}} \cdot f_{ij}(\theta) \tag{1}$$

  where $f_{ij}(\theta)$ is an angular function;

- the relationship between the external global stress σ_g and the mode-I intensity factor K_I

$$K = \sigma_g \sqrt{\pi a} \cdot F_\sigma(a/W) \tag{2}$$

  where $F_\sigma(a/W)$ is a geometry correction factor, "a" is the size of the defect;

- the relationship between the critical strain energy release rate G_{Ic} and the critical stress intensity factor K_{Ic}

$$G_{Ic} = \frac{K_{Ic}^2}{E'} \tag{3}$$

  where $E' = E$ Young's modulus in plane stress, $E' = E/(1-\nu^2)$ in plane strain, and ν is Poisson's ratio.

The major difficulty when applying EPFM comes from the existence of a very large number of fracture criteria. A clear comprehension of the elastoplastic fracture mechanics requires a classification of fracture criteria. In addition, it is important to define the domains of validity of these criteria and a comparison between these criteria is mandatory. A first classification of the fracture criteria into four classes is proposed here which are related to stress, strain and energy, and to a method of interpolation between the two well-defined limiting cases of brittle fracture and instability.

An improved description of all aspects of EPFM can be achieved by using the classes of fracture criteria shown in Table I.

Table I: Fracture criteria

local stress criteria	local strain criteria	local energetic criteria
global stress criteria	global strain criteria	global energetic criteria
	interpolation criteria	

The seven classes indicated in Table I comprise 30 major elasto-plastic fracture criteria [2].

2 Examples of Practices in Fracture Mechanics

2.1 Application of a Formula

Explicit application of a practice is given in the case where a formula derived in the course is frequently applied in practical exercise problems with the purpose of stably memorizing the knowledge taught. This kind of practice is generally performed during the lectures and going through the calculations of the exercise problems demonstrates the order of magnitude and the units of the results.

Figure 1 shows the results of computing the opening displacement of a crack in a steel plate with a yield stress of 1500 MPa subjected to a 300 MPa global stress

2.2 Treatment of Data

Rough data as obtained from testing can be used in practical exercise problems. A typical example is the determination of the fracture toughness, K_{Ic}, or the determination of the critical crack opening displacement from the load-displacement diagram. This procedure must be performed according to the appropriate standards which will be studied at the same time (Fig.2).

2.3 Application of a Design Route

The application of a design route (e.g. the determination of the critical defect size using the RCC-MR A16 Route) is a very practical problem. Many engineers are

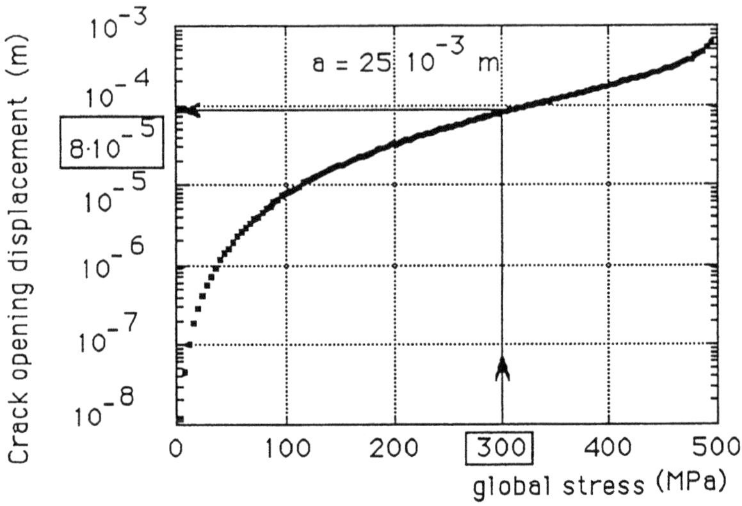

Figure 1. Evolution of crack opening displacement versus critical global stress (global stress 300 MPa)

concerned with this kind of problems. As much time is required for the calculation of the solution, therefore this kind of practice is more suitable for home training.

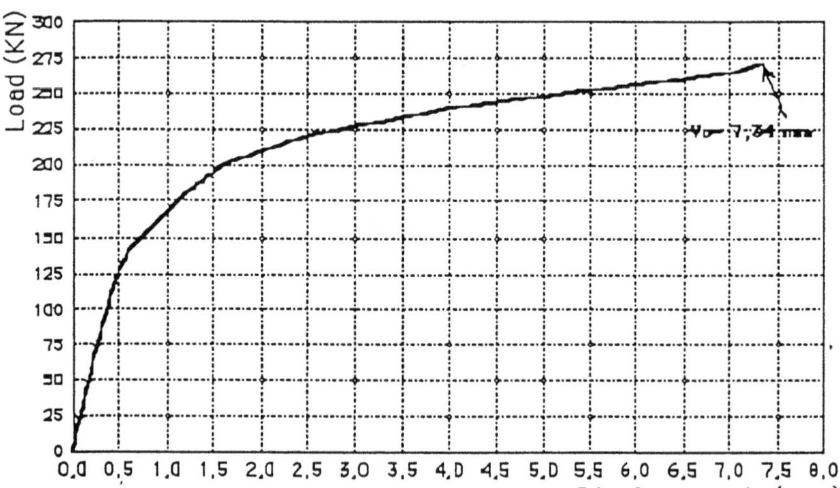

Figure 2. Load-displacement diagram for the determination of the critical crack opening displacement

Example:

A pipe with an annular defect is made of stainless steel (J_{Ic}: 78 KJ/m^2, Re: 141 MPa) and subjected to a global tensile stress of 180 MPa.

The RCC-MR A16 Route assessment diagram is given by the following equation:

$$k_r = \left(1 - 0.14.(S_r)^2\right).\left[0.3 + 0.7.\ \exp\left(-0.65\ .\ (S_r)^6\right)\right] \quad (4)$$

It is shown, in Fig.3, that for the given input values the integrity of the structure is secured.

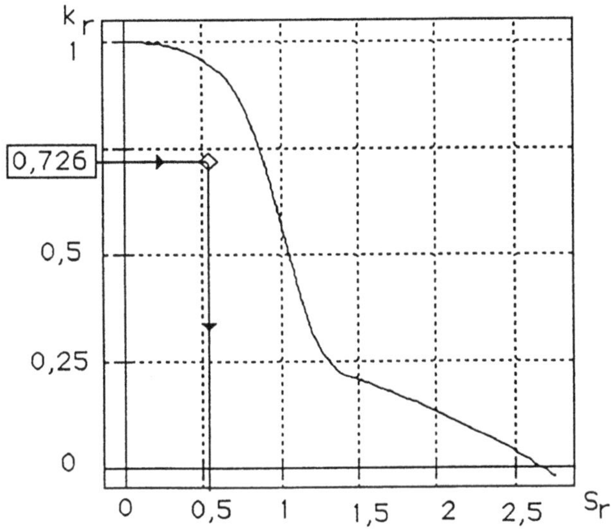

Figure 3. The RCC-MR A16 Route assessment diagram

2.4 Search for an Optimum

Engineers should be sensibilized about the influence of mechanical and metallurgical parameters on the fracture toughness. A fracture mechanics controlled optimum is required. This can be done by deriving formulas. An example is given by the derivative of the Mac-Clintock formula for critical failure strain with respect to the strain hardening exponent n.

$$\frac{\partial \varepsilon_l^C}{\partial n} = \frac{\ln\left(f_v\right)^{1/3}}{sh\left[(1-n)\sqrt{3}.\beta\right]}\left[\frac{\sqrt{3}.\beta.\ln\left(f_v\right)^{1/3}}{th\left[(1-n)\sqrt{3}.\beta\right]}.-1\right] \quad (5)$$

Here, β is the triaxiality of stresses and f_v is the volumetric fraction of inclusions. It can be seen that the critical strain increases with increasing strain hardening exponent.

2.5 Modelization of Physical Phenomena

In the general case, this type of practice is very difficult to master on the undergraduate level and, therefore, is reserved for a series of Master's lectures. The use of the concept of a micro-specimen at a crack tip to determine the influence of the grain size on the fracture toughness is a good example for this kind of modelization. The students' attention must be drawn to appreciate the influence of stress triaxiality on the fracture toughness (Fig.4)

3 Prerequisites

It is interesting to notice that fracture mechanics needs an interdisciplinary approach of practices. Knowledge in the fields of mathematics, statistics, strength of materials,

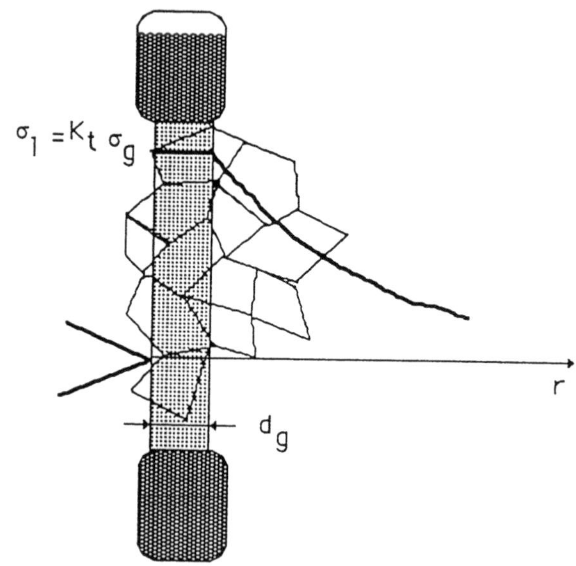

Figure 4. The concept of micro-specimen at a crack tip

solid mechanics, plasticity, metallurgy and materials science is required to understand and solve fracture mechanics problems.

The following examples of fracture problems are designed in order to demonstrate the importance of having acquired an appropriate level of knowledge of the subjects mentioned above.

Example 1: *Mathematics*

Integration and derivation are the most frequently used mathematical tools for solving fracture mechanics problems.

In this example problem the stress intensity factor shall be calculated for a crack type defect located in a beam which is submitted to a constant bending moment (see Fig.5).

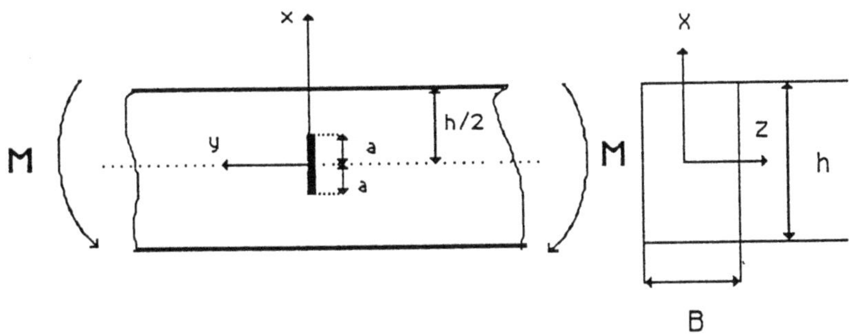

Figure 5. Beam submitted to a constant bending moment

The calculations can be performed by using the Green's function approach:

$$K_I = \frac{1}{\sqrt{\pi a}} \cdot \int_{-a}^{a} \sigma_{yy} \sqrt{\left(\frac{a+x}{a-x}\right)}\, dx \tag{6}$$

After a change of parameters: $x = a \sin \alpha \Rightarrow dx = a \cos \alpha \, d\alpha$, one obtains:

$$K_I = \frac{12\,M}{B.\sqrt{pa}.\,h^3} \int_{-\frac{p}{2}}^{p/2} \frac{a \sin a(a + a \sin a)}{\sqrt{a^2 - a^2 \sin^2 a}}\, a \cos a \, d a = 6\,\frac{Ma}{B\,h^3}\sqrt{pa} \tag{7}$$

Example 2: *Strength of Materials*

Principles and results of strength of materials are widely used in fracture mechanics. Simple beam theory shall be employed to find the value of the J-integral in the case of a DCB specimen (see Fig.6).

The J-integral is defined in the form

$$J = -\frac{1}{B}\frac{\partial \Pi}{\partial a} = \frac{1}{2B}\frac{\partial d}{\partial a} \tag{8}$$

where the half-beam displacement is given by

$$d = \frac{2}{3}\frac{Pa^3}{EI} \tag{9}$$

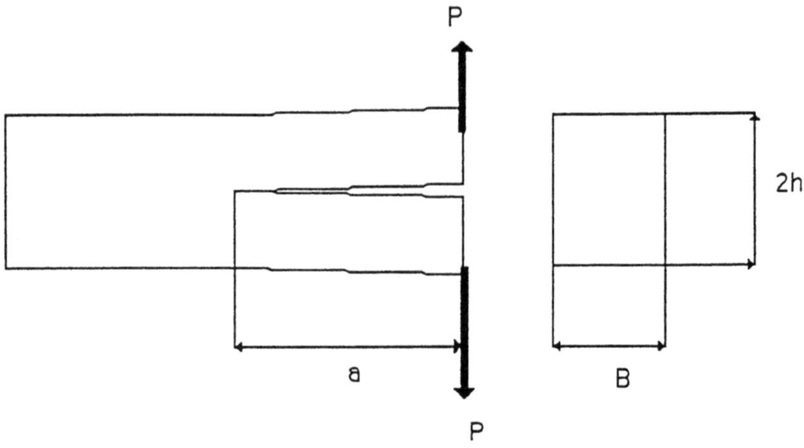

Figure 6. DCB specimen loaded by forces P

which leads to

$$J = \frac{6\,P^2}{6\,B\,EI} \,.\, a^2 \tag{10}$$

Example 3: ***Solid Mechanics***

Solid mechanics, in particular the theory of linear elasticity is employed. The following example pertains to the computation of the strain energy density W* in the vicinity of a crack tip which is subjected to mode-III loading. The stress distribution is given by:

$$\sigma_{xz} = \left[- \frac{K_{III}}{\sqrt{2\,\pi\,r}} \,.\, \sin\theta \right] \qquad \sigma_{yz} = \left[- \frac{K_{III}}{\sqrt{2\,\pi\,r}} \,.\, \cos\theta \right] \tag{11}$$

Using the definition of the strain energy density

$$W^* = \frac{s_{ij}}{2} \,.\, \frac{1}{2m}\left(s_{ij} - \frac{n\,.}{1+n} s_{kk} d_{ij} \right) = \frac{1}{4m}\left(s_{ij}.s_{ij} - \frac{n}{1+n} s_{kk} s_{ii} \right) \tag{12}$$

one obtains

$$W^* = \frac{1}{2m}\left[\left(- \frac{K_{III}}{\sqrt{2\,\pi\,r}} \,.\, \sin q \right)^2 + \left(- \frac{K_{III}}{\sqrt{2\,\pi\,r}} \,.\, \cos q \right)^2 \right] = \frac{K_{III}^2}{4\,\pi\,m\,.\,r} \tag{13}$$

Example 4: *Plasticity*

Plasticity, specifically instability criteria and limit analysis, are suitable tools for the evaluation of the J-integral and global strain. A well-known example of the use of a limit analysis is the computation of the limit load for a CT specimen (Fig.7) by means of the lower bound solution by Merkle and Randall.

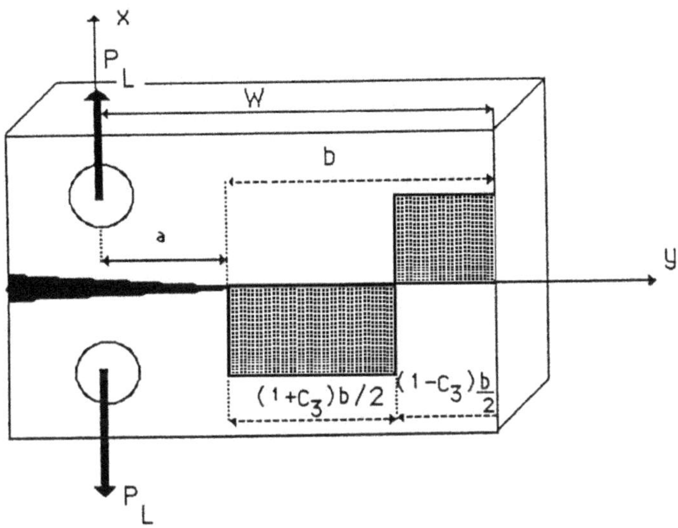

Figure 7. CT specimen, loading and lower bound solution by Merkle and Randall

The constant C_3 is geometry-dependent

$$C_3 = \frac{\sqrt{2a^2 + 2W^2} - (W-a)}{(W-a)} \tag{14}$$

and the limit load, P_L, is given by

$$P_L = Re\, B \,.\, C_3 \frac{b}{2} \tag{15}$$

Example 5: *Materials Science*

Metallurgical laws, particularly relationship describing the influence of parameters such as grain size, loading rate and temperature on the yield stress are of primary importance. The relationships developed by Hall-Petch, Ryvkina and Yarowzewitch will be introduced into fracture mechanics models to describe the influence of metallurgical parameters.

The yield stress is given by:

$$Re = \sigma_{\mu} + \left(Re^{o} - \sigma_{\mu}\right) \exp\left(- CT \log \frac{A_F}{\dot{\varepsilon}}\right) \quad (16)$$

where σ_{μ} is the athermal stress, Re^{o} is the yield stress at zero temperature, C is a constant, and A_F is a frequency factor.

4 Conclusions

Five kinds of practices employed in fracture mechanics have been identified. These are: the application of formulas, treatment of data, application of procedures, search for an optimum, and modelization.

These practices correspond to two situations. It is proposed that some of the practices be employed in training only, while the more synthetic ones shall serve as a model to be followed in an actual situation. The educator may utilize the practices of the first category for "classical" examinations (two hours in a class room), while those of the second category are more useful for home examinations.

It is also interesting to note that the fact that the technical subject of fracture mechanics requires an interdisciplinary approach, the "case studies" method of teaching seems to be particularly suitable.

References

[1] G.Pluvinage, 120 Exercises de Mécanique Elastoplastique de la Rupture, Editions CEPADUES, 1994.

[2] G.Pluvinage, Mécanique Elastoplastique de la Rupture, Editions CEPADUES, 1989.

Regional Differences

Teaching and Education in Fracture and Fatigue. Edited by H.P. Rossmanith. Published in 1996 by Taylor & Francis, 2 Park Square, Milton Park, Abingdon, Oxon, OX14 4RN. ISBN 0 419 20700 7.

17 ADVANCED STUDIES DIPLOMA IN MECHANICS AND MATERIALS

D. FRANCOIS
Laboratoire de Mecanique, Ecole Centrale Paris, France

Abstract

Industry requires engineers with a strong knowledge of solid mechanics and materials science. In this respect the general education of engineering schools in France needs to be completed by more specialised courses. A community of scientists exists which is able to teach the links between microstructures of materials and their mechanical behaviour. Thus was created an Advanced Studies Diploma in Mechanics and Materials. It is organised jointly by top level laboratories in the Paris area. A Council including representatives from industry looks after its continued relevance. The programme deals with the main classes of mechanical behaviour, going from microscopic mechanisms to macroscopic behaviour. A common core is followed by three more specialized options, concerned with forming by plastic deformation, damage and fracture mechanisms and mechanics. Training exercises include laboratory work and a detailed case study based on PhD. thesis works. As an example, the students were asked to work on the fracture of zinc, including doing compression tests on single crystals and reassessing papers concerned with cleavage fracture and mixed mode fracture of textured plates.

1 Introduction

A few years ago mechanics of solids and materials science used to be fairly dissociated fields. Fracture mechanics was an important factor in revealing the need of bridging the gap.

Teaching and Education in Fracture and Fatigue. Edited by H.P. Rossmanith. Published in 1996 by Taylor & Francis, 2 Park Square, Milton Park, Abingdon, Oxon, OX14 4RN. ISBN 0 419 20700 7.

Whereas the brittle -ductile transition of ferritic steels was well understood and studied by metallurgists, it was not until the mechanics of cracked parts was rigorously established that safe predictions of fracture could be attempted.
But it was soon realized that the ideal elastic solutions needed to be amended in view of the real behaviour of materials which is so intrinsically linked with their microstructure. Nowadays designing against fracture, predicting its occurrence, analysing broken parts require engineers with a strong background both in materials science and in solid mechanics.

To speak of the situation in France, training of engineers with a large background had been efficient for a long time in top engineering schools (grandes écoles). However in view of the large number of subjects in the curriculum it was out of question to go into advanced descriptions of each of them. Now the relations between materials science and solid mechanics are arduous, particularly so because they involve such large and frequent changes of scale.

It is only learned students who can fully grasp these intricacies. Special courses were needed at a high level, at the end of engineering studies.
The need to better understand how the microstructure of materials influence their macroscopic mechanical behaviour was felt more and more in the seventies both by industry and by academics. The National Centre for Scientific Research (Centre National de la Recherche Scientifique CNRS) launched a special programme to promote such research: it was the "GRECO" "large deformations and damage", which was very efficient in creating a common language between scientists and engineers in the fields of materials science and of solid mechanics. This created a community of scientists well prepared to disseminate their knowledge.
In the early eighties strong incentives came from the nuclear, aeronautical, spatial, steel and aluminium industries and others as large programmes in these areas needed well trained engineers in the mechanics of materials.
In France Engineering Schools were, and still are in most cases, disconnected from Universities and research did not occupy a strong position in most of them. However it became more and more obvious that research ought to play an important role in the training of engineers, and the situation has now profoundly changed.
In this respect the advanced studies diplomas (diplomes d'études approfondies DEA) which lead the way to thesis work are now not only widely opened to engineering students, but are in many cases considered in themselves as useful supplements to their studies. DEA are one year programmes which begin graduate studies.Most engineering students can be allowed to take them at the same time as their last year of engineering studies (five years after the "baccalauréat").

Taking advantage of these evolutions, pushed forward by industrial needs, André Pineau, André Zaoui and Dominique François decided, at the beginning of the eighties, to create a DEA in Mechanics and Materials aimed at promoting research at the border between these fields and at giving engineering students a special training by research in this direction. Because it has so much to do with fatigue and fracture, it seemed legitimate to explain how it works in this conference. Explanations will be given about organisation features which ensure continued relevance, some details about the programme will be given and emphasis will be put on special training exercises which are somewhat original.

2 The conditions for relevance

Several features avoid deviating to a teaching far from industrial needs or out-of-date. The DEA is organised in co-operation by several laboratories which are considered among the best in the field.
They are the following:
-Laboratoire de mécanique des solides. Ecole Polytechnique
-Laboratoire des matériaux. Ecole Nationale Supérieure des Mines de Paris.
-Laboratoire de Mécanique (Sols, Structures, Matériaux). Ecole Centrale de Paris
-Laboratoire de mécanique et technologie. Ecole Normale Supérieure de Cachan.
-Laboratoire M3. Ecole Nationale Supérieure d'Arts et Métiers.
Ecole Nationale Supérieure d'Arts et Métiers.
-Laboratoire des propriétés mécaniques et thermodynamiques des matériaux.
Université Paris-Nord - Villetaneuse.
This ensures that the teaching staff is at the top level and provides for a large enough renewal to avoid sclerosis.

A council meets twice a year to follow the evaluations of the DEA. It includes representatives of numerous industries and is chaired by a leading personality of a large company. Thus the industrial needs keep being accounted for.

3 The programme

The mechanical behaviour of materials must be well understood for the sound design of equipments and for improving their properties. This involves grasping the relations between processes which take place at microscopic scales and the macroscopic mechanical behaviour. They are studied according to the typical classes of mechanical behaviour, namely elasticity, plasticity, viscosity, and to the one which results from damage. Lastly fracture mechanics comes as a sort of synthesis, all these phenomena taking place at crack tips.
The detailed programme of the courses is given in annex I. They occupy Thursdays and Saturday mornings from September to January for 175 hours. They are followed by more specialised topics in three options taking place in February and March: mechanisms and mechanics of mechanical behaviour, damage mechanisms and mechanics.These correspond to 60 hours of lectures but project work can easely add 100 hours more. In these options again a balance is kept between the study of micro mechanisms and of macroscopic behaviours. Based on the courses Dominique François, André Pineau and André Zaoui wrote two books on the mechanical behaviour of materials. [1] They include problems which help to handle the various concepts. The DEA finally includes a project in a laboratory which lasts for at least three months.
As the hired students do not possess the same background, equalising courses are given in solid mechanics and in materials science before the programme itself begins.
Materials science being strongly rooted in metallurgy, metallic alloys occupy a large part of the programme. However the courses bear as much as possible on other materials as well, and special courses deal with polymers. As an example the courses on damage mechanisms include microcracking in concrete.

Annex II gives at list of topics covered by the laboratory projects.

4 Training exercises

The evaluation of the students progress includes usual examinations with problem solving. However it was felt that exercises requiring more initiative should be an important part of this evaluation.
First the students are required for a limited time, to do some laboratory works, helping Ph.D. students in their experiments. They must not keep confined in a single technique, but should, on the contrary, learn to use various instruments: microscopes, testing machines, Xrays diffraction...They are then asked to present their results on a poster, which can be seen by other students and which is examined by professors. Questions are asked about the experimental techniques and about the reliability and precision of the measurements. At this stage the interpretation and the discussion of the results are avoided. The Ph.D students who headed the work are asked to give their appreciations. The marking tries to bear essentially on the experimental ability.
Secondly, the students are given a case study (or what we like to call a "long exercise"). Usually based on Ph.D thesis, it consists in the step by step exploration of a subject. It requires the reading of various papers. It involves a certain amount of calculations, and critical appraisal of models. This work is achieved at home and the students are not discouraged to work together. However each one must write a complete answer. The next paragraph describes such a case study which was given this year and which deals with the mechanical behaviour and fracture of zinc. Annex III gives the list of the subjects covered by these case studies.

5 Example of a case study on the fracture of zinc

The case study included first a general paper about the properties and uses of zinc. It gave most of the data needed for solving the various questions.
In the first part the students were given specimens of zinc single crystals and were asked to determine their orientation. They then had to test them in compression and to explain their observations. According to the orientation they could obtain glide or twinning. The second part consisted in the calculation of the elastic properties of polycrystals given the elastic constants of the single crystal. This was to be done according to given pole figures of zinc textured rolled plates. The third part of the case study dealt with the anisotropic properties of the zinc textured rolled plates, which had to be discussed in view of the pole figures.
In the fourth part the students were asked to reanalyse the famous paper by Stroh [2] in which he gave the conditions for cleavage fracture of zinc single crystals triggered by the separation of a kink [3]. It could be shown that a slightly different criterion from the one given by Stroh resulted from rigorous calculations of the stress intensity factor induced by a split dislocations wall. The experimental cleavage stress measurements of Deruyterre and Greenough [4] had then to be compared with this cleavage criterion.
Finally the last part consisted in analysing mixed mode fracture toughness results for textured zinc plates which had been obtained by Lemant and Pineau [5]. This involved the computation of the stress intensity factor in an anisotropic material. Use was made of the results obtained on the elastic constants in the second part of the exercise. Annex IV gives a more detailed outline of this particular exercise. The questions to be solved are a mixture of classical problems found in text books, and of more involved ones including reappraisal of published models or of interpretations given in a thesis work.
It is believed that such work which gets the students much interested and involved, gives them the opportunity to exercise their initiative, their ability for critical appraisal, for searching for relevant informations. A few go beyond what is strictly required and they are able to produce magnificent reports.

6 Conclusion

An average of thirty students every year have now taken the DEA in Mechanics and Materials which has been in operation for nine years. Many of them found positions in industry. Quite a few succeeded in achieving a Ph.D work (70 % about). A number of them were hired in Universities or in CNRS. Apparently what they have learned in the DEA helps them in their career, not only because they acquired a solid background on the mechanical behaviour of materials but also because of the training they got on research methodology.

It must be stressed that the DEA welcomed a number of foreign students. It would be nice if this trend could be strengthened. It might be good idea to include it now in a wider European project.

References

1. D. François, A. Pineau and A. Zaoui. Propriétés mécanique des matériaux. Elasticité et plasticité. Hermes. Paris (1991, 1992) Viscosité, endommagement, mécanique de la rupture, mécanique du contact. Hermes. Paris (1993).

2. A.N. Stroh. "The cleavage of metal single crystals". Phil. Mag.3 (1958) 597-606.

3. J.J. Gilman "Mechanism of ortho kink-band formation in compressed zinc monocrystals" - Trans. AIME . J. of Metals. 64 (1964) 629

4. A. Deruyterre and G.B. Greenough. "Cleavage of zinc single crystals" Nature 172 (1953) 170- 171

5. Lemant and A.Pineau. "Mixed mode fracture of a brittle orthotropic material. Example of strongly textured zinc sheets" Engng. Fract. Mech. 14 (1981) 91-105.

6. J. Levasseur. "Le zinc, propriétés et applications". Matériaux et Techniques. 6-7 (1993) 43-52.

7. J. Friedel. Les dislocations. Gauthier-Villard. Paris (1956) - Dislocations. Pergamon Press (1964).

Annex I
DEA - Mechanics and Materials

Programme

1. Introduction - main types of materials and of mechanical behaviour- scales- observations and testing techniques.

   Mechanical modelling - anisotropy and heterogeneity - homogenization.

2 Elastic behaviour
   - elastic potential - linear elasticity - anisotropy - waves - variational formulation - finite elements - heterogeneous materials - bounds - inclusion - porous and composite materials.

3. Plastic behaviour: threshold - criteria - strain hardening - maximum plastic work - normality - limit load
   Dislocations: perfect and imperfect dislocations - twinning- obstacles = Pearls force - dislocations forest - grain boundaries - surfaces.
   Recrystallisation.
   Solid solution and precipitates hardening

4. Viscosity.
   -thermal activation of deformation -creep mechanisms, -dissipation potentials, -mechanical behaviour of polymers.

5. Damage.
   -cleavage and ductile fracture - brittle ductile transition - statistical aspects,
   -nucleation, growth and coalescence of cavities,
   -fatigue mechanisms,
   -creep damage,
   -damage mechanics.

6. Fracture mechanics.
   -stress intensity factor,
   -strain energy release rate,
   -fracture toughness.

Annex II
DEA Projects - examples

-Fatigue crack growth in an austenitic stainless steel at high temperature.
-Experimental and numerical study of stresses in the grains of a polycrystal.
-Thermomechanical behaviour of a cast iron - ceramic assembly.
-Ageing and damage of an austenitic stainless steel.
-Study of a cyclic elastoplastic constitutive equation.
-Mechanical behaviour of a dual phase single crystal.
-Influence of strain hardening on the service behaviour of surgical implants.
-Behaviour of hard layers.

-Fatigue resistance of an elastomeric material.
-Microstructure and mechanical behaviour of Al-SiC composites.
-Fatigue cracking of brake disks.
-Influence of thermal treatment on the fatigue resistance of a carbon steel.
-Fretting of titanium.
-Brittle-ductile transition of polyepoxy.
-Characterisation of the surface by its fractal dimension.
-Modelling of machining by cutting tool.
-Drawing behaviour of aluminium alloys.
-Local texture of a fine grains steel.
-Mechanical behaviour of aluminium - alumina composites.
-Elastic and viscoelastic properties of expanded polystyrene.
-Influence of elaboration on the mechanical properties of organic matrix composites.

Annex III
list of case studies

87/88	Alluminium alloys and composites. Thermomechanical behaviour
88/89	Aluminium - Lithium alloys - Mechanical properties including fatigue
89/90	a) Nickel alloy single crystal. Anisotropy b) Fracture tougheness of fibers reinforced comerate
90/91	a) Ceramic materials sintering - Mechanical behaviour of porous media b) Polymeric materials
91/92	Plasticity and damage under multiaxial loading in austenitie stainless steels
92/93	Mechanical behaviour of zirconium

Annex IV
Case study on the mechanical behaviour of zinc

1 General information

A general paper on Zinc = "Le zinc, propriétés et applications" by Jean Levasseur [6].

Physics and mechanics data.

composition of Zn GOB
melting temperature - surface tension - crystallographic parameters.
Young's modulus - shear modulus - Poisson ratio - elastic compliance of single crystals - critical shear stress.
-references.

2 Experiments

2.1 Description of specimens (Each student received a single crystal specimen).

2.2 Crystallography

-give the value of the angle Ø between the basal plane and K_I and the value of γ for twinning. [K_I being $(10\bar{1}2)$]

-calculate the maximum associated dilatation

-is twinning expected in a compression test of a zinc single crystal ?

2.3 Orientation of the single crystal using the Laue method. Give the orientation on a stereographic projection.

2.4 Compression test.

-Record the stress - strain curve - take a photograph after the test.

-Measure the Young's modulus and the critical shear stress.

-Did you observe slip lines, twins, cleavage ?

-Explain your results.

3 Elasticity

3.1 Single crystal properties.

3.1.1 Write the matrix of elastic compliance's.

3.1.2 Give the Cauchy relations between the elastic constants.

3.1.3 Show that they are not observed by metals.

3.1.4 Demonstrate that in the case of hexagonal symmetry there are only five elements in the matrix of elastic compliances.

3.2 Influence of anisotropy

3.2.1 Give the tensor to transform the local axis in the axis of the sheet.

3.2.2 Give the elastic compliances in the axis of the sheet.

3.2.3 Calculate the Young's modulus of a zinc single crystal, the c axis of which makes an angle θ with the tensile axis.

3.2.4 Calculate the shear and the compressibility moduli.

3.3 Zinc rolled sheet.

The Young's moduli in the longitudinal and transverse directions are given as a function of the rolling ratio, as well as pole figures.

3.3.1 Explain qualitatively the pole figures observed. How a rolled Zr rolled sheet would differ ?

3.3.2 Consider the sheet as a pseudo single crystal. Calculate the evolution of the Young's modulus in the longitudinal direction during rolling and compare with the given results.

3.4 Polycristal.

3.4.1 Give the average property of the localization tensor, considering the pole figures.

3.4.2 Define the effective compliance tensor of the homogeneous equivalent material.

3.4.3 Determine the compliance tensor in the case of Voigt's approximation.

3.4.4 Determine the compliance tensor in the case of Reuss's approximation.

3.5 Young's modulus of a textured zinc sheet.

3.5.1 Give the Reuss bound for a given function of the crystalline orientation density.

3.5.2 Propose an approximate function of the crystalline orientation density from the pole figures given.

3.5.3 Calculate the Young's modulus using Reuss's approximation. Compare with the given experimental results.

4 Plasticity

4.1 Plasticity criterion.

4.1.1 Show that Schmid's law corresponds to plastic incompressibility.

4.1.2 Calculate the elastic limit of a zinc single crystal whose c axis makes an angle θ with the tensile axis.

4.1.3 Show that the principle of maximum plastic work applies for materials which obey Schmid's law.

4.1.4 Give the loading function for a zinc single crystal. Show that it is a standard material.

4.1.5 Show that Tresca's criterion is only macroscopic.

4.2 Elastic limit of textured zinc.

4.2.1 What are the most favourable orientations to activate various glide systems ?

4.2.2 The evaluations of the elastic limits in the longitudinal and transverse directions are given for a rolled sheet as a function of the rolling ratio. Discuss these results.

5 Cleavage of zinc

5.1 Examine the micrograph of cleavage in zinc published bu Gilman [3].
Study the models of Friedel [7] and of Stroh [2].

5.2 The cleavage mechanism.
Looking at Gilman's micrograph explain the cleavage mechanism.

5.3 Kink band.

5.3.1 Draw what would be the free deformation of a zinc single crystals in a compression test. Calculate the glide as a function of the vertical displacement of the faces.

5.3.2 Calculate the density of geometrically necessary dislocations when the faces cannot move horizontally.

5.3.3 Show that the arrangement of these dislocations in a wall forms a kink band and give the angle of disorientation.

5.3.4 Show that this arrangement is favorable.

5.4 Separation stress for a kink.
Calculate the effective shear stress needed to split a kink in two parts by slip as a function of the distance of separation (replace the dislocation wall by a continuous distribution).

5.5 Nucleation of cleavage cracks.
Calculate the disorientation angle of the kink band needed to nucleate a cleavage crack.

5.6 Propagation of cleavage cracks.

Calculate the stress intensity factors K_I and K_{II} at the tip of the cleavage crack and deduce the strain energy release rate as a function of the cleavage length and of the applied stress. Show that the cleavage criterion can be written

$L(\tau - \tau_i)(\sigma_n + t) = C^{te}$

Where L is the size of the crystal

$\tau - \tau_i$ the effective shear stress on the basal plane

σ_n the normal stress on the basal plane and

$t^2 = \sigma_n^2 + \tau^2$

- 5.6.1 Compare this criterion with the results published by Deruyterre and Greenough [4].
- 5.6.2 How is this criterion transfered to a polycrystal ?

6 Mixed mode fracture of a zinc sheet

6.1 Results obtained by Lemant and Pineau [5] are given.

6.2 Plastic zone size.

- 6.2.1 Check that the condition of plane strain is fulfilled.
- 6.2.2 Give an approximate expression of the plastic zone size in mixed mode.
- 6.2.3 Calculate the plastic zone sizes at fracture for the given results. What should have been the precracking ΔK value to avoid preloading effects ?

6.3 Erdogan and Sih maximum stress criterion.

- 6.3.1 Calculate the matrix of elastic constants at liquid nitrogen temperature.
- 6.3.2 Study the characteristic equation used in the calculation of the stress and strain fields at the tip of the crack in the anisotropic material.
- 6.3.3 Give the stress field at the tip of the crack in cylindrical co-ordinates.
- 6.3.4 In mixed mode loading. Calculate the normal stress on an inclined plane at the tip of the crack as a function of its orientation. Check whether the maximum stress criterion applies in the case of zinc.
- 6.3.5 As in the experiments the crack propagated without deviation, check if the normal stress on the plane of the crack can be used as a criterion.

6.4 Energetic criterion.

- 6.4.1 Give the crack opening displacement in modes I and II for the anisotropic material.
- 6.4.2 Calculate the stresses on the plane of the crack in mixed mode;
- 6.4.3 Calculate the work needed to close the crack tip along a small length in mixed mode for the anisotropy material.
- 6.4.4 Calculate the strain energy release rate for the anisotropic material.
- 6.4.5 Calculate the critical strain energy release rate for the experiments reported on zinc. Check whether the criterion applies.

6.5 Local criterion.

6.5.1 Check whether Stroh's cleavage criterion or the criterion calculated in 5.6. apply for fracture stress data obtained for zinc at 77°K.

6.5.2 What relation should relate K_{IC} and K_{IIC} if one of these criteria applied ? Check their validity for the tests on textured zinc.

18 TEACHING FRACTURE MECHANICS IN CIVIL ENGINEERING EDUCATION: THE SPANISH EXPERIENCE

J. TORIBIO
Department of Materials Science, University of La Coruna,
La Coruna, Spain

Abstract
This report is devoted to the Spanish experience in teaching fatigue and fracture mechanics in the framework of civil engineering courses. Formative and cognitive objectives of teaching are mentioned, as well as different teaching methods. Emphasis is placed on the degree of application of fracture mechanics techniques in civil engineering design, particularly in structural mechanics. Some open questions concerning specific problems in teaching fatigue and fracture mechanics are discussed in the final section.
Key Words: Teaching fracture mechanics, civil engineering, Spanish experience.

1 Introduction

In Spain, civil engineers have been reluctant to use fracture mechanics techniques, either considering them as too complicated tools or classifying fracture mechanics approaches as unnecessary in the design of safe and reliable structures. Designing on the basis of fracture mechanics principles has been considered exotic and thus by-passed in engineering practice in favour of too conservative stress analyses according to classical elastic-plastic methods, which are well known by traditional civil engineers. Such a gap between fracture mechanics principles and civil engineering practice is even wider in specific fields such as bridges design and construction or dams engineering.

The reasons for the afore-mentioned situation can be found in the fact that fracture mechanics was not included in the civil engineering educational plans until 1976, when the enthusiastic and forward-looking action of Professor Manuel Elices led to the introduction of a fracture mechanics subject in the School of Civil Engineering in Madrid [1], although this was mandatory only for those students specializing in foundations and structural mechanics.

Teaching and Education in Fracture and Fatigue. Edited by H.P. Rossmanith. Published in 1996 by Taylor & Francis, 2 Park Square, Milton Park, Abingdon, Oxon, OX14 4RN. ISBN 0 419 20700 7.

This remained so until the modification of civil engineering educational plans in 1991, when fracture mechanics was introduced as a mandatory subject in all civil engineering schools in Spain [2], as part of a wider course including continuum mechanics, elasticity, plasticity and materials science. The first plan according to these new ideas is now being developed in the School of Civil Engineering at the University of La Coruña, in which the author of the present essay is now working. The paper presents the main features of these fracture mechanics courses as well as questions of teaching fatigue and fracture mechanics to civil engineering students.

2 Fracture mechanics in engineering and technology

Fracture mechanics plays an increasingly important role in engineering and technology to guarantee structural integrity of the whole engineering construction or of any structural members. Since no material or structure can be considered perfect, and defects such as cracks or notches can pre-exist in the material or appear in the structure after construction (due to loading and environment), *damage tolerant* design offers many advantages over classical continuum mechanics-based design procedures [3].

As an example, one can mention two important innovative engineering areas in which fracture mechanics plays an important role: aerospace and nuclear technology. Aerospace engineering requires structural design under damage tolerance codes to assure a correct service life with a minimum of repair and maintenance under very severe and complex loading conditions, which emphasizes the importance of fatigue and fracture mechanics techniques [4]. Nuclear technology is related to environmentally-assisted fracture mechanics problems due to neutron irradiation or the presence of hydrogen isotopes (in the case of fusion nuclear technology), which may affect the safe structural behaviour under such severe environmental conditions [5]. These examples demonstrate that continuing education in fracture would undoubtedly improve engineering practice, and ideas along this line are suggested in [6].

The role of fracture mechanics in engineering and technology is extended over the three basic stages of an engineering work: project, construction and maintenance of the structure. At the project level, the fracture criterion provides the engineer with a design criterion. In the construction stage, fracture mechanics allows one to control possible defects. In maintenance, fracture mechanics techniques provide an estimation of the service life and insight into the real necessity of repair work.

As a consequence for civil engineering in the near future, it seems to be necessary to project structures of special risk under damage tolerance conditions, an approach which has shown to be valid and adequate in aircraft and aerospace engineering. Then the continuum mechanics-based elastic and elastic-plastic approach —on which traditional structural engineering is based— is not enough to project safe and reliable structures. This limitation, however, is classically overcome in civil engineering with a too conservative design that requires a more expensive structure.

A final reflection regarding the importance of fracture mechanics from the structural integrity point of view. According to Prof. Keith Miller [7], it is estimated by various national and international agencies that up to 8% of GNP is lost annually due to failures in engineering plants, especially in advanced-technology countries.

3 The use of fracture mechanics in civil engineering

The increasing importance of fatigue and fracture mechanics in civil engineering in particular is a consequence of the general improvement in structural engineering due to technological development, and this can be understood by taking into account the following facts:

• Improvement of properties of traditional structural materials, which makes them more resistant from the point of view of classical elastic or elastic-plastic design (since they posses higher yield strength than conventional materials) but, on the other hand, less resistant from the fracture mechanics point of view in the presence of defects such as cracks or notches.

• Evolution of computational methods for determining stress and strains, with better and cheaper hardware and more developed engineering software. Thus, the engineers can use a really powerful tool for accurate and reliable analysis of stress and strain states at any point of the structure. On solving the problem of analysis of stress and strain, the material itself and its properties —not yet fully understood— become the only open issue and the main objective from the engineering point of view.

• New operating conditions for civil engineering structures, as in the case of civil engineering constructions working in hostile or harsh environments or under dynamic loading, e.g., offshore structures suffering corrosion-fatigue, or criogenic prestressed concrete tanks, which may fail due to the decrease in fracture toughness at very low temperatures, with the subsequent risk of brittle failure.

• Higher demand for structural safety and reliability. Technological development requires the construction of engineering structures whit a high safety margin, due to the catastrophic consequences of failure. In these cases, the damage tolerance design on the basis of fracture mechanics principles is not only recommended, but is unavoidable.

• Increasing use of new structural materials, e.g., composite materials, widely used in aircraft and aerospace engineering, and increasingly employed also in civil engineering (apart from traditional composite materials such as the well known reinforced or prestressed concrete). The possibilities of these new materials in engineering design make them very interesting in structural engineering.

4 Fracture mechanics and civil engineering in Spain

The Spanish Group on Fracture was founded in 1984 by Prof. M. Elices (Polytechnic University of Madrid) and Prof. M. Fuentes (CEIT-University of Navarra). Since then, the Spanish Conference on Fracture has been held every year at different locations during the spring season, and it has become the main forum for Spanish scientists and researchers in the field of fracture mechanics and structural integrity. In addition, a new official degree —*materials engineer*— was created in 1994 in Spain [8], devoted to materials science and engineering. It is a two year course which can be followed by students coming from different previous degrees on physics, chemistry and engineering, civil engineering being one of the possible previous studies to reach the materials engineering degree.

This favourable environment for fracture mechanics advance, together with the afore-mentioned reasons showing the increasing importance of fracture mechanics in civil engineering, brought two consequences of a different nature in Spain, one related to teaching and education in civil engineering and the other linked with civil engineering professional practice. From the educational point of view, this led to the introduction of mandatory courses devoted to fracture mechanics and materials science in the new civil engineering educational plans in Spain, as described in further sections of this paper, thus allowing the students to acquire basic knowledge of this matter of primary importance.

From the professional point of view, fracture mechanics techniques are now included —if not explicitly, at least implicitly— in civil engineering codes such as the European EUROCODE 2 [9] or the Spanish RPM-95 [10]. The EUROCODE 2 [9] (European pre-standard) includes concepts of cracking, fatigue and stress corrosion, but in a rather simple manner. Cracking is profusely mentioned as a phenomenon in concrete, but no mention is made of cracking of prestressing wires and tendons, a more critical problem from the engineering point of view. Furthermore, no critical crack is considered in the concrete, except classical multi-cracking. Fatigue of prestressing steel is treated on the basis of the classical S-N curves and no comment is made regarding fracture-mechanics approaches such as the Paris law. Fatigue cracking is not considered in steel, since no mention appears devoted to cracks as a result of the fatigue phenomenon; it is a purely phenomenological treatment of the fatigue phenomenon or a "black box" approach. Stress corrosion is mentioned very briefly in the requirements for reinforcing and prestressing steel. Thus it is assumed that concrete can crack (in the form of non-critical multi-cracking) while reinforcing or prestressing steel is unable to crack, but can suffer fatigue.

The Spanish RPM-95 [10] is a document containing recommendations for the project of metallic bridges. It implicitly includes fracture mechanics approaches when it gives minimum fracture toughness values in the form of Charpy energy, and recommends direct measuring of more reliable fracture toughness values by means of pure fracture mechanics techniques. The well known effect of temperature on toughness is included in design charts.

In spite of this progress —small but not null— in standards through the explicit or implicit inclusions of fracture mechanics principles, engineering practice continues using classical structural mechanics techniques. This is a kind of elusive approach in which crack-like defects —a part of reality— are obviated, and the engineer designs and projects the structure as if cracks did not exist and could not appear or grow during service life. Therefore traditional civil engineering work is based on the two following conjectures. First assumption: cracks do not exist in civil engineering structures; second assumption (if the first fails): if cracks did exist in any structure, they would never grow; third assumption: all environments are inert from the electro-chemical point of view. Therefore, the only accepted possibility is the existence of sub-critical non-growing cracks in inert environments which would not imply a risk for the structure. This is of course a mistaken way of thinking, since all cracks can grow if the stress level is sufficiently high, especially in an aggressive environment, but one can remember an old Chinese proverb which says: "None are as blind as those who do not want to see".

5 Educational framework for teaching civil engineering in Spain

The orientations of civil engineering educational plans have been deeply analyzed in recent years to see if a future civil engineer must receive education in matters of a more proper scientific nature or whether his training must be focused only on pure technical matters of —apparently— more practical interest for his ulterior professional job. The answer to this question has been traditionally the same. On the basis of the existence of real engineering problems, it was usually assumed that a solution to such problems requires only the application of previously-known recipes, with virtually no creative work of thinking (with the exception of design, a traditional field of interest for engineers).

However, more modern approaches to the problem consider also that a solid scientific foundation is fundamental in engineering knowledge, since it permits a more flexible adaptation to technical changes and a more creative way of thinking. Changes in technical fields are so fast that engineers should be able to assimilate tremendous amounts of scientific and/or technical information in increasingly shorter periods of time, and this should be achieved in an ever-demanding and competitive environment. This is the general orientation of new civil engineering educational plans in Spain, according to the new regulations of 1991 [2], whose main advance is the inclusion of fracture mechanics and materials science as mandatory subjects for all civil engineering educational plans in Spain.

6 Civil engineering educational plan at the University of La Coruña

The civil engineering educational plan of the University of La Coruña (Spain) is the first in Spain according to the new legal requirements for civil engineering education. It is organized in five academic courses and includes a mandatory subject entitled *Continuum Mechanics and Materials Science* (including explicitly fracture mechanics as a part) in the third course. It is the first civil engineering plan in Spain to include the fracture mechanics item as a mandatory subject for *all* students.

The philosophy of the plan tries to unify concepts of continuum mechanics, fracture mechanics and materials science. In this regard, it is useful to analyze previous subjects supplying physical and mathematical bases, which are:

- Physics: *Applied Physics* and *Mechanics*
- Mathematics: *Differential and Integral Calculus, Mathematical Analysis, Differential Geometry, Ordinary Differential Equations, Partial Differential Equations, Field Theory* and *Tensor Calculus.*

A wide set of later courses are based on continuum mechanics and/or materials science . The main big fields are the following:

- *Soil Mechanics and Foundations*
- *Hydraulic Engineering*
- *Structural Engineering*

There is a previous matter related to materials science: *Construction Materials* in the first course, of a descriptive nature. Some simplified models of mechanical behaviour are considered in other courses, e.g. strength of materials on the basis of the Navier hypothesis. Other subjects are associated with a specific construction material, e.g., reinforced and prestressed concrete and metallic structures. They usually insist very little on the constitutive equations of the correspondent materials and on the causes of their mechanical behaviour.

It is important to emphasize here that students of civil engineering receive or can receive teaching in structural analysis in the following ten specific subjects:

- *Structures I-II-III*
- *Reinforced and prestressed concrete I-II*
- *Metallic structures and composite construction*
- *Structures under dynamic loading*
- *Bridges I-II*
- *Dams*

a large amount when compared with lessons of fracture mechanics and materials science, since engineering design involves two key values: the stress-strain state in the solid or structure (computed by means of strength of materials or structural analysis) and material strength to guarantee structural integrity against catastrophic failure (obtained with fracture mechanics and materials science tools). To have a reliable design, both have to be properly determined. If not, too high safety factors —and consequently more elevated costs— are required. It is useless, therefore, to improve structural analysis (very well solved with the help of modern computers) if there is only a superficial knowledge of material properties and damage tolerance of the structure. Furthermore, even in stress-strain analysis, a constitutive equation of the material is required, and this topic is far from being totally understood when material behaves beyond the elastic range (in the case of metals) or in other cases such as concrete, a widely used example of rough ceramic material in civil engineering, whose behaviour is better understood with the help of fracture mechanics.

7 Teaching programme for fracture mechanics and materials science

A 150 hour mandatory course entitled *Continuum Mechanics and Materials Science* is taught to the students of civil engineering at the University of La Coruña, including the following main topics:

- Continuum mechanics
- Constitutive equations
- Elasticity and viscoelasticity
- Plasticity and viscoplasticity
- Fracture mechanics
- Materials science
- Composite materials

It results in a general evolution of knowledge from the simply-connected continuum media (continuum mechanics), including the particular theories of elasticity and plasticity, through the multi-connected continuum media (macro-aspects of fracture), to the discontinuous or disconnected media (micro-aspects of fracture and materials science), and finally materials made with other materials, which can be named *meta-materials* (composite materials).

Firstly, continuum mechanics analyzes the mechanics of continuous solids, which from the topological point of view are simply-connected continuous media. Students should assimilate the conceptual abstraction of the continuous material medium as a mathematical model representing the real material when the minimum volume of analysis is sufficiently greater than the material grain size, to be considered then as continuous and overcoming microscopical discontinuities.

Secondly, general constitutive laws are analyzed: the elastic behaviour (including viscoelasticity) and the plastic behaviour (including viscoplasticity). Elasticity lets the students know how materials behave when the external load is low enough to maintain stresses well below the yield strength, thus avoiding irreversible strains. This is the most common situation in civil engineering, since design and safety factors are chosen to permit a global elastic behaviour in the structures.

Plasticity theory extends the validity domain of elasticity equations. This includes an important idea: a new model or theory does not necessarily make another invalid, but it may complete the previous one. Conversely, it is important to think that any scientific theory has certain limits according to its basic hypotheses, and the equations do not represent universal laws, but correspond to *that* specific theory.

The next logical step is the study of multi-connected continuum media, i.e., fracture mechanics dealing with cracked solids, in which a certain part of the solid has a crack which produces a disconnection between two portions of the body. The analysis of the macroscopic aspects of fracture is, however, that of continuum mechanics theory.

A reflection on the conceptual similarity between plasticity theory and fracture mechanics. Plastic and fracture behaviour represent opposite kinds of behaviour prior to the failure of engineering materials. However, they present a conceptual similarity, since irreversible damage from the thermodynamic point of view appears in the form of yielding (irreversible plastic strain) or cracking (irreversible damage).

Disconnected media are taken into account when micro-aspects of fracture (physics of fracture) are studied. Physical micro-mechanisms producing different macroscopical behaviours (elastic, plastic and fracture behaviour) are considered. Given the scale of analysis, media are treated as discontinuous (crystalline lattice, dislocations theory) and disconnected (fracture micro-mechanisms: cleavage-like, micro-void coalescence or dimpled fracture).

Finally, composite materials are considered, i.e., materials made with other materials or *meta-materials*. From the civil engineering viewpoint, composite materials are structures made with other materials, which permits the designer to choose the desired properties, an action very close to the typical work of structural civil engineers, who have to design the material when they use reinforced and prestressing concrete, materials which may be classified as fiber-reinforced ceramic matrix composites from the materials science point of view.

The most important challenge is to define fracture mechanics and materials science courses as a rational and coherent part of a more general training in strength of materials and structural mechanics. Furthermore, it would be desirable to connect the general ideas of continuum mechanics with those of soil– and fluid–mechanics, since all of them are basic for the future civil engineer. Fracture mechanics and related fields should be considered as general matters useful in all structural engineering branches (aerospace, mechanical, civil engineering and so on), in the conceptual framework of damage tolerance analyses.

The above description is of a general introductory course at the undergraduate level, which should be completed with some lessons on the relationship between microstructure and typical material behaviour for metals, ceramics and polymers. In addition, some higher level courses of advanced fracture mechanics, elastic-plastic fracture, computational methods and fracture of concrete are now being planned for the near future.

8 Formative objectives of teaching

It is possible to distinguish between the formative and cognitive objectives of teaching. Formative objectives are based in the development of general student capacities (personal, scientific and technical), and they are of a more general nature, whereas cognitive objectives are linked with the specific matters (knowledge acquisition) and they have precise attributes.

The first formative objective has a *personal* character, and it consists of continuing the intellectual training of students, developing intellectual habits and improving their deduction and synthesis capacity. In addition, the improvement of the working ability of students is another objective of primary importance.

The second formative objective is *scientific*. Students should acquire a scientific knowledge of the matter (fracture mechanics and materials science in this case). To this end, the subject must be taught in the most rigourous possible manner, the basic hypotheses being clearly stated. With regard to this objective, and considering the double role of the University as an Institution devoted to both teaching and research, it is important to introduce students to the scientific field and discover those scholars more qualified for scientific work.

The third objective has a *technical* nature, and consists of giving the students the capacity for applying the knowledge of fracture mechanics and materials science to real engineering problems. This is achieved by showing in class the technological applications of theoretical ideas as a tool for solving civil engineering problems. Exercises could be based on case studies of real civil engineering problems: beams, bridges, structural elements, etc. with crack-like defects and subjected to static or fatigue loading in inert or corrosive environment.

There is another important training objective which consists of stimulating a creative way of thinking —which is essential not only for the scientist, but also for the engineer— and a critical discussion of the solution obtained for the problem, eliminating non realistic solutions. To this end, solving open problems and questions (with no definite solution or with a set of possible solutions) appears the best way.

9 Global cognitive objective of teaching

The global objective of this teaching programme in fracture mechanics and materials science is that the student become aware of the idea that materials can be considered as *live entities*, susceptible to the effects of their external environment (in general sense, including both mechanical and electro-chemical actions), and that knowing this dependence is essential for a proper use of the specific material.

Paraphrasing Ortega y Gasset —the Spanish philosopher— one could say: *the material is itself and its circumstance*, which indicates, firstly, that the material is intrinsically imperfect, and therefore, that surface or internal defects are inherent in it; and secondly, that there is an evolution of its mechanical properties throughout the service life, due to the combined effect of the mechanical load history and the surrounding physico-chemical environment.

This consideration of the material as a live entity immersed in the surrounding environment has an important consequence: the concept of material is strongly linked with the existence of superficial or internal defects or geometrical flaws such as cracks or notches (from the macroscopical point of view) or imperfections in the microstructure such as lattice defects, dislocations, micro-voids, etc (from the microscopical point of view).

Thus the classical approach in civil engineering in which material is totally defined by its constitutive equation —or, even worse, merely by its elastic properties— turns to a new approach in which materials behaviour depends not only on intrinsic characteristics of the material itself, but also on the circumstance, i.e., on extrinsic factors such as load history (load magnitude, kind of loading, loading rate, etc) and environment (temperature, humidity, corrosive agents, etc.) which make previous defects grow. It is important to notice, therefore, that a given material does not have a behaviour *per se*, but can exhibit one or another behaviour depending on the specific working conditions (mechanical and pysico-chemical environment).

10 Specific cognitive objectives of teaching

The specific cognitive objectives of teaching continuum mechanics, fracture mechanics and materials science in civil engineering educational plans are the answers to the following questions:

- How do structural materials behave?
- What are the causes of that behaviour?
- How can one modify this behaviour?

The answer to the first part is given in the realm of continuum mechanics (including elasticity and plasticity) and fracture mechanics, whereas the answers to the second and third questions can be found in the field of materials science (including the study of composite materials), through the consideration of the crystalline lattice (elastic behaviour), dislocation theory (plastic behaviour) and microscopic mechanisms of fracture (fracture behaviour). Possibilities of modifying the behaviour of materials are presented in the chapter of hardening mechanisms (materials science) or materials design (composite materials).

This approach has two main advantages for teaching in civil engineering schools. Firstly, teaching is focused on modes of material behaviour and not on types of materials, which implies a didactic economy, since a given behaviour may correspond to different materials, e.g., elastic-plastic or fracture behaviour. Secondly, students become aware of a fundamental idea: materials do not always exhibit the same behaviour, but this depends on the specific mechanical and physico-chemical environment (the afore-mentioned *circumstance*).

11 Teaching methods and aids

Generally speaking, the first question is the choice of the didactic method: inductive or deductive. Inductive logic starts from experimental observations in various materials in different circumstances and tries to establish general laws of mechanical behaviour. Deductive logic starts from behaviour laws and gives predictions of such behaviour. Both inductive and deductive methods will be used in class.

With regard to types of groups for teaching, they may be classified as follows:

- Individual work (1 student)
    - Autonomous study
    - Directed study
    - Programmed study
- Work in groups (2 to 25 students)
    - Team: Autonomous
Directed
    - Discussion group: Open (Round table)
Directed (Seminar)
- Work in classroom (more than 25 students)
    - Lecture
    - Conference

Autonomous study must be the result of a personal decision of the student, which may be suggested, but not imposed. Directed and programmed learning is associated with tuition and, in a certain sense, with laboratory activity. Work in teams will be developed to prepare small projects or to solve extended problems. Round tables and seminars will be held under the direction of a professor of the Department. Standard teaching in classroom will be performed in the form of lectures and exercises. Finally, invited lectures given by specialists in fracture mechanics and materials science should be welcome.

Apart from using the conventional blackboard (or whiteboard), other relevant media should be used such as overhead projector, slide projector and didactic video. The slide projector is very important in materials science to show micrographs and other images. Didactic video is also a fundamental teaching aid, and excellent examples can be found in didactic collections (see, for example, [11]).

12 Ways of teaching

Theoretical lectures will follow a scheme in three phases: *input, output* and *feedback*: first, the professor explains the theme and suggests open questions (*input*); second, students ask questions (*output*); third, the professor answers questions and receives information about the degree of understanding (*feedback*). There are three improvements over the traditional approach which can be used: i) using audio-visual aids; ii) referring to real engineering problems; iii) using bibliographic sources.

The specific objectives of exercises are primarily three: i) to apply theoretical concepts to real situations, ii) to enhance the student's creative way of thinking, iii) to stimulate his critical sense of the student. In this framework, there is a real necessity of books containing worked examples in fracture mechanics. Although for the moment they are very scarce, some examples can be found, as mentioned in further sections.

Laboratory practice is a task of conceptual and practical complexity. Conceptual complexity lies in the fact that experimental magnitudes which can be directly measured —forces and displacements— are related to more complex entities —not directly observable— such as stresses (components of the stress tensor) or strains (components of the strain tensor) or stress intensity factors. This means that students should know the theoretical bases very well before starting laboratory testing. Practical complexity appears because most laboratory measurements require the use of very complex experimental devices: load cells, strain gages, extensometers, computer control, etc. In addition, there is a real physical danger in mechanical testing due to the very high loads required to perform a fracture test on a material sample.

The first decision is whether or not students must learn how to use laboratory testing facilities. The answer depends on the specific test required. If testing facilities orientated to teaching are available, the answer should be affirmative, but this possibility seems remote because of the increment of costs. If, as usually happens, testing facilities are directed to scientific research, then there is no real possibility of direct handling of testing machines by students. In this case, students can observe the test (e.g. standard measurement of fracture toughness) and work later on the results to obtain final values (e.g. fracture toughness from load displacement measurement and crack profile) and determine if the test is valid.

The final aim of laboratory practice is a better understanding of material behaviour and fracture phenomena, as well as an introduction to experimental techniques, with emphasis on the interpretation and evaluation of results in a critical manner. In this case, as in many others, *knowing why* is far more important than *knowing how*.

Seminars and conferences on specific topics of great interest should be enhanced, focused on fracture aspects of civil engineering problems or case stories. The level of seminars and conferences can vary from the purely technical to the scientific level, but in the latter case the practical applications of the theory should be emphasized.

Students can develop monographic works of a voluntary character, consisting of a small technical project or a limited scientific work. Two approaches can be suggested:

• Damage tolerant design of an engineering structure or an structural element (pressure vessel, pipe, tendon, ...) under constant or variable loading, in inert or aggressive environment. This way seems to be adequate for students orientated towards professional practice. The aim is to show the engineering applications of fracture mechanics. One example of a very simple engineering approach is given in Ref. [12], which allows an estimation of the fatigue life of a pressure vessel under thermal cyclic loading. In this framework, another interesting possibility is the analysis of case histories involving fatigue and fracture mechanics.

• Performing a small scientific project —of limited scope— in fracture mechanics or materials science. It could be a simple part of a bigger scientific project, or a specific aspect of it. Topics would be included in the main research lines of the Department. This way would be adequate for students orientated towards scientific research in the future. The aim is to introduce them to the techniques of scientific research.

Tuition will be designed to help in the aforesaid tasks and motivation is the key element for its success. It will be organized by means of periodical interviews between students and a professor of the Department to discuss specific problems and to suggest scientific and/or technical references in the field of the project or scientific work developed by the scholar.

The best system to evaluate students is the continuous evaluation by means of the following activities:

- Attendance at theoretical lectures.
- Solving practical exercises.
- Small exams or control tests.
- Laboratory practice.
- Monographic works.
- Tuition.

All these activities will give sufficient information to evaluate the students in a continuous manner. Nevertheless, a final exam is another possibility for those students who failed with the previous system.

13 Selected text books

There are very good texts on fatigue and fracture mechanics which can serve as a basis for lectures in class. A detailed analysis of most of them is beyond the scope of this essay, but a very interesting comparison of the content of selected textbooks on fracture and fatigue (as well as the complete references) can be found in [13]. In the preference of the author of the present paper, the best basic book on engineering is the classical by Broek [14], which can be completed with that of Kanninen and Popelar [15] for more advanced questions. An interesting book fully devoted to worked examples is that by Knott and Withey [16]. In Spanish there is an excellent work in several volumes by Elices [17], covering different aspects of fracture mechanics and including worked examples. Handbooks of stress intensity factor (SIF) solutions [18] are the basic tool to solve fracture mechanics problems.

In engineering education, solving case histories involving fatigue and fracture mechanics is a very interesting task for future engineers. References [19, 20] contain numerous examples, and others can be found in the recent journals devoted to engineering failure analysis [21]. Case studies are very useful in engineering education, but they are generally rather complicated, since ususlaly involve different problems.

With regard to civil engineering, no basic text book can be found in the scientific literature. However, advanced teaching may be carried out on the fracture mechanics of reinforcing and prestressing steel [22] or concrete [23], materials of the highest interest in civil engineering construction. In the matter of fracture mechanics applications to the whole structure, a chapter on bridge failures is given in Ref. [24], including case studies, failure prevention and full-scale testing. A recent handbook on fatigue crack propagation in metallic structures [25] contains specific chapters on fatigue of steels for concrete reinforcement and cables, fatigue propagation of surface cracks in round bars, fatigue behaviour and testing of offshore structures and joints, and fatigue phenomena in railway rails and wheels, all matters of *specific* interest in civil engineering, apart from other chapters on reliability and structural integrity. Finally, an interesting paper [26] should be mentioned, since it presents a method for calculating the SIF in cracked beams, with applications to frames containing cracks.

14 Open questions and problems

This section deals with some open questions regarding teaching fracture mechanics, with emphasis on civil engineering education. They are only briefly sketched due to space limitations, and most of them do not have a clear answer or solution, which would require further discussion. The following ideas may be suggested:

- The name *fracture mechanics* could be changed to *structural integrity*, following the change from *European Group on Fracture* (EGF) to *European Structural Integrity Society* (ESIS). However, the term *fracture mechanics* is adequate for teaching. In engineering practice, *damage tolerance* and *structural integrity* are very adequate.
- A global terminology in fracture mechanics should be welcome [27], towards a unambiguous and precise nomenclature. In this framework, a multi-lingual dictionary of fracture mechanics terms would also be required.
- The problem of transferability of laboratory test results to engineering structures (closely related to constraint and triaxility conditions in 3D states) remains open, and solutions are too complex to be used in practical engineering.
- Environmentally assisted cracking problems represent one of the most difficult fields in fracture mechanics education, since they involve not only mechanical, but chemical problems, and dependence on time.
- Many engineering problems have a 3D nature, thus increasing their difficulty. In civil engineering, the problem of a cracked cylindrical bar (associated with prestressing steel wires, cables and tendons) is typically 3D.
- Cold drawn prestressing steel is a highly anisotropic material from the fracture mechanics point of view. As a consequence, cracks frequently change their trajectories and a mixed mode problem appears.
- Fracture process in concrete is rather complicated, and present developments in research are far from being applied in practice, in spite of the effort made in this promising area.
- Application of fracture mechanics to reinforced and prestressed concrete is even more difficult for the engineer, since it includes problems of concrete and steel, as well as combined problems, e.g., the interaction between the two materials.

15 Conclusions

If crack-like defects did not exist or could not appear, i.e., in the case of perfect structures, fracture mechanics concepts would be unnecessary. However, in the more realistic case of defects in the material and/or structure (either previously existent or grown after construction), fatigue and fracture mechanics principles have to be used in civil engineering in the framework of damage tolerant design.

An increasing use of fracture mechanics in civil engineering is the basis for an effective teaching of these subjects in educational plans. Conversely, enthusiastic and challenging teaching of fatigue and fracture in civil engineering education is the key for stimulating the students and making them conscious of the usefulness of fracture mechanics in civil engineering.

Acknowledgements

The author is indebted to Prof. Manuel Elices, whose enthusiastic and forward-looking action allowed the introduction of fracture mechanics in Spanish civil engineering. Acknowledgement is also given to Prof. Andrés Valiente, for helpful suggestions and comments to this paper.

References

1 B.O.E. (1976) Plan de estudios de la Escuela Técnica Superior de Ingenieros de Caminos, Canales y Puertos de la Universidad Politécnica de Madrid. *Spanish Official Bulletin*, 30-10-76.
2 B.O.E. (1991) Título universitario oficial de Ingeniero de Caminos, Canales y Puertos y directrices generales propias de los planes de estudios conducentes a su obtención. *Spanish Official Bulletin*, 10-10-91.
3 Wnuk, M.P. (1992) Fundamental concepts of damage tolerant design, in *Reliability and Structural Integrity of Advanced Materials-ECF9* (ed. S Sedmak, A. Sedmak and D. Ruzic), EMAS, West Midlands, pp. 687-716.
4 ESA (1986) *Damage Tolerance of Metallic Structures–ESA TRP Contract 6978/86/NL/PH*, European Space Agency, European Space Research and Technology Centre, Noordwijk, The Netherlands.
5 CEC (1985) *NET Status Report*, Commission of the European Communities, Directorate General XII-Fusion Programme, Brussels.
6 Faria, L. (1990) Engineering requires continuing education on fracture, in *Fracture Behaviour and Design of Materials and Structures-ECF8* (ed. D. Firrao), EMAS, West Midlands, pp. 1719-1723.
7 Miller, K.J. (1993) Management and editorial comment, *Fatigue and Fracture of Engineering Materials and Structures*, Vol. 16.
8 B.O.E. (1994) Título universitario oficial de Ingeniero de Materiales y directrices generales propias de los planes de estudios conducentes a la obtención de aquél. *Spanish Official Bulletin*, 6-9-94.

9 EUROCODE 2 (1995) European pre-standard, *Design of Concrete Structures -Concrete Bridges*, European Comittee for Standardization, Brussels.

10 RPM-95 (1995) *Recomendaciones para el proyecto de puentes metálicos para carreteras*. Spanish Ministry of Public Works, Madrid.

11 BBC (1992) *To engineer is human*, Horizon, BBC videos for education and training, London.

12 Toribio, J. and Valiente, A. (1993) An engineering approach to determine the fatigue life of sheet structures under cyclic loading. *ASME Journal of Engineering Materials and Technology*, Vol. 115, pp. 106-108.

13 Rossmanith, H.P. (1990) How to teach fracture mechanics? In *Fracture Behaviour and Design of Materials and Structures-ECF8* (ed. D. Firrao), EMAS, West Midlands, pp. 1703-1717.

14 Broek, D. (1982) *Elementary Engineering Fracture Mechanics*, Martinus Nijhoff Publishers, The Hague.

15 Kanninen, M.F. and Popelar, C.H. (1985) *Advanced Fracture Mechanics*, Oxford University Press, New York.

16 Knott, J.F. and Withey, P.A. (1993) *Fracture mechanics. Worked examples*. The Institute of Materials, London.

17 Elices, M. (1995) *Mecánica de la Fractura* (in Spanish), Polytechnic University Press, Madrid.

18 Murakami, Y. (1985) *Stress Intensity Factors Handbook*, Pergamon Press, Oxford.

19 Hudson, C.M. and Rich, T.P., Eds. (1986) *Case Histories Involving Fatigue and Fracture Mechanics, ASTM STP 918*, American Society for Testing and Materials, Philadelphia, PA.

20 Esaklul, K.A., Ed. (1992) *ASM Handbook of Case Histories in Failure Analysis; Vol. 1*, American Society for Metals, Metals Park, OH.

21 Jones, D.R.H., Ed. (1995) *Engineering Failure Analysis*, Pergamon-Elsevier Science Ltd., Oxford.

22 Elices, M. (1985) Fracture of steels for reinforcing and prestressing concrete, in *Fracture Mechanics of Concrete: Structural Application and Numerical Calculation*, (ed. G.C. Sih and A. DiTommaso)., Martinus Nijhoff, Dordrecht.

23 Elfgren, L. and Shah, S.P., Eds. (1991) *Analysis of Concrete Structures by Fracture Mechanics*, Chapman & Hall, London.

24 Rossmanith, H.P., Ed. (1993) *Structural Failure, Product Liability and Technical Insurance, IV*, Elsevier, Amsterdam.

25 Carpinteri, A., Ed. (1994) *Handbook of Fatigue Crack Propagation in Metallic Structures (2 Vol.)*, Elsevier, Amsterdam.

26 Valiente, A., Elices, M. and Ustáriz, F. (1994) Stress intensity factors in statically indeterminate cracked beams, in *Structural Integrity: Experiments–Models–Applications–ECF10* (ed. K.H.Schwalbe and C. Berger), EMAS, West Midlands, pp. 781-786.

27 Kausch, H.H. (1990) Is now the time for a global terminology in fracture mechanics? In *Fracture Behaviour and Design of Materials and Structures-ECF8* (ed. D. Firrao), EMAS, West Midlands, pp. 17-22.

19 TEACHING STRATEGIES AND METHODS FOR STIMULATING THE INVENTIVE ABILITIES OF ENGINEERING STUDENTS

V. BERINDE
Department of Mathematics, University of Baia Mare, Baia Mare, Romania

Abstract

The aim of this paper is to present some basic principles and strategies for developing the creative thought and inventive ability of engineering students. These principles and strategies can be introduced inside or alongside conventional lectures and can be adopted to form a style of work at any problem solving session in class. This style of work consists in a gradual questioning and answering on the main ideas, notions and facts involved in the given problem, which is in fact the typical working behaviour of any researcher. Our approach is directed mainly at the engineering mathematics and it consists in shaping metacognitive skills and inventive abilities by training students to learn to solve problems in a creative manner. Such active and creative learning can be adapted to various other technical and non-technical subjects in the engineering curriculum.

A partial evaluation of the application of this approach to the non-ferrous metallurgy, electromechanical and mechanical engineering programmes of our university is also given.

Key words: active learning, creative teaching and learning, case studies, problem based learning, modelling steps.

1 Introduction

A problem which faces the engineering educator today is to give an adequate answer to the continuously increasing technical, economic and ecological needs of industry

Teaching and Education in Fracture and Fatigue. Edited by H.P. Rossmanith. Published in 1996 by Taylor & Francis, 2 Park Square, Milton Park, Abingdon, Oxon, OX14 4RN. ISBN 0 419 20700 7.

and society and to ensure the development and improvement of the education and training of engineers by adapting it to these needs. In the contemporary technical world we especially need good engineers, but a twenty-first century engineer will be a competent engineer if and only if he can be an inventive engineer.

We are concerned here with an attempt to form and develop specific and general creative and inventive skills and abilities by means of mathematical courses. There are, of course, conflicting opinions concerning the position of mathematics in relation to other technical or non-technical subjects in the engineering curriculum. There are many people involved in engineering education who consider that in the contemporary world - when computers are intensively used in teaching and education- mathematics is not necessary or not as important as before in the engineering curriculum (see references) and that consequently it may be replaced by appropriate software packages.

In our opinion, teaching mathematics as well other non-technical subjects to engineering students is necessary not only for its importance at certain levels of various technical subjects, but also as an intellectual process. However, some European universities of technology report that engineering faculties have plans to decrease or have already decreased the number of hours of mathematics in the last five years. The effect of this change will be felt in the future.

In our opinion, mathematics still forms a significant part of engineering degree courses because mathematics develops not only manipulative skills but can be seen as a vehicle for creative and analytical thought alongside its role in developing logical thought. Educators must therefore be convinced of the importance of encouraging students to regard mathematics as a natural part of their package of engineering skills, as a fundamental engineering subject, in spite of the increasing penetration of computers and mathematical software packages in education. Of course, we do not intend to contest the role of computer packages in engineering education. On the contrary, we are amongst the strong advocates of injecting computer-based exercises into the teaching of mathematics. This new technology is intended to make teaching more effective, to make the mathematics material seem more relevant, to equip students so that they can recognise and cope with situations in engineering where mathematics is appropriate, but we do not expect to succeed in a total substitution of the subject of mathematics by computer packages, however user-friendly and exhaustive they may be. Therefore, from a pedagogical and didactic point of view, the teaching of mathematics must keep pace with new technologies, but in a temperate manner, because computers can not replace the essential metacognitive role of *mathematics learned by pen and paper*. Without an appropriate mathematical knowledge - learned by pencil and paper - our engineering students will tomorrow be poor at employing basic mathematics to link theory and the physical world and will use software packages without an understanding of the fundamental principles on which the software is based. Even though in practice engineers seldom need to apply mathematics at the level taught at a technical university, the study of mathematics must at least be seen as a training in logical thought and creative and inventive skills. These important shaping functions of teaching mathematics can be accomplished only if mathematics is taught in a creative style, in a manner which is appropriate to users and maintains students interest by demonstrating the relevance of mathematics to engineering needs and by emphasising its significance to engineers in developing inventivity, logical analysis and creative thought. If the teaching of mathematics is non-creative, namely if it is reduced to computing without thinking, applying formulae without deducing and reasoning, without reflection on what is being done, then the desired side of the educational process is not attained.

But if the teaching of mathematics is directed toward this creative way, then we can expect at least that our mathematics instruction will help students to think both inside and outside the mathematics subject. Our experiments are however based on a strong argument: in all Romanian schools, beginning with primary schools, continuing with secondary and high schools and finishing with undergraduate and postgraduate courses in universities, mathematics is taught at a very high level. Or, more exactly, it was taught at a high level, because, unfortunately, in the last five years almost all Romanian technical universities have decreased the number of courses in mathematics.

Although, there are many people who argue that a good engineer is not necessarily a good mathematician, we have established a correlation between the high level of mathematics attained at the Romanian technical universities until 1990, on one hand, and the highly appreciated Romanian inventors [14], on the other hand. By means of such examples as the following, we are trying to convince educators in engineering of the metacognitive relevance of mathematics in shaping creative and inventive thought.

Example:
In the course of the year 1994 Romanian inventors obtained [14]: 99 gold medals, 44 silver medals, 23 bronze medals and 15 special prizes, at five international competitions (held in Switzerland, Croatia, U.S.A., Bulgaria and Romania). At the international competition EUREKA (Brussels, 9-16 Nov. 1994), alone the Romanian Institute for Invention obtained no less than 60 medals (40 gold medals, 12 silver medals, 2 bronze medals and 6 special prizes). (Unfortunately, there is not enough money and too little effort is invested in Romania to apply many of these inventions).

In our opinion, students will only become competent Twenty-first Century engineers if they are inventive engineers. So, we think that mathematics - which plays an important role in shaping a creative and inventive thinking - is indeed essential in the engineering curriculum.

2 Basic creative principles for solving a problem

The basic idea of our attempt to develop a creative manner in solving any problems (mathematical, technical and so on) is to try to follow the specific behaviour of a research worker, that is a gradual questioning and answering on the main notions, facts and ideas involved in the problem and its solution. It is far better if we begin to train our engineering students in such a style of work by mathematical problems, because in the early part of their course the students are unfamiliar with the engineering topics. Therefore, this training may be seen as a metacognitive background of their engineering inventive ability, as a preparation of those logical tools which are to be used in their possible future research activity in engineering. In our opinion, in order to equip students with such inventive skills, we must promote a creative behaviour in teaching and learning mathematics.

By creative behaviour we mean, first of all, the style of teaching. Indeed, engineering mathematics could be taught as a vehicle for creative thought and analysis rather that an academic subject which seeks for the most rigorously correct explanation. An active teaching is to be adopted so that the teacher can encourage students in a continuous logical questioning and answering on the facts he teaches and on the reasoning he uses. He will often be willing to reduce the content of courses and to increase the intellectual challenge which he offers. By a creative behaviour in teaching, we further mean a permanent care to equip students with a firm base of mathematical knowledge and a clear understanding of its main concepts together with a constant effort to demonstrate its relevance to engineering needs. Modelling with mathematics is an area where the relevance can be demonstrated most satisfactory, but having in view the fact that students have little or no engineering background at the start of their course, the early examples of modelling should be the simple and realistic.

Following these modelling exercises it is desirable that any engineer should be able to understand the formulation of an engineering problem, to construct a model, to recognise limitations inherent in the model, to follow mathematical arguments in a critical and constructive way and to interpret results from the model in terms of the given engineering problem. In order to train students in a creative learning of mathematics we have to direct the ordinary problem solving session towards an active and creative one. To this end one will follow some general creative principles and suitable strategies in solving problems. These basic principles - as stated in our paper [3] - are the following three ones: 1) the algorithmicity principle; 2) the generality principle; 3) the generalisation principle.

To solve a problem by the applying the *algorithmicity principle* we mean to construct a well ordered solution which includes all essential steps in a gradual succession. The *generality principle* applied in solving a problem assumes to construct or to adopt and to retain only those methods which could be applied to other similar or kindred problems. In short, the generality principle selects the general methods for solving a given (class of) problem(s). By applying the *generalisation principle* in solving a problem, we try to obtain a more general statement of this problem, if possible, the most general statement of it, replacing all particular data or assumption by a general one. It is important to stress the fact that not all mathematical problems can be tackled in such a creative way. Therefore we need a collection of selected problems which can be taken or adapted, by a careful analysis, from several textbooks and collection problems [9]. These exercises and problems can be taken from different levels of mathematical knowledge and can have different degrees of difficulty. By systematically applying one, two or all three of these principles in solving such creative problems, we expect to train students in an active and creative approach to any problem they meet, we hope to discover and develop their inventive abilities and skills inside and outside mathematics courses. To attain this goal we have to choose between initiating special case study exercises inside conventional lectures and problem solving sessions in class on the one hand, and organising special seminars alongside ordinary lectures, on the other hand. Our experiments have been concentrated mainly on combining these two approaches, laying stress on the second. Regarding these special problem solving sessions, we try to solve as many accessible creative problems as possible stressing the clear application of each creative principle, in order to help students to learn to:

- analyse carefully the statement of the problem,
- construct logically an algorithmic solution,
- analyse in detail the critical steps and reasonings in the obtained solution,
- discover the reasonings we can improve or generalise,
- detect and replace any particular data and assumptions by a general one,
- endow with generality attributes the method(s), algorithms and techniques used,
- discover as many questions arising from the given problem and its solution as possible,
- answer these questions in a critical way and to formulate new problems arising from the given problem,
- summarise and report conclusions.

After five years of experience and in view of our partial evaluation we are convinced that, if we solve together a sufficient number of such problems and then the students solve themselves other similar or kindred problems, we discover, form and develop a creative manner and reasoning in our students. We think this methodology can be easily adopted and adapted in teaching other fundamental, technical or non-technical subjection the engineering curriculum.

3 Case studies

A collection of case studies is being prepared by the author and it is hoped that it will soon be published. It would be desirable that the book should be accompanied by a software library suitable for computer aided lessons, as in [5] and [2]. From a didactic point of view, in solving these problems we will simultaneously stress the main steps in modelling a physical problem, because a mathematics course specially for engineers must clearly show students that the application of mathematics to an engineering problem consists essentially in the following three phases:

- translation of the given physical information into a mathematical form, that is a construction of a mathematical model of the physical situation;
- treatment of the model by mathematical methods and solving of the obtained equations by analytical or numerical methods. This furnishes the solution of the given problem in mathematical form;
- interpretation of the mathematical results in the physical terms.

All three steps seem to be of equal importance in modelling, but it is generally very difficult to interest students in solving such problems. As it can be seen [2], [8], our creative approach is rather convenient for illustrating these modelling phases too and to encourage students to practise it.

Each principle has its own relevance, the most important seems to be the *generalisation principle* which allowed us to initiate explorations in many directions [4], [8]. If we succeed in emphasising so many aspects starting from a given problem, we form and train a creative manner of thinking and reasoning, we develop inventive skills in our students. But, how many problems can we solve in such a creative manner during the mathematical instruction of our students?

To be honest: very few. However, we could use parts of this creative style of teaching (or we simply use it, consciously or not) in our activity, instead of the whole approach which is very difficult to implement. In any case, I am convinced that success in training such creative skills and abilities depends in a large measure on the existence of such creative moments in our teaching activity.

I occasionally inserted them in my lectures and frequently in problem solving session class. A correlation between this style of teaching and the inventive abilities in which we train our students, has been studied on a sample of graduates (three engineering programs) and reported elsewhere (see section 4). After these experiments we are convinced that, if we solve together a sufficient number of such problems and then the students solve by themselves other similar or kindred problems, we contribute to the development of their metacognitive inventive skills. but it is important to stress the fact that not any problem or exercise can be tackled in such a creative way. During the last ten years, I have made a collection in a special problem solving session - The Seminar on Creative Mathematics - intend to stimulate the research abilities of the students from the mathematics and some engineering programs at our University.

This activity usually covers 6 weeks, but we are going to extend it to 12 weeks. Any weekly session consists in

- a problem solving session in class,
- home work,
- reports on the individual (or team) activity.

4 Evaluation and final conclusions

Following a certain measure the creative style described in section 2 and 3 we have taught a course of lectures to the mechanical engineering programme and to the electromechanical engineering programme. We have also taught two courses of lectures to the metallurgy programme following in a large measure the principles described. Inventive abilities and skills have been evaluated by the activity of the Seminar on Engineering Invention (directed by Professor Gh.Volcovinschi). His report is the following [6]:

- Metallurgy programme (1990 - 1995): from 28 graduates we have 6 promising inventors;

- Electromechanical programme (1990 - 1995): from 25 graduates we have 1 promising inventor;
- Mechanical engineering programme (1990 - 1995): from 50 graduates we have 2 promising inventors.

In spite of our particular experiments - on a sample of metallurgy, electromechanical and mechanical engineering, from a single series of graduates - the results obtained encourage us to extend and to recommend this creative teaching and learning to other basic subjects in the engineering curriculum and to other programs. Furthermore, a clear correlation between the mathematical knowledge of the students and aptitude in invention (eight of nine promising inventors obtained high marks at all mathematical examinations) it is shown by our experiments.

References

[1] Barry,M.D.I. and N.C.Steele (1992) A core curriculum in mathematics for the European engineer, Document 92.1 SEFI Bruxelles

[2] Berinde,V. (1995) How to stimulate engineering invention by means of mathematical approaches, Proceed.Int.Conference "Teaching Mathematics for industry", 18-20 Sept.1994 (Ed. J.Cerny), CTU Publishing House, Prague, pp.19-29

[3] Berinde,V. (1994) Basic principles concerning the creative solution of problems (in Romanian), *Lucr.Sem.Creativ.Mat.*, vol. 3 (1993-1994), 89-98

[4] Berinde,V. (1994) Strategies for solving "following problems" (in Romanian), *Lucr.Sem.Creativ.Mat.*, vol. 3 (1993-1994), 29-42

[5] Berinde,V. and O.Cosma (1994) An elementary geometric construction performed by a computer, *Bul.St.Univ.Baia Mare*, vol.X, 99-108

[6] Berinde,V. and Gh.Volcovinschi (1995) About the relevance of developing the creative and inventive abilities to engineering students by means of nontechnical courses, Inter.Conf. "Educating the whole engineer", Cracow Univ. of Technology, 4-6 May 1995, to appear in SEFI Document No. 15, 1995

[7] Berinde,V. (1995) Active problem based learning for developing the inventive abilities of engineering students, Proceedings of SEFI/CDWG Conference 1995, 1-3 November 1995, Univ. of Twente, The Netherlands (to appear)

[8] Berinde,V. (1995) A collection of creatively solved problems for simulating the inventive abilities of students, ICTMT, Napier University Edinburgh, 4-7 September 1995

[9] Clayton,B.R. (1995) Mathematics for engineers: an integrated approach in Mustoe, L.R. and S.Hibberd (eds) Proceed. of Conf."Mathematical education of engineers", Oxford University Press

[10] Craggs,I.W. (1978) Rigorous engineering mathematics, *Bull.I.M.A.*, 1979, pp.307-308

[11] Grandsard,F. (1989) Problem solving for first year university students, in Modelling, Applications and Applied Problem - Solving, Blum,W., Niss,M. and Huntley,I.(eds), Ellis Horwood

[12] Grandsard,F. (1990) Can a local problem solving contest help to develop problem solving skills?, *Mathematics Competitions*, Vol. 3, No 1, 53-56

[13] Mustoe,L. (1994) Industry experts - but who should determine the curriculum?, SEFI Math. Working Group Newsletter, Nov.1994, 2-25

[14] Plahteanu,B. (1995) 1994 - A new horizon of the Romanian invention (in Romanian), National Institute for Invention, Iasi, Romania

[15] Polya,G. (1965) Mathematical discovery. On understanding, learning and teaching problem solving, John Wiley & Sons, New York

[16] Pope, I.A. (1978) The mathematical education of engineers - where next?, *Bull.I.M.A.*, 274-278

[17] Rade,L. (editor) (1993) Teaching of Modern Engineering Mathematics, Studentliteratur & Chartwell - Bratt, Lund, Sweden

[18] Shercliff,I.A. (1978) Can mathematics form the heart of the engineering curriculum?, *Bull.I.M.A.*, 279-281

20 THE PROBLEMS OF FRACTURE AND FATIGUE IN EDUCATION AND TRAINING AT THE TECHNICAL UNIVERSITY OF SOFIA

D.M. DIMOV and K.V. VESSELINOV
Department of Strength of Materials, Technical University, Sofia, Bulgaria

Abstract

This paper informs about the educational situation at the Technical University of Sofia in terms of teaching and training of engineering students in the field of fatigue and fracture.

Keywords: Fracture, fatigue, strength of materials, testing, Bulgaria.

1 Introduction

At present the application and solving problems in fatigue and fracture of materials and structures are taught as part of several courses offered to students at the Technical University of Sofia.

The significance with respect to engineering education is the main reason to include issues of fracture and fatigue into the basic course on "Strength of Materials". This basic course is attended by all first level students of mechanical engineering. A short course on "Strength of Materials" is presented also to students in management and electrical engineering.

We think that by means of these courses mechanical engineering students are offered the possibility to become aquainted with the basic problems in fatigue of materials and structures and gain some engineering experience in strength of materials subjected to variable loads. At the same token these courses nearly exclude the subject of fracture mechanics.

Teaching and Education in Fracture and Fatigue. Edited by H.P. Rossmanith. Published in 1996 by Taylor & Francis, 2 Park Square, Milton Park, Abingdon, Oxon, OX14 4RN. ISBN 0 419 20700 7.

2 Teaching Fatigue

Let us concentrate on the course on fatigue of materials and structures. During the lecture span of 6 hours of fatigue the students will get an idea about fatigue of materials, strength and endurance under cyclic loading, fatigue curve and fatigue limit. The student will also learn about the main factors influencing the fatigue limit, such as: load conditions, mean stress, stress concentration, size, surface roughness factor and hardening factor. The students are shown how to determine the safety factor under uni- and multiaxial state of stress. Having worked their way through these 6 lectures the students should have required appropriate knowledge about accumulation of fatigue damage (hypotheses by Palmgren-Miner and by Serensen-Kogayev).

Time limitations preclude these lectures to concentrate on some particular questions addressing strength endurance estimation under low cycle fatigue, thermal and random loading, methods of classification and experimental determination of the life time of structures.

Three hours of lecture time are currently dedicated to the teaching of introductory fracture mechanics including the following items: classification of cracks, stages of their evolution, initiation and propagation, and stress intensity factors.

Time limitations prevent the teachers from addressing recent research problems, but even more crucial there is not enough time for demonstrating how to put to work all these ideas which they have been exposed to. The importance of fracture under extreme conditions (nuclear radiation, low temperature, etc.) and the increase of research and development in this field support the necessity to include these subjects into the curriculum. Students, particularly the ones who are involved in structural design must be offered the possibility to study these kinds of problems in a profound manner.

In the seminar "Strength under variable loading" of the course "Strength of Materials" the students are supposed to calculate the fatigue limit and assess the safety factor.

In the course, the students are provided with the textbooks and manuals they need for their work. These textbooks include recent methods of evaluation, reference books and standards which are valid in Bulgaria but usually based on the standards of CMEA, the former USSR and Germany.

During the last two years the students were familiarised with laboratory work and exercises on fatigue testing of engineering materials. In this lab work they are expected to determine the endurance limit of steel specimens under various amplitudes of loading and to determine a few data points on the fatigue curve. This way they become aquainted with fractography, i.e. the fracture surfaces after fatigue rupture and with fatigue crack propagation.

Some aspects of problems of fracture and fatigue are also considered in the following courses:

- Structural parts - estimation of the strength of beams and shafts subjected to cyclic loading;
- Reliability of vehicles and building equipment - determination of the life timeMechanical technology - weld assessment and optimisation of the mechanical and heat treatment conditions.

3 Requirements for teaching design students

Given the current situation, a mechanical engineering student's current exposure to teaching and education in fracture and fatigue and the knowledge derived therefrom is, in general, considered to be sufficient as the students usually become technologists or managers in mechanical engineering plants. However, the knowledge acquired on the basis of these courses taught will prove to be completely insufficient for designers of modern equipment such as structures and components in the transport, chemical, aircraft and nuclear industry. These specialists should be in a position to estimate the life time and the reliability of such complicated, expensive and reliable equipment.

Using a probabilistic approach to the load history and the material properties the students are expected to come up with an adequate statistical evaluation of the reliability of the structure and be able to carry out theoretical and experimental investigations of the ability of the structure to resist fatigue and fracture.

From the preceding one may conclude that design students need additional training in:

- Experimental methods for the investigation of stress and strain in structures;
- Estimation of strength and reliability of structures subjected to complicated service loading.

We propose that a new curriculum at the Technical University of Sofia should contain separate courses in problems in fracture and fatigue from among the students can then freely select. In this way they can most effectively be trained and gain profound knowledge in contemporary research in fracture mechanics. This also aims at the students to get better aquainted with recent results about estimation of fatigue strength of structures which were obtained at the leading institutions and laboratories all over the world.

Over the last two decades a large amount of work in research and development has been carried out at the Department of Strength of Materials of the Technical University if Sofia particularly with regard to fatigue life prediction criteria on the basis of inelastic material behaviour under cyclic and random loading. Because of the importance of fatigue and fracture for practice a fatigue oriented software package for the prediction of fatigue life and probability aspects was developed at the Department of Strength of Materials. This general purpose software can be utilised on an undergraduate level, in postgraduate studies and also for design purposes.

4 Conclusions

From a Bulgarian point of view it will in the future be important to establish a collaboration between all European countries both in teaching as well as in research. The authors advocate that in the field of engineering education the curricula contain courses where the most important problems of fracture and fatigue be discussed and techniques for their solution be presented. New courses should

- illustrate and present more practical examples of damaged structures,
- be specifically designed to teach testing methods, fracture and fatigue for mechanical engineers with design of structures as their major subject.

In addition, the laboratory equipment of the universities should be modernised at the same time.

This will give the students a chance to deepen their knowledge in the above mentioned research areas and methods. Progress in modernising Eastern European university laboratories without the help of the Western European countries seems to be impossible at present.

As teaching goes hand in hand with research the authors propose a network between European universities be established which will foster the exchange of results of research projects and assist mobilisation of students and lecturers. A possibility would be to develop a Joint European Project inside the TEMPUS action of the European Community.

21 STRENGTH CALCULATION OF CONSTRUCTION ELEMENTS UNDER CONDITIONS OF FATIGUE AND IN THE PRESENCE OF CRACKS IN THE SUBJECT OF 'STRENGTH OF MATERIALS'

A. ZAKHOVAYKO, S. SHUKHAYEV and N. BOBYR
Kiev Polytechnical Institute, Kiev, Ukraine

Abstract

The paper presents the methodology of training on the resistance of the elements of constructions under the fatigue conditions and the presence of cracks. The approach is based on the independent research of strength of bars with gradual complication of the calculated model.

Using the example of statically indeterminable bar system the elastic elements of which are subject to the definite tension-compression the calculation algorythms for different stages of problem solving are presented. There was also given an example of some versions of the problem according to which one can formulate the conditions of the level to solve the problem.

1 Introduction

Today, in the higher educational institutions of many countries, they study the strength of materials — one of the main themes of the general engineering — according to the programs which in some way use the concepts proposed by professor Timoshenko S.P. at the beginning of the century. At the same time the development of sciences and technology requires to implement into the engineering practice the new methods of calculations based on the latest achievements of the mechanics of solids. And perhaps the most up-to-date problems here are those related to the fatigue and the longevity of the construction elements. The rapid growth ofnumber of the experimental and theoretical researches in the field of fatigue and fracture mechanics permitted to develop quite simple and reliable methods of evalution of the strength reliability and the life of the construction elements under the indicated conditions and to implement them into the practical desinge. And our objective is to give students the possibility to learn them at a proper level.

Teaching and Education in Fracture and Fatigue. Edited by H.P. Rossmanith. Published in 1996
by Taylor & Francis, 2 Park Square, Milton Park, Abingdon, Oxon, OX14 4RN. ISBN 0 419 20700 7.

We have chosen the way of creating sets of calculations combining two different levels of evaluation of the strength and the bearing ability of constructions — from the classic calculations of the tolerances permitted, the definition of the marginal loading-up to the evaluation of the longevity of this construction under the cyclic loading. This provides the students with the possibility to carry out the comparative evaluation of the calculation results and to visually demonstrate to them the possibilities of the calculation methods used.

2 The essense of the problems at the different stages of their solution

In Kyiv Polytechnical Institute "The Resistance of Materials" course is lectured at the departments with the specialization in mechanics. The curriculum at foresees department here the maximum number of lectures — 104 hours; practice work — 36 hours; laboratorial studies — 36 hours; individual work with students — up to 36 hours for each academic group. For the students at the departments with the no-specialization in mechanics 'The Resistance of Materials' course is lectured as either a short original course or a chapter of "Applied mecanics". The curriculum foresees the maximum number of academic hours according to the specializations. All syllabae of instructions (full and short program) contain much information about the fatigue strength. Full course programs and some short programs (according to specialization) raise issues of mechanical fracture and fatigue life of constructions with the crack of materials. Further, we'll consider the maximum number of hours in these section only in the framework of the full "The Resistance of Materials" course.

The section of fatigue strength consist of 8–10 hours of lectures and 3–4 practice hours. It means 6–9 % of total number of lecture hours. The curriculum of the "The mecanics of fracture and the steadfastness cracks" course includes the lectures — 6–8 hours, and practice studies — 2–3 hours. It means 4–6 % of total number of academic hours.

There is no doubt that this volume of lectures is no sufficient for such serious sections. At the same time, there is no reserve of lectures in the curriculum for these specializations. That's why we give consideration for individual student work. The carrying-out of calculation of this problem according to given version is the result of the work.

The multistage approach was used for the statement of these problems. We propose to students the complex of problems of principals sections of this course: tension-compression, torsion, bending, and as subjects of calculations — bars and bar systems. The essence of different stages are following.

Stage I: Elastic desiegn calculation of given construction under the external load.
Stage II: Unelastic desiegn of given construction.
Stage III: Fatige desiegn of given construction.
Stage IV: Design of construction under the presence of cracks in its element and calculation of thier life under cycleloading.

The scheme design is given on the drawing 1. To reveal the essence and the volume of problems we shall consider the problem from the section "Tension – compression" and expose the principal steps of its solution at different levels.

In the Table 1 we see the proposed version of bar systems for calculations. Materials and specifications of bars are presented in the Tables 1 and 2.

Table 3 contains working drawings of bars and types of cracks. Table 4 contains geometric dimensions of the construction and its elements, other initial data.

Drawing 2 presents the scheme of one of version. It contains the necessary data for calculating: materials and drawings of bars, type of cracks, changing of temperature etc. We'll examine briefly the essence of each stage of this problem.

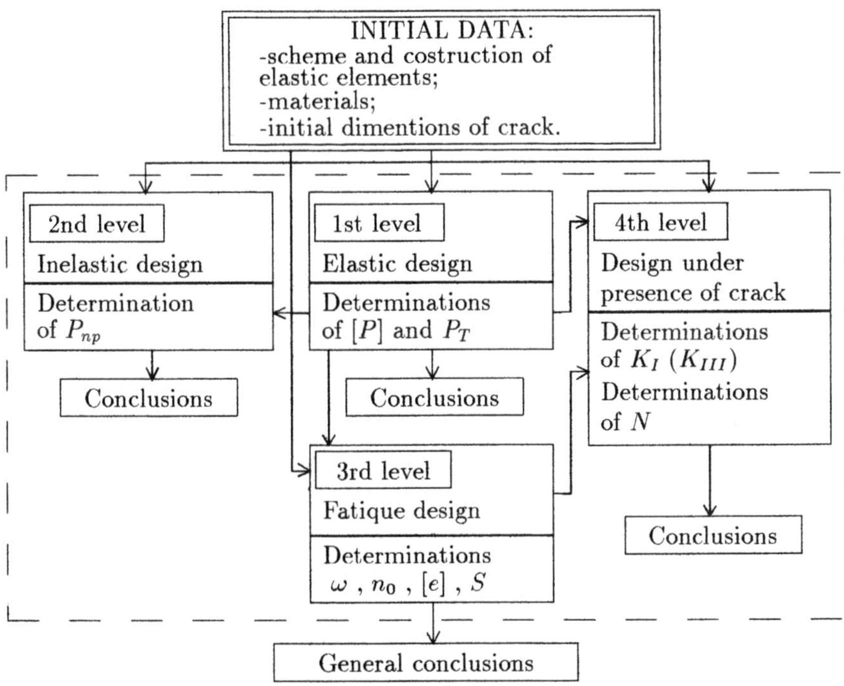

Drawing 1. Scheme of solution of the problem.

Tabl. 1 Design schemes.

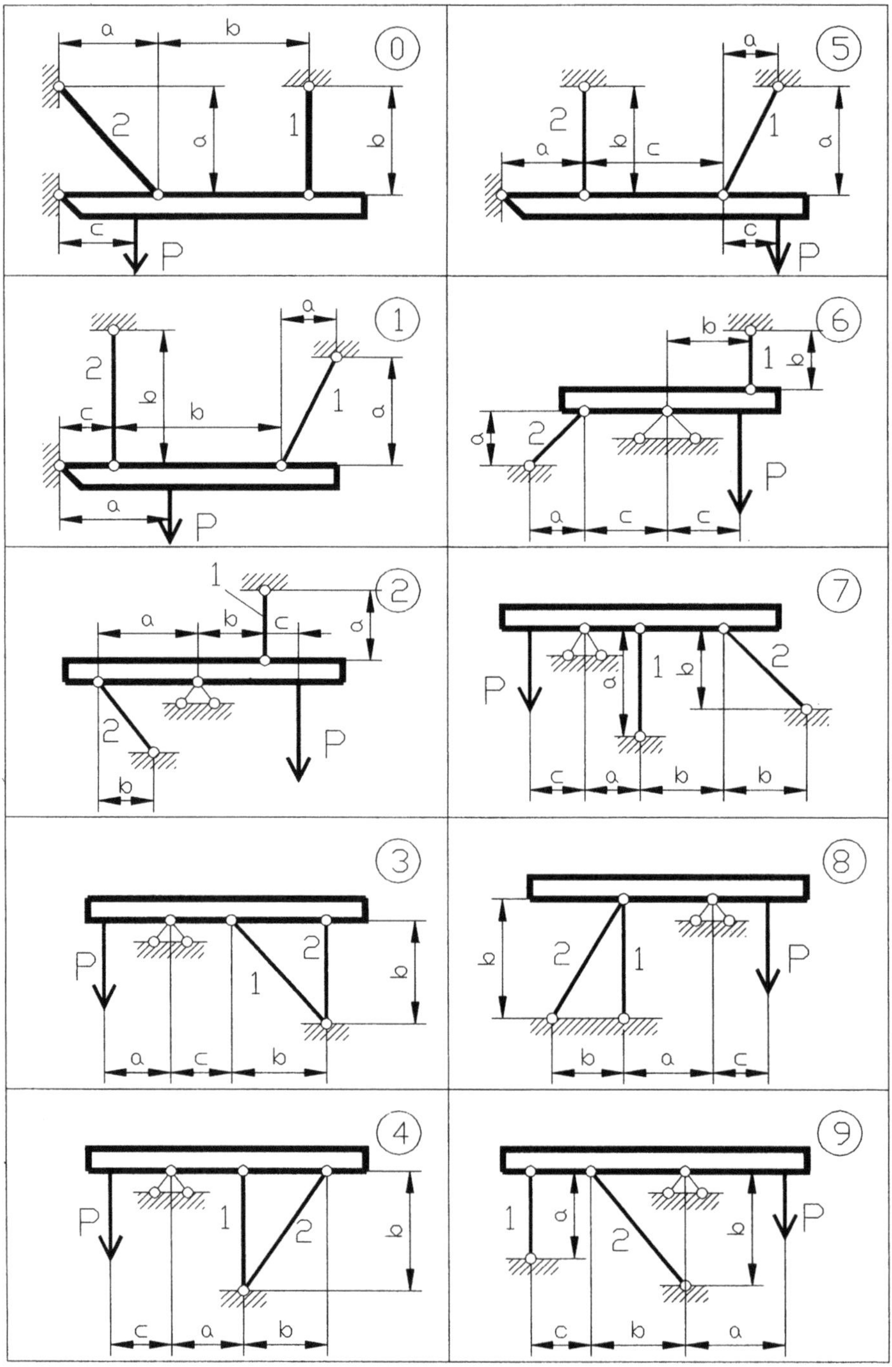

Table 2.1. Mechanical characteristics of materials.

No var	Materials	Charact of strength		Charact of plast		$E * 10^5$	$G * 10^4$	σ_{-1}
		$\sigma_{0.2}$ MPa	σ_B MPa	δ %	$\frac{\varepsilon_B}{\delta}$	MPa	MPa	MPa
1.	Сталь 45	350	590	16	0.5	2.1	8.1	250
2.	КП 60	410	690	12	0.5	2.1	8.1	330
3.	КП 80	950	1100	10	0.7	2.1	8.1	340
4.	КП 100	981	1069	15	0.7	2.1	8.1	450
5.	40Х	460	815	17	0.5	2.1	8.1	330
6.	20Х13	430	858	33	0.6	2.15	8.1	405
7.	15Х2МФА	584	700	21	0.7	2.2	8.1	315
8.	25Х1М1ФА	560	770	20	0.6	2.2	8.1	220
9.	27ХН3М2ФА	615	817	18	0.6	2.2	8.1	376
10.	ЭП428	580	710	21.4	0.6	2.2	8.1	520
11.	ВТ 3-1	1040	1100	18	0.8	1.1	4.2	336
12.	ВТ 9	1030	1090	16	0.8	1.1	4.2	350
13.	ВТ 20	930	940	5	0.8	1.1	4.2	345
14.	СЧ 18-36	170	180	-	-	1.2	4.5	60
15.	СЧ 24-44	230	240	-	-	1.2	4.5	78

Table 2.2. Mechanical characteristics of materials.

No var	Materials	Stress intensivity factors, $MPa\sqrt{m}$				n	C
		K_{Ic}	K_{IIIc}	K_{fc}	K_{th}		
1.	Сталь 45	73	-	-	3.3	3.3	$3.03 * 10^{-12}$
2.	КП 60	-	241.25	-	-	-	-
3.	КП 80	-	324.9	-	-	-	-
4.	КП 100	63.7	-	52.6	8.1	4.12	$2.33 * 10^{-13}$
5.	40Х	63.3	168.4	-	2.1	3.6	$1.18 * 10^{-12}$
6.	20Х13	42	-	-	5.45	4.16	$2.05 * 10^{-13}$
7.	15Х2МФА	93.3	-	121.3	9.3	2.55	$2.05 * 10^{-13}$
8.	25Х1М1ФА	-	260.8	-	-	-	-
9.	27ХН3М2ФА	-	348.9	-	-	-	-
10.	ЭП428	-	259.8	-	-	-	-
11.	ВТ 3-1	59	-	52.8	-	4.15	$6.82 * 10^{-13}$
12.	ВТ 9	76	-	73	-	4.0	$1.09 * 10^{-12}$
13.	ВТ 20	75	-	71.5	-	3.01	$2.08 * 10^{-11}$
14.	СЧ 18-36	20	-	-	7.0	6.6	$4.9 * 10^{-12}$
15.	СЧ 24-44	18.7	-	-	10.9	7.0	$0.7 * 10^{-12}$

Tabl.3 Forms of bars and types of cracks.

No var	Construction of bar	No var	Type of crack	Formula for K_I
1	2	3	4	5
1	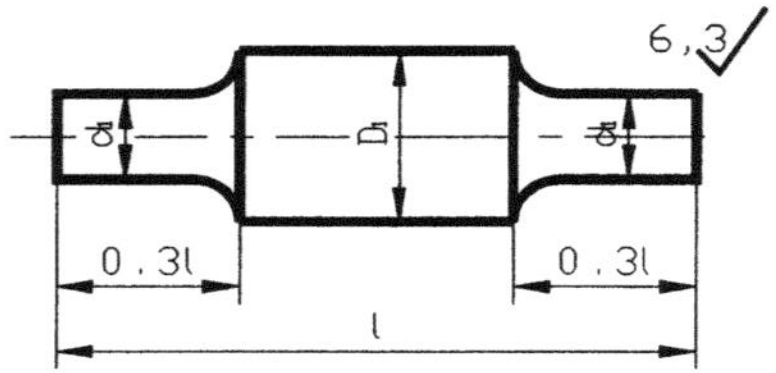 Carburizing	1	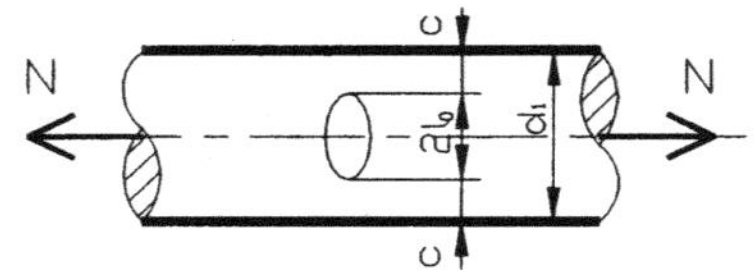	$K = YN \frac{\sqrt{2c/d}}{\pi(\frac{d_1^2}{4} - l_0^2)}\sqrt{\pi l_0}$ $Y=\frac{2}{\pi}(1+\frac{\lambda}{2}-\frac{5}{8}\lambda^2)+0,268\lambda^3$ $\lambda = \frac{2l_0}{d}$
2	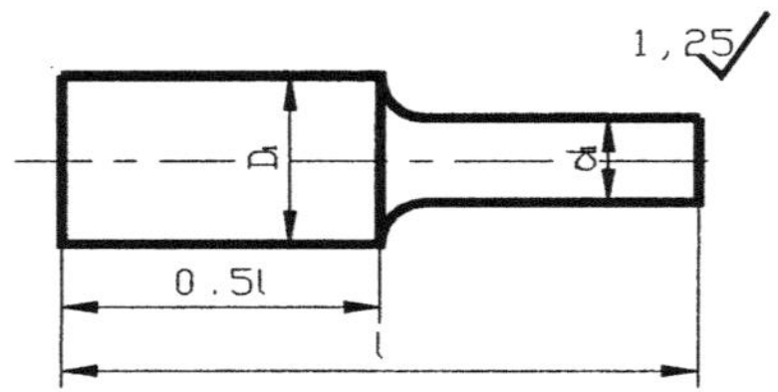 Cold work	2	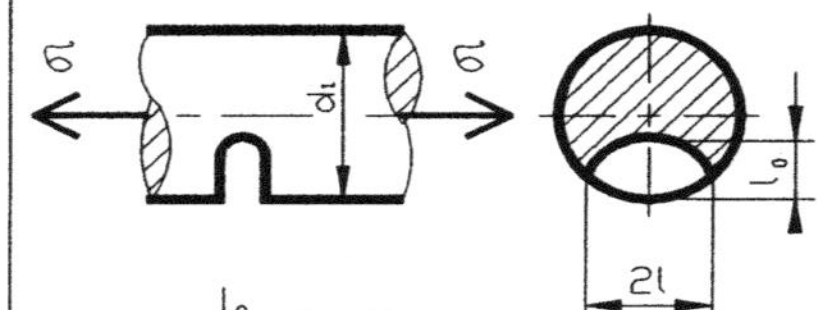 $\frac{l_0}{l_0}=0,5$	$K_I = Y\sigma\sqrt{\pi l_0}$ $\frac{2l_0}{d}$ 0,125 0,25 0,375 0,5 Y 0,89 0,92 0,976 1,064
3	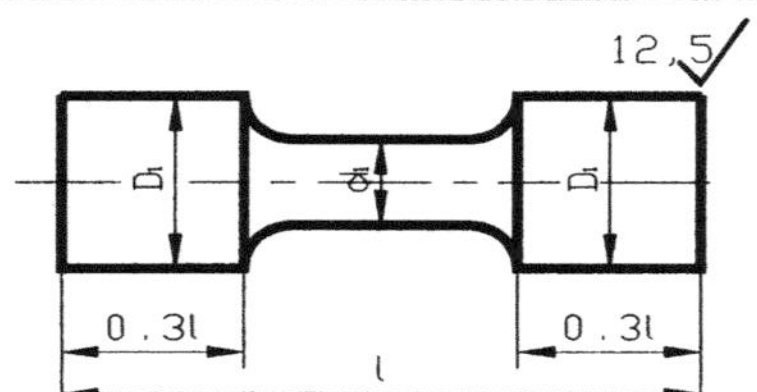 Self hardening	3	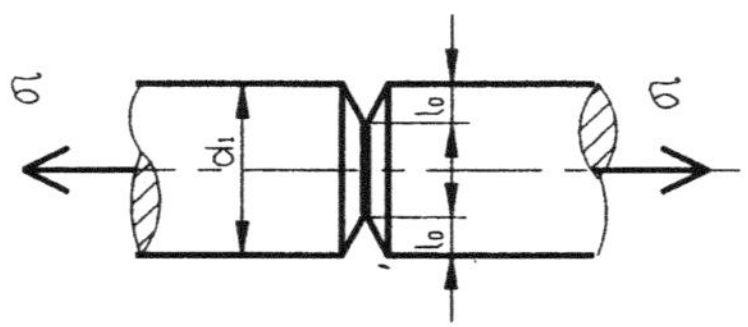	$K_I = Y\sigma\sqrt{\pi l_0}$ $\frac{2l_0}{d}$ 0,03 0,05 0,1 0,2 0,4 Y 1,88 1,82 1,66 1,41 1,01

Tabl.3 (Cont.)

1	2	3	4	5
4	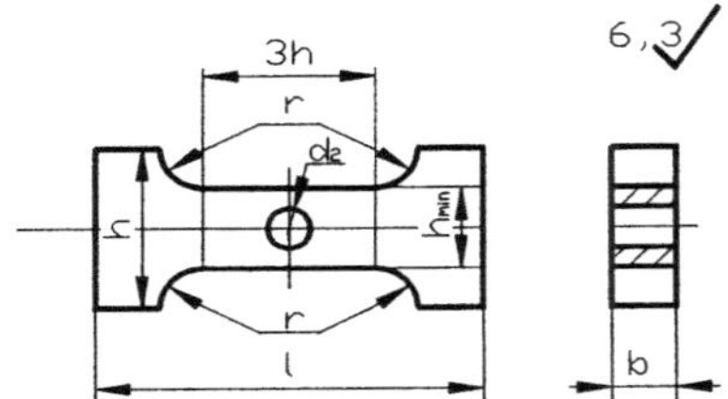 Induction hardening	4	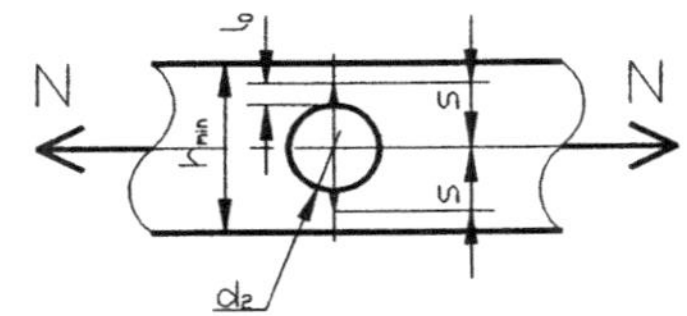	$K_I=\frac{N}{b\sqrt{h_{min}}}(\lambda)^{1/2}(1,77+$ $0,454\lambda-2,04\lambda^2+$ $+21,6\lambda^3)$ $\lambda=\frac{s}{h_{min}}$
5	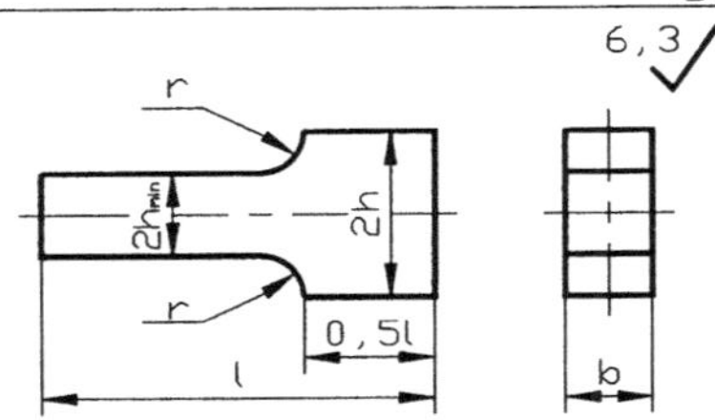 Carburizing	5	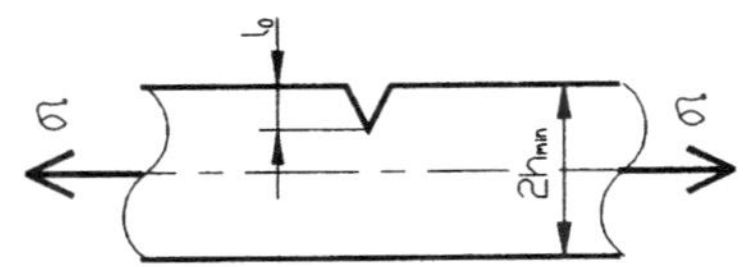	$K_I=Y\sigma\sqrt{\pi l_0}$ $Y=1,99-0,41\lambda+18,7\lambda^2-$ $-38,48\lambda^3+53,85\lambda^4$ $\lambda=\frac{l_0}{2h_{min}}$
6	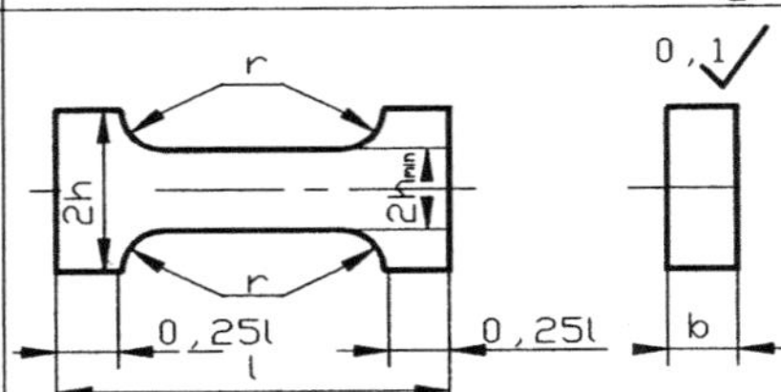 Normalizing	6	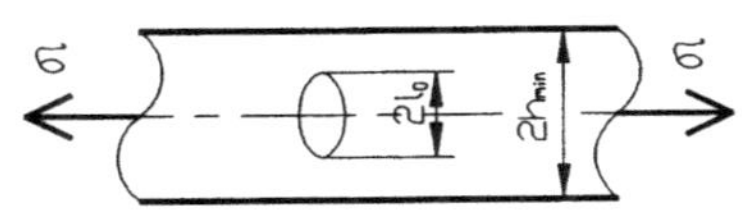	$K_I=Y\sigma\sqrt{\pi l_0}$ $Y=1+0,128\lambda-0,288\lambda^2+$ $+1,525\lambda^3$ $\lambda=\frac{l_0}{h_{min}}$ l_0

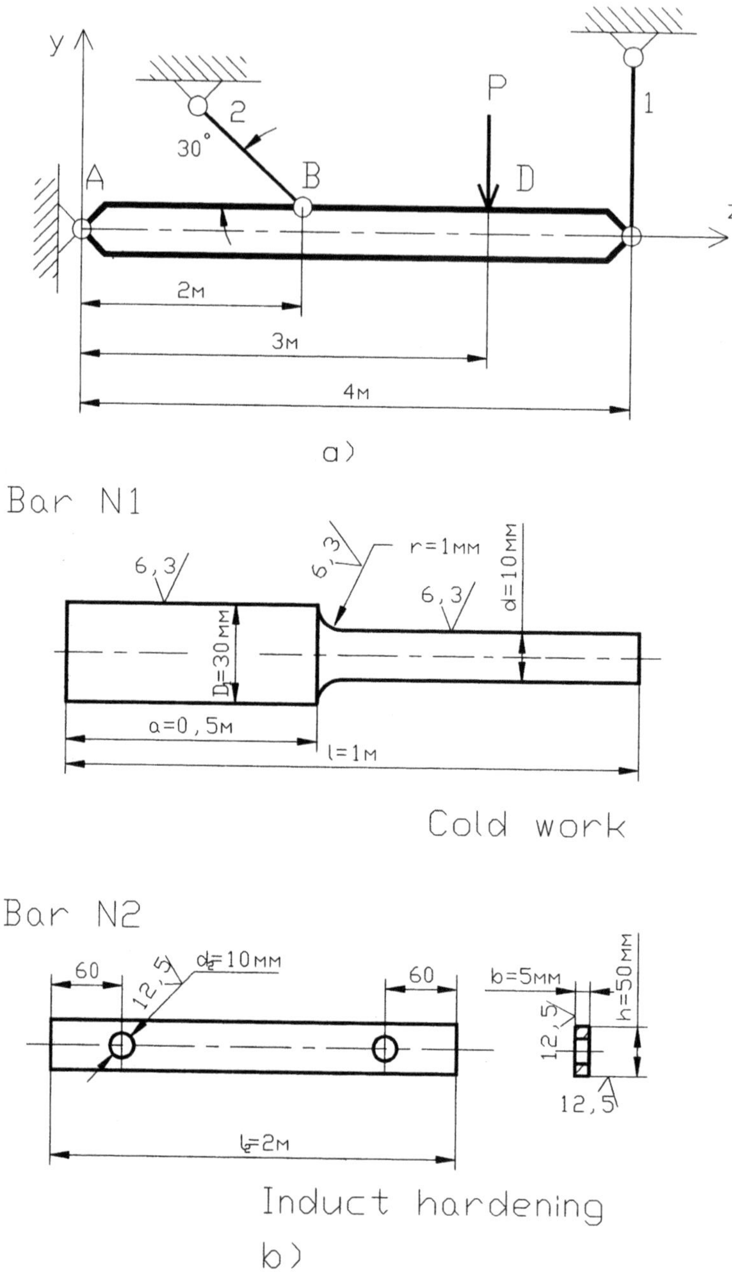

Drawing 2.1 Scheme of loading and construction of bars

MATERIALS OF BARS

Version 1	Bar No.1	*Steel CT.3*
	Bar No.2	*Steel CT.3*
Version 2	Bar No.1	$C\text{ч}12-28$
	Bar No.2	*Steel CT.3*

MECHANICAL CHARACTERISTICS OF MATERIALS

Steel CT.3	$\sigma_{0.2} = 250MPa;$	$\sigma_B = 420MPa;$
	$E = 2.1 * 10^5 MPa;$	$\delta = 25\%;$
	$\varepsilon_B = 0.5\delta;$	$\sigma_{-1} = 120MPa$
	$K_{IC} = 2.2MPa\sqrt{m};$	
$C\text{ч}12-28$	$\sigma_B = 120MPa;$	$E = 1.15 * 10^5 MPa;$
	$\sigma_{-1} = 45MPa;$	$K_{IC} = 3.16MPa\sqrt{m};$

TEMPERATURE CONDISIONS: $t = 293K$, $\Delta t = 0$

THE TYPE OF A CRACK

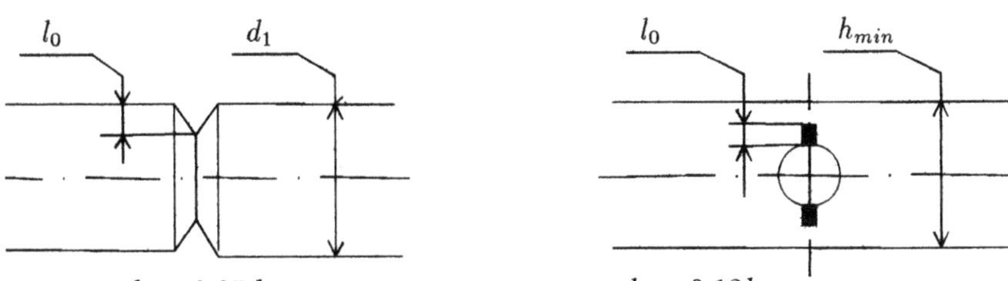

$l_0 = 0.05d_1$ $\qquad$ $l_0 = 0.12h_{min}$

Drawing 2.2.

Stage I. Determination of maximum admitted external force with the given strength factor of material.

The absolutly rigid square beam AC is supported by two bars and is loaded by the force P. Determine the admitted value of forse $[P]$ based upon the given strength factor of bar materials. Determine also the forse P_t, given the begining of creation of plastic deformation in the most loaded bar.

Drawing 3 contains calculation diagram of the problem at the first stage. Calculation is made for one or two version of bar materials. The received data are used at the third and forth stages of solution of this problem.

Table 4. Datas on tension.

No. var.	0	1	2	3	4	5	6	7	8	9
a, m.	0.8	1.5	1.2	1.8	2.0	1.6	1.4	1.7	2.0	2.2
b, m.	1	1.5	1.8	1.2	1.5	2.0	1.4	2.1	2.4	1.8
c, m.	0.9	1.0	1.0	0.8	1.2	1.2	1.0	1.5	1.0	1.5
Bar N1	1	3	2	2	1	3	5	4	6	5
Bar N2	5	6	4	6	5	4	3	2	1	2
d1, mm.	15	10	20	22	30	40	32	20	25	28
D1, mm.	25	30	30	44	60	60	58	60	50	60
d2, mm.	-	-	9.3	-	-	7.6	-	6.0	-	-
h, mm.	80	40	50	60	65	45	48	52	58	45
b, mm.	16	12	10	15	20	10	12	15	12	10
h/h_{min}	1.6	1.5	1.6	1. 25	1.3	1.25	1.6	1.4	1.2	1.5
r, mm.	10	5	3	4.5	7	3	6	4.5	12	5
e, mm.	2.2	1.5	1.6	1.0	1.2	2.0	1.0	1.8	0.8	1.5
n, min^{-1}	970	1500	600	1500	800	1000	970	1200	500	1200
No. bar	1(2)	1(2)	1(2)	1(2)	1(2)	1(2)	1(2)	1(2)	1(2)	1(2)
Type of cracks	2(6)	3(6)	1(4)	3(5)	1(6)	2(4)	6(3)	4(1)	5(2)	6(3)
l_0/d_1	0.12	0.05	0.1	0.08	0.15	0.2	0.22	0.12	0.1	0.2
l_0/h_{min}	0.1	0.12	0.05	0.2	0.1	0.1	0.1	0.18	0.25	0.1

Stage II. Determination of limited load during the deformation of bars outside of the range of elasticity.

Here, based upon calculation data we determine the most loaded bar. Using the diagram of tension (a - idealy plastic material; b - material with linear hardening) from the condition of stable equilibrium we determine the limited loading at the time when fluidity begins in all bars or one of them begins to destroy.

Drawing 4 contains the calculation diagram at the second stage.

Stage III. Determination of fatigue strength factor of bar system under the variable loading.

At this level the loading at the point D we propose to interpret as a power mit with a weight $G = 0,6...0,8$ [P], because it is the vibration source. The main shaft weights $G_b = 0,4G$ and the excentricity of mass e and speed n_p are given. Here, the student determines the condition of formation of resonance in this construction, the coefficient of increasing of forced amplitude, and for the given excentricity of the rotor mass a student determines the average σ_m and amplitude stress of cycle σ_a, based upon the conditions:

$$\sigma_m + \sigma_a \leq [\sigma]; \tag{1}$$

If it is necessary a student corrects the admitted allowable excentricity [e]. Then the student determines the fatigue strength factor of the construction and a student draws conclusions for different materials.

The diagram of calculation is presented on the drawing 5.

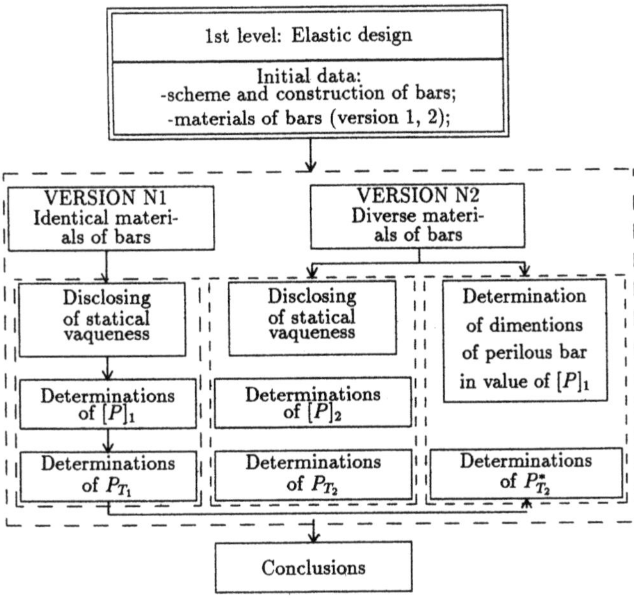

Drawing 3. Design of the problem at the first level.

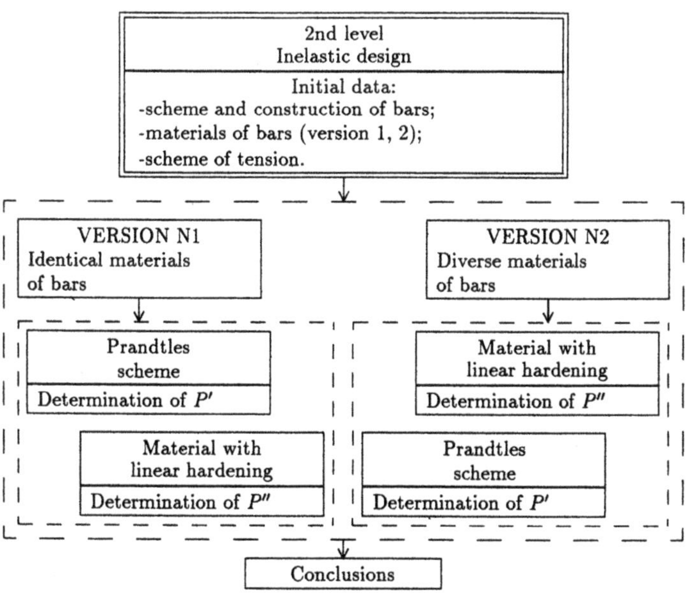

Drawing 4. Design of the problem at the second level.

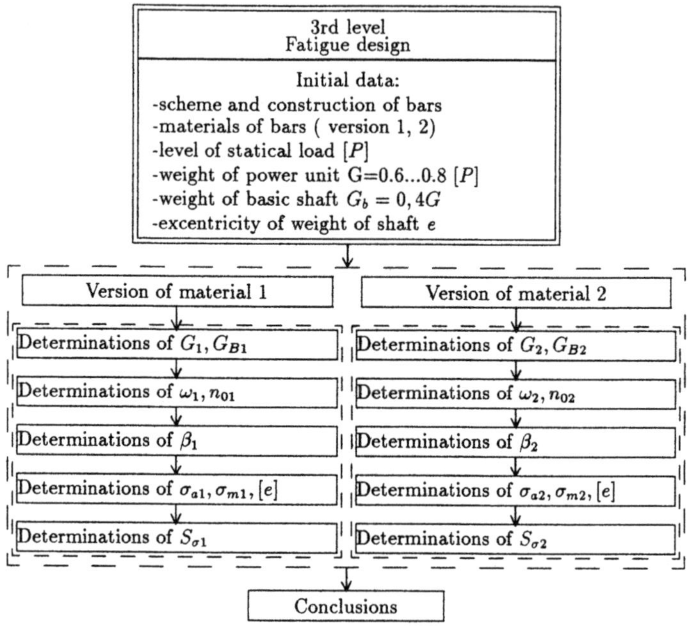

Drawing 5. Design of the problem at the third level.

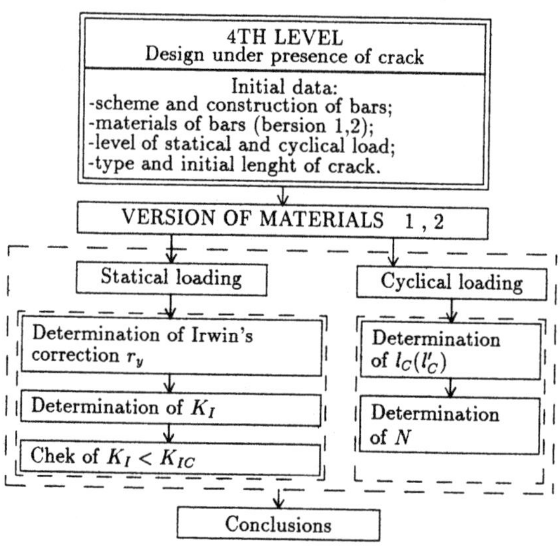

Drawing 6. Design of the problem at the forth level.

Stage IV. Check the security of exploitation of bar system under static and cycle loading, if the crack has appeared in one of the bars.

Drawing 6 contains the design scheme at forth level. The problem is solved in two ways: the static and the cycle one. We receive the initial data at the first and the third level. If the statement of this problem corresponds theoretically the first and the third level, to traditional student problem for opening the essence of the fourth stage we need some comments concerning to program of theoretical course of this theme.

It consists of:

- main knowlege of the mecanics of fracture;
- fragile fracture and Greaffits' ideas;
- appreciation of plastic zone at the top of crack;
- methodology of experimental determination of formation of cracks in construction;
- cycle of crack life.

The linear mechanics of fracture should be given the main consideration. The problem of unlinear mechanics of fracture are also present, but at the level of acquaintance. In the framework of theoretical course the student recevies the sufficient information about criteria of a fragile fracture choice of the stress intensity factor depending on kinds of loading, crack and shape of bodies to solve real engineering problems.

The forth stage contains the following steps:

1.Static loading.

- according to given specification of material of bars with a crack we find the amendment of Erwing for the value of plastic zone r_y;
- determination of value of stress intesity factor K_I;
- checking of carrying-out of these conditions

$$K_I < K_{Ic};$$

2. Cycle loading.

The Paris law is used during the cycle loading:

$$\frac{dl}{dN} = A\,(\Delta K_I)^n\,; \tag{2}$$

The determination of critical length of crack of given bar is necessary for solving of this problem. We can find it from:

$$l_c > l'_c; \tag{3}$$

Here

$$l_c = \left(\frac{K_{IC}}{Y \cdot \sigma_{max}}\right)^2 - r_y; \tag{4}$$

and l'_c – the length of crack, when the bar section is reduced until the dimensions when the stress reaches the limitation of strength:

$$\sigma_{max} = \sigma_B; \tag{5}$$

After the integration of this ratio we received the relation (2):

$$N\left(\frac{n}{2}-1\right)A(\Delta\sigma\cdot Y)^n = l_0^{\left(1-\frac{n}{2}\right)} - l_c^{\left(1-\frac{n}{2}\right)}; \qquad (6)$$

when
$\Delta\sigma$ – range of stress into the cycle;
l_0 – initial length of a crack.

Instead of K_{IC} during the cycle loading we recommend students to use the viscosity of fracture under the cycle loading K_{fc} if its known for this material. For the material as a cast iron instead of K_{Ic} we use the coefficient of intensity of stress K_{th}. For this material, as well known, the stage of creation of crack is absent, because the presence of graphite disturbs the contunuality of material, what is mean an existence of crack [1].

From the ratio (5) we determine the residual rating life N until the fracture of bar. We change the materials to find the maximum vitality of construction.

3 Conclusion

Given structure of calculation for problem solving for students, has certain advantages:

a) it permits students to skill the methology of appreciation of strength construction under several eniteria breaking conditions.

b) the numder of levels is not limited by the considered ones. This problem can be used as a basic one for a special course under the low-cycle fatigue conditions, creeping etc.

c) during the formation of calculaton problems the teacher can choose the stage levels and the essence of problems, which mostly correspond with the specialization program and with the number of academic hours.

References

[1] *Механика разрушений и прочность материаллов. Справочное пособие: В 4-х т./ Под общей ред. Панасюка В.В.*– К.: Наукова думка, 1988.

[2] Трощенко В.Т., Сосновский Л.А. *Сопротивление усталости металлов и сплавов. Справочник.*– К.: Наукова думка, 1987. – Ч.1. – 512 с.

[3] *Справочник по коэффициентам интенсивности напряжений: В 2-х т.*– Пер. с англ. /Под редакцией Ю. Мураками.– М.: Мир, 1990.

Advanced Teaching Aids

Teaching and Education in Fracture and Fatigue. Edited by H.P. Rossmanith. Published in 1996 by Taylor & Francis, 2 Park Square, Milton Park, Abingdon, Oxon, OX14 4RN. ISBN 0 419 20700 7.

22 ON UNDERSTANDING FRACTURE MECHANISMS THROUGH COMPUTER SELF-STUDY AND TESTING

I.I. KOKSHAROV
Computing Center of Russian Academy of Sciences, Krasnoyarsk, Russia
N.A. MACHUTOV
Institute of Machinery of the Russian Academy of Sciences, Moscow, Russia

Abstract
A method to gain experience on strength, structural integrity, engineering safety of carrying structures by computer testing and study is presented. The course 'Fundamentals of structural strength and safety' is intended for practical engineers and senior students of technical universities. The aim of the learning is to strive to understand fracture mechanisms. To realize the aim, special computer textbook and tests have been created.
Keywords: computer test, computer self-study, fatigue, fracture mechanisms, hypertext.

1 Introduction

Methods of computer self-study applied in teaching and training in engineering are rapidly progressing [1]. Let's consider pecularities of teaching fatigue and fracture. We suggest two principal ways for training : 1) intensive course delivered in lectures, examination of cases at seminars and testing by special computer programs; 2) learning by reading "pages" on computer screen, reading paper textbooks cited on computer "pages" and doing numerous control tasks. While the former provides a professor ready to help understand and examine the theoretical problems and practical cases the latter makes possible to study the subject independently at the working place.

Questions in the computer tests differ from tasks we solved in classical textbooks on theory of elasticity [2], fracture mechanics [3] : they are more simple - less time is spent to solve each problem (correct answer must be

Teaching and Education in Fracture and Fatigue. Edited by H.P. Rossmanith. Published in 1996 by Taylor & Francis, 2 Park Square, Milton Park, Abingdon, Oxon, OX14 4RN. ISBN 0 419 20700 7.

chosen from several ones or inserted). The feature allows to try the student's knowledge on a great number of various questions (more than 40 for one-hour long test). Much depends on the quality of the questions. The classical style of essence concentration by simplest explanations peculiar to J. Gordon's books [4,5] may be considered to be the best. It is precisely the questions that lead to clear insight of the subject. It is also very important to use the experience of exploitation and accidents study described in a simple and understandable language[6,7].

2 Computer textbook and self-study reference

For tests and self-study reference preparations we have developed program-shell PAGES (1995). The shell allows to create hypertext with built-in tools for testing. The hypertext allows to move and go back from one page to many other according to programmed cites on the page. The students answer two types of questions: 1) 'choose one from several items' (from 2 to 8, usually 4); 2) 'insert a correct text line or value' (alternative answers are acceptable). The reference includes content, chapters, paragraphs, bibliography, examples of accidents study, control tasks (Fig.1). An ordinary test includes a fixed number of cites (about 50) on "pages-exercises" chosen according to the identification number (ID). The shell gives the score table both for each student and for the entire group (rating list). The figure is not a commercial advertisement. It shows the structure of the self-reference and examples of its pages.

The new form of the textbooks gives new advantages : 1) a specialist with a personal computer can make his own tests without help of publishers; 2) time for preparing, transmiting information and learning may be shorter; 3) transmission of information by computer networks allows to study independently without gathering under 'one roof'.

3 On understanding fracture mechanisms

Attention is concentrated on the principal items - fracture causes, mechanisms of failure, driving force, main environment impacts. Step-by-step thorough analysis of different factors (temperature, loading mode, environment, material fragility, ...) and their effect on limit loads, fracture mechanisms is briefly explained in the textbook. After it a student does appropriate control tasks. Various approaches to the idea help him memorize and master the material under study.

For education in fatigue and fracture it is convenient to use simple graphical schemes, force lines drawings [8] (Fig.1, pages 8, 4.1.6), plastic zones representation (page 3), dangerous zones of structure (page 4), best versions of designer's decision (Fig. 2, page 2).

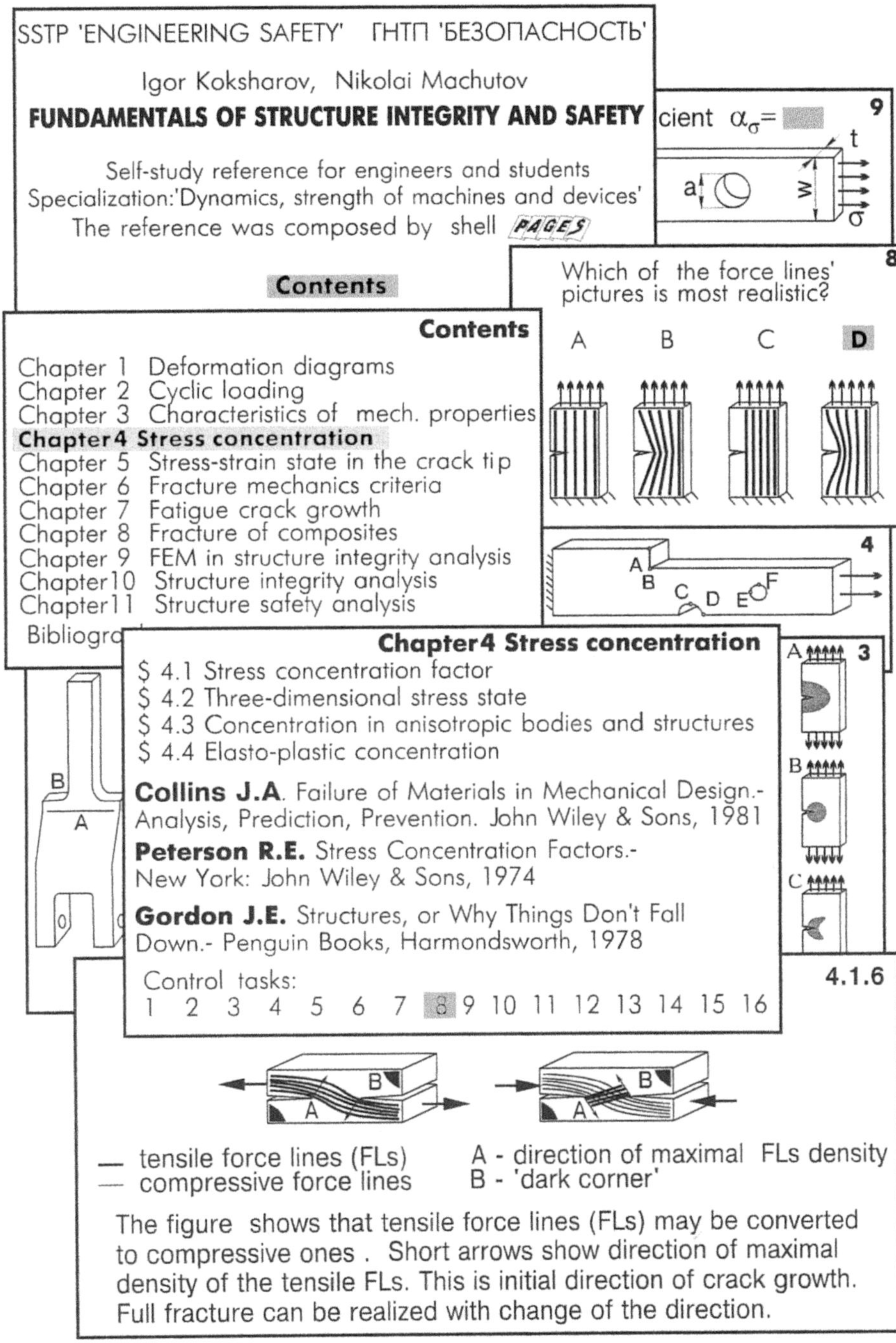

Fig.1 "Pages" of self-study reference.

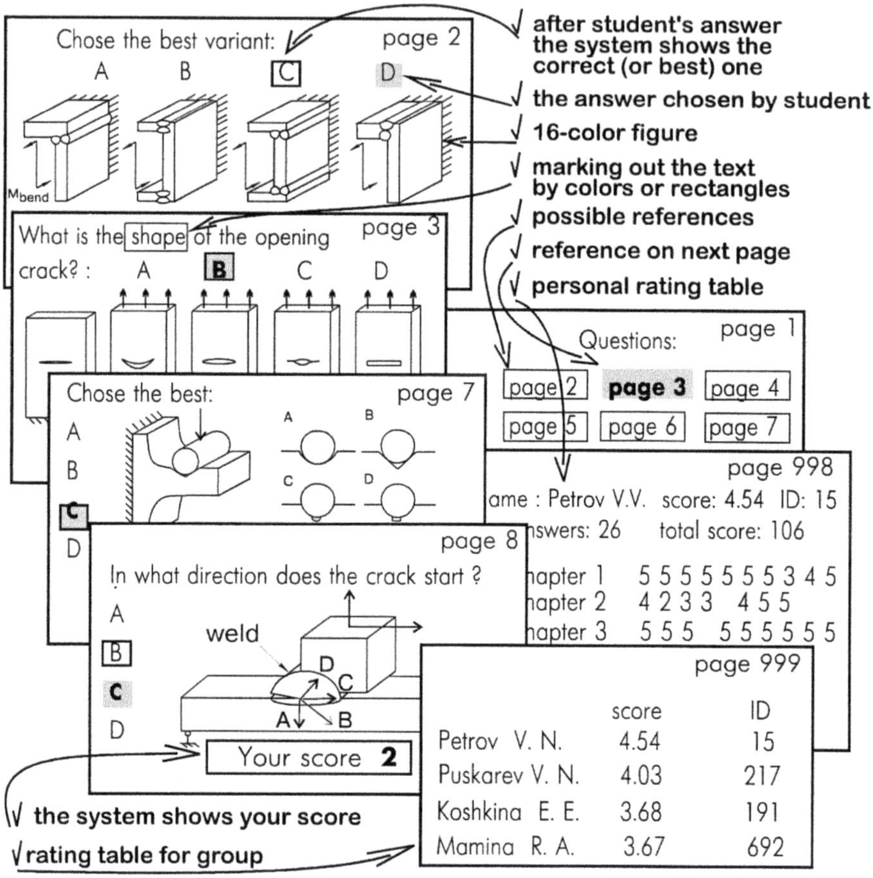

Fig.2 "Pages" from the test.

To understand fracture mechanisms it is useful to answer such questions:

Will the structure destroy or not?
From what point will the fracture initiate?
In what direction will the crack initiate?
Which type (mechanism) of fracture is most possible?
What is the decisive factor for this type of fracture?
How does the factor affect strength?
What is form of 'strength parameter-factor' dependence?
Which structure is most reliable?
Which structure is most dangerous?
Which structure is more rational?
Deformed shape of which structure is proper?
Which stress (force lines) distribution is proper?

Sensible answers promote understanding of causes, mechanisms, driving forces and consequences of failures.

4 Discussion

1) **Is it possible to study fracture mechanics independently by computer self-references and cited books only?** Can students or engineers understand principal fracture mechanisms by accident cases described in the computer reference? Our experience gives the positive answer to this question. Short, clear explanations of subject's essence with numerous graphical examples help to achieve the result.

2) **Is it enough to pass the computer exam to consider students trained specialists in fracture mechanics?** We think that the computer study by student himself is enough for primary engineering training. Profound work with a professor and/or practical consultants is necessary to prepare top-class experts.

3) **Is it worthwhile to create unified tests?** Should a special equivalent of IQ on fracture problems be used? Professors or universities can prepare their tests, computer references. However, joint work of the specialists with different experience, knowledge and viewpoints essentially improves the quality of the tests. And such tests will be very useful for many engineers and students. The work is not only to collect information on all known facts, but also to derive the essence of practical case - main causes of the failures.

4) **Is it possible to convert current textbooks on fracture mechanics into 'electronic encyclopaedia', self-control tests?** Simple page-to-page copy is inadmissible. The information must be compressed, the essence of problems must be pointed out, "screen pages" must be interesting to keep a student at the computer. The tests may be like computer games.

5) **Should the tests to be introduced in the design offices, the engineering firms?** It is obvious that such a work would promote both knowledge on fracture and prestige of the universities, engineering societies making these noble efforts. The tests give an impetus to spreading knowledge about fracture and help improve the professional skills of the engineers at large.

References

1. Wallace P. (1995) Virtual classes teach a real sense of community, *Computerworld-Moscow*, N 1, p.36.
2. Timoshenko S.P., Goodier J.N. (1970) *Theory of Elasticity*, McGraw-Hill, New York.
3. Broek D. (1988) *Practical Use of Fracture Mechanics*, Kluwer.
4. Gordon J. (1968): *The new science of strong materials or why we don't fall through the floor*, Harmondsworth: Penguin Books.
5. Gordon J. (1978): *Structures, or Why Things Don't Fall Down*, Harmondsworth: Penguin Books.
6. Baker R.D. Peter B.F. (1971) *Why Metals Fail*, Gordon and Breach Science Publishers.
7. Fisher J.W. (1984) *Fatigue and Fracture in Steel Bridges - case studies*, John Willey & Sons.
8. Koksharov I.I., Burov A.E. (1991) Force lines drawing for deformed body. An analysis of solutions for plate with hole or crack, Preprint of Computing Center of the RAS, N 6 (in russian).

23 EDUCATION: THE INTERFACE BETWEEN RESEARCH IN FRACTURE MECHANICS AND ENGINEERING PRACTICE

S.E. SWARTZ
Department of Civil Engineering, Kansas, State University, Kansas, USA
R.J. O'NEILL
Department of Civil and Mechanical Engineering, United States Military Academy, West Point, New York, USA

Abstract
Getting the word out is the basic premise of this paper. A tremendous amount of research has been conducted in the field of fracture mechanics of concrete and it is time to get this information to the practicing engineers. This paper discusses the reasons for providing this information to the practicing engineer and the format of its presentation. In this age of inexpensive personal computers most, if not all, professional engineers have immediate access to and are familiar with the use of personal computers. The information is based within an interactive, multimedia presentation operating on a personal computer within Microsoft Windows. The information and presentation are structured in such a way that the software can be used either as a tutorial or a reference. It provides information about basic fracture mechanics concepts, materials testing, and structural applications. Additionally, it can access an extensive bibliographical database. The interactive multimedia allows for the user to approach the information in a non linear manner while applications of machine intelligence assist the user in determining the depth of the knowledge desired.

1 Introduction

Since the early 1940's the methods of fracture mechanics have been used in analysis of structural failures, in design to prevent failures, in life assessment plans, on development of inspection schedules and other applications. These have included structures subjected to static loads (short duration and creep), time-varying loads (fatigue), and short duration events (impact loads, blast loads). Textbooks such as Broek [1] or Barsom and Rolf [2] describing basic theory are available and material testing standards have been developed, for instance for metals, ASTM [3]. Proposed material testing standards for rock, Ouchterlony [4], and concrete, RILEM [5] [6] [7], are currently being evaluated. There is

Teaching and Education in Fracture and Fatigue. Edited by H.P. Rossmanith. Published in 1996 by Taylor & Francis, 2 Park Square, Milton Park, Abingdon, Oxon, OX14 4RN. ISBN 0 419 20700 7.

quite literally an "explosion" of research activity investigating fracture mechanics of metals, plastic composites, wood, cementitious materials and aggregative materials at the macro-, meso- and micro-levels of behavior. An important problem related to this research activity is the timely transfer of this rapidly-expanding knowledge base into the realm of design practice.

One way of achieving technology transfer is through conferences and symposia. Another way is through committee reports, RILEM [8][9] and ACI [10]. Another approach is through educational workshops. All of these methods are basically intended to supplement formal educational programs.

Over the years, these educational efforts in the areas of metal and composite materials have resulted in acceptance of the methods of fracture mechanics in material characterization, structural design and performance evaluation by a large - and growing - community of engineers.

The same cannot yet be said to be true for the practice of structural engineering in concrete and in geotechnical engineering. This is partially due to a lack of consensus by researchers on the basic issues as well as a lack of standardized material testing procedures. However, these problems are rapidly being resolved. The main difficulties lie in

- lack of cross-fertilization between disciplines, i.e. the vast body of knowledge of fracture mechanics applications to steel structures is unknown and/or ignored by those working with reinforced concrete;
- to date there is only limited published evidence of practical applications in the failure analysis and design of R/C structures;
- as a result of the above there is limited - or no - understanding of the basic ideas of fracture mechanics by many designers of R/C structures.

It is thus incumbent upon the researcher to provide useful, readily accessible, educational programs to a large body of potential users at a reasonable cost. This may be achieved through the use of a microcomputer-implemented, educational tools based on concepts of machine intelligence (MI).

The first attempts at such an approach are described in Swartz and Kan [11]. While this effort was directed to implementation both in design practice and as an educational tool, it became apparent that the initial applications would be primarily as a stand-alone system to explain fracture mechanics concepts in general and their utilization with concrete structures in particular.

Thus, the emphasis has changed somewhat and the focus of this research is the development of an educational tool that incorporates multimedia, a graphical user interface (GUI), and MI concepts into a complete package to explain to the practicing structural engineer the need, theory, recent developments, design parameters and applications of fracture mechanics. It is intended that this be implemented on a 486 level multimedia PC operating within Microsoft Windows 3.1.

2 Interactive Multmedia / Hypermedia

The tremendous amount of information concerning fracture mechanics of concrete has to be presented in a manner that is both educational and user friendly. The structure chosen for this presentation is Interactive Multimedia.

2.1 Definitions

At this point some brief definitions of important terms are required. Computer Based Training (CBT), Computer Assisted Instruction (CAI) and Computer Aided Learning (CAL) are all names for software programs that allow a user to receive instruction about a particular topic by accessing a computer terminal. Hypertext is a tool used within a computer program that allows the user some control over the sequence of events in a given application. "The idea of hypertext is simple: You link related information together, regardless of its location and medium, and links lead to other links in a nearly endless chain from text to graphics to video to applications and back again. Leaping through information this way makes reading an interactive, three-dimensional activity" [12]. "Thus, the word 'hyper' in hypertext would give 'text' an extended or generalized form. In hypertext, the information is not presented in a single layer of text. It is a multilayered environment. What you see on the computer screen is only one layer: but there is 'more' than what you see" [13]. A program that uses hypertext , thus becomes interactive with the user. Originally, hypertext was only capable of displaying textual information, but as the definition above suggests, it now goes beyond just text sequences and encompasses graphics, animation, audio and motion video displays. Because of this evolution, a new word was created, hypermedia. Interactive multimedia can be considered synonymous with hypermedia.

2.2 Multimedia Personal Computer (MPC)

What is an MPC? Recently a group of hardware and software vendors joined together to create the Multimedia OC Marketing Council. Their goal was to develop a consistent set of standards that software developers could use in order to develop multimedia applications. The current MPC standard, shown in Fig. 1 [14] , requires:

1. an 80386SX or better processor;
2. VGA (Video Graphics Array) graphics capabilities;
3. support for MIDI (Musical Instrument Digital Interface) and waveform audio;
4. CD (Compact Disk) ROM (Read Only Memory) Drive;
5. Multimedia Windows (Microsoft Windows with Multimedia Extensions 1.0);
6. 4 MB (Mega Bites) RAM (Random Access Memory);
7. 100 MB hard drive.

The 80386SX processor is the minimum required due to speed requirements for playback of audio and video. Older processors like the 80286 are not sufficiently fast enough to provide adequate and realistic playback.

The standard VGA graphics with a screen resolution of 640x480 pixels is the bare minimum. Most PC's sold today come with graphics adapters and monitors capable of displaying 800x600 or 1024x780 pixels for clearer, high quality pictures.

Support for MIDI and waveform audio usually comes from a plug-in sound card such as Creative Lab's Sound Blaster [15] or Media Vision's Pro Audio Spectrum [16]. Waveform audio is a form of digital audio understandable by the PC.

The CD ROM provides over 600 MB of storage. The CD ROM disks (similar to an audio CD) are cheap and easily distributed [13]. The storage requirements for interactive multimedia grow quickly as audio, animation and full motion video are added to applications. One CD ROM is capable of replacing over 400, 3.5" HD (high density) floppy disks.

Multimedia Windows is a set of software drivers that support waveform, MIDI and CD audio and animation.

2.3 Interactive Multimedia Applications

Some examples of interactive multimedia applications are shown in Fig. 2 [14]. In this figure are two different applications currently available for multimedia. The first is called "Multimedia Beethoven: The Ninth Symphony". "This program, created by Winter and the Voyager Company, features an interactive learning session where you can read about Beethoven's world, explore the structure of the symphony, and see and hear the symphony in detail" [14]. The second is called "Compton's Multimedia Encyclopedia for Windows". "Both sound and animation make [this] more than just a disk version of a print encyclopedia. You can hear famous phrases, see animations of how things work, and jump easily from topic to topic" [14].

2.4 Why Interactive Multimedia?

Melhem, in a presentation at a videoconference on multimedia technology [17], listed six advantages of Computer Based Training (CBT) and Computer Aided Instruction (CAI). These were:

1. Cost Effectiveness
2. Standardization of Learning
3. Individualized Attention
4. Flexibility: Self-paced Training
5. Trainee is in Control
6. Training Readily Available (24-hour Access)

"The use of multi-media computer assisted learning (CAL) has grown quite considerably over the last few years. This growth has been primarily due to three major factors:

- the relatively low cost of the terminals and other types of resource needed to implement it;
- some degree of dissatisfaction with conventional approaches to computer assisted learning; and least of all,
- the effectiveness of the instructional strategies that can be produced using multimedia CAL "[18].

Additionally, Miller [19] lists ten learning benefits of interactive technology. Some of these include:

- Reduced learning time by as much as 50 percent due to various reasons that include easily understood formats, personalized instruction that accommodates different learning styles and constant, highly-effective reinforcement of concepts and content.

- Privacy that allows users to explore and ask questions that they may not be comfortable with in a group setting.
- Strong reinforcement is gained by the interactive process with the material that provides increased content retention.
- The responsive feedback and individual involvement increases student motivation.
- Learners enjoy interactive learning.

3 Interactive Multmedia Software

This work is a continuation of the efforts described by Swartz, Kan and Hu [11] and while the medium of the expert system is not included, many of the reasons for adopting a CBT approach are still viable. These are:

1. Software is portable, relatively inexpensive and may be disseminated readily. It is envisioned that the final product of this work will be a self-sufficient computer program that can be distributed to practicing engineers and engineering students alike.
2. Aspects of the software can be made to interact with present computer aided design systems in a logical way.
3. The most important reason is that the massive amount of information presently available has been sorted, culled and distilled into a manageable form to be presented to the user. This is provided by structuring the information in a way that allows the user to be selective about how much or how little the user wants to know about a particular subject. The linking of the information within the program is thus crucial to the overall performance of the system.

3.1 Scope of the Proposed System

The objective of this research is to develop an interactive, multimedia, microcomputer system in the domain of fracture mechanics of concrete structures which can be used as a Text/Reference "Book". This system would be constructed such that it would provide the user with the ability to learn about fracture mechanics starting with the very basic and progressing along a logical path to more complex theories. In this manner the system would be operating as the "textbook" in a stand-alone tutorial. This tutorial assumes that the user's competency is equivalent to a Bachelor of Science degree in Civil Engineering. Additionally, the system would be a handy reference to the more experienced user as a way to quickly relearn or investigate a particular topic. The system is being constructed using an interactive multimedia structure that allows the information to be accessed in a nonlinear manner somewhat like using the index of a book to seek out pertinent information throughout the book.

3.1.1 Knowledge/Data Base

The information that will be contained in this system will include the following:

1. Basic fracture mechanics -- localization concepts, linear elastic fracture mechanics, nonlinear fracture mechanics (fictitious crack model, size effect law, R-Curve (Fracture Resistance) approach, failure classification, why fracture mechanics, domain of applicability compared to-say - plasticity, etc.);
2. Materials testing -- usual tests for properties of hardened concrete, tests to obtain fracture parameters, closed loop testing, data reduction/interpretation, explanations for preferences supported by experimental evidence in the literature;
3. Structural applications -- failure analysis case studies, design, software availability and utilization. Current application areas include:
   - bar pullout, bond
   - anchor bolts
   - shear
   - minimum reinforcement
   - fatigue
   - creep
   - plastic hinge formation
   - A tool to characterize advanced construction materials
4. Access to an extensive bibliography of fracture mechanics of concrete references;
5. Access to a glossary of fracture mechanics terms through hypertext.

3.1.2 Operating System and Hardware

This interactive multimedia system will operate within Microsoft Windows 3.1 [20]. The software will be written in C++ using the Borland C++ 3.1 compiler [21]-[23]. A number of other references were used in the software development [24]-[32] while Refs. 33-45 were used in subject matter development. Microsoft Windows is currently becoming the operating system of choice for personal computers. Its Graphical User Interface (GUI) makes it an easy system to operate. The user is usually presented with action type icons that must be pointed to with a mouse and clicked on. The selection of an icon then causes an action to occur within the program. This system has the ability to interact with our concrete fracture mechanics bibliography which is contained in dBase IV [46]. Additionally, the system can operate external DOS programs.

In order to make this system easily available to the practicing engineer, it is important to provide it on a common platform. That platform is IBM PC compatible with at least an 80386SX processor, 2 MB of memory, 40 MB hard drive, VGA monitor and Microsoft Windows. Due to the large files created by sound and graphics, it is envisioned that this software will eventually be available for distribution on CD ROM. If the user has multimedia hardware available, CD ROM and soundcard, then the performance of the system will be enhanced with animation and digitized audio and video and the mass storage available on the CD ROM.

3.1.3 Features

Features to be included in the system:

1. Browse - move through the material forward and backward;
2. Search - search for specific words or topics;
3. Bookmark - allows user to mark a specific item for later use;
4. Help - On-line help to assist the user in using the system;
5. History - Keeps track of where the user has been in a current session to allow backtracking;
6. NotePad - A pop–up notepad for writing notes during a session for later printing;
7. Digitized sound where appropriate;
8. Graphics and scanned graphic images.

Features to be considered for inclusion:

1. Animation and motion video;
2. Operation of finite element program demonstrating fracture mechanics modeling;
3. Voice recognition using neural nets to allow use of voice commands to replace some mouse commands;
4. Demonstrate use of Fuzzy Arithmetic/Logic in fracture design and analysis;
5. Context sensitive help;
6. CD ROM distribution disks.

3.2 FRACMECH

The organization of the major areas of the program are shown in Fig. 3. This is an actual reproduction of the user screen. Each item in the table of contents is in blue on the user screen to designate it as a hypermedia item. Therefore, the user merely points the mouse pointer at a particular choice and clicks the mouse button to select that topic. Fig. 4 shows the sub-table of contents that appears when the user selects

"2 Learning Fracture Mechanics".

The development of this program is in progress and is expected to be completed early next spring. Figs. 3-29 show a sequence of screens that display several of the topics that are currently being constructed. The sequence in which they are shown is selected by the user. By following the mouse cursor, the reader can determine how the user moves from one screen to another. Each screen is actually constructed as an ASCII text file with codes that determine which graphics files to display along with the text. Once the multimedia hardware is attached to the PC, additional codes can be added to the text files for executing audio and video files.

The ability to create each screen as a separate text file allows the program to be modular and easily changeable in the future. The actual information available and presented can be changed without changing the executable code. An example of a text file that would display what is shown in Fig. 16 is shown in Fig. 30. Everything after the "\" is code to tell the program what Bitmap graphics to display and where to display it on the screen, and what, if any, additional screens are linked to this screen.

Refer to Fig. 30:

- **3.1.c Displacement Test, RILEM Method`** - Screen title.
  This is a load point displacement (LPD) test. The displacement is measured indirectly as the movement of the of the beam surface near the crack as opposed to directly measuring the displacement at the load.
- \ - everthing before this symbol is printed as text.
- **{Bitmap31c}100x250y** - display Bitmap31b (drawing of test setup) at pixel position 100,250.
- | - the following are connections to other locations in the program.
- **{cont_arw}13105**- display the MORE button - 13105 is the label of the screen that is linked to the MORE button. If the user wants to continue along this path the MORE button is clicked on.
- **{back_arw}13102**- display the BACK Button- 13102 is the label of the screen that is linked to the BACK Button. This returns the user to the previous screen.
- **{camera1}13104**- Enables the PHOTO Button to tell the user that a scanned image is available to view - 13104 is the label of the screen that contains the scanned image.
- * - The "*" tells the program that there are no more codes.

3.2.1 Learning About Fracture Mechanics

The initial element of FRACMECH is a Fracture Mechanics Text/Reference "Book". The construction of this reference is in its early stages. Fig. 4 Shows the Table of Contents for the coverage of this topic. Since each item in the Table of Contents is a hypertext item the user simply clicks on the topic desired and that particular subject area is presented. Figs. 4-11 demonstrate a short session. The user has selected **"2.1 Introduction"** in Fig. 4. This causes the screen in Fig. 5 to be displayed. In Fig. 5 the user elects to move on to the next subject, **"2.1.1 Why Fracture Mechanics"**, by clicking on the ">>" symbol. This causes the screen shown in Fig. 6 to be displayed. This continues on until the user reaches the screen shown in Fig. 8. In the lower left portion of this screen the MORE button becomes enabled that indicates that there is additional information concerning this topic. The user selects this and the screen shown in Fig. 9 appears. The MORE button remains enabled which means there is still more. Additionally, the BACK button is now enabled to allow the user the ability to return to the previous screen. Figs. 10-11 continue this process until the MORE button is no longer enabled, signifying the end of this particular path. The user can either backtrack or merely select ">>" to go on to the next subject, **"2.2 Linear Elastic Fracture Mechanics (LEFM)"**. This demonstrates the flexibility of this program. The user controls the way in which information is presented.

3.2.2 Displacement Measurements for the RILEM Method

The method of determining the energetic parameter fracture energy, G_F, is a proposed RILEM test standard which requires the measurement of energy of deformation of a beam in three point bending (TPB) [5]. Although error limits on displacement are specified, a preferred method to measure the displacement is not given. FRACMECH supplies a recommended procedure as shown in Figs.13-19. Explanations are available as to why this procedure should be used in order to mitigate against "punching" errors [39], errors associated with indirect measurement [40] and others, including rolling friction [41]. The logical basis for this test is given by Hillerborg [42].

As experience is gained in implementing this method - either by the user of the program or by others - these screens can be updated and/or augmented in a logical, and reasonably simple way.

3.2.3 Equivalent Crack Models

One of the problems in applying LEFM directly to crack growth and fracture in concrete is the non-uniform nature of the crack front [43] which implies the need for three-dimensional modeling. In order to overcome this problem in a practical way and to obtain size-independent fracture properties, the concept of the "equivalent crack" is used. The basic idea is that the combination of open crack and process zone ahead of it can be replaced by an equivalent, open crack for which the fracture mechanics parameters can be evaluated from LEFM. Approaches using compliance comparisons are given by Jenq and Shah (J/S) [44] and Karihaloo and Nallathambi (K/N) [45]. An approach using peak load-crack length calibration is given by Go, Refai and Swartz (G/R/S) [43]. The J/S method has been proposed as a RILEM standard test [6]. This is displayed on FRACMECH screens in Figs. 20-25 and 27-28. The determination of effective crack length a_e requires solution of a non-linear equation. A suggested algorithm for rapidly-converging, trial and error solution is presented in Fig. 26. The effectiveness of this method is described in Ref. 9.

4 Conclusions

The work presented here represents the start of a program to provide practical information to the engineering profession on the concepts of fracture mechanics and how these can be implemented in the design of reinforced concrete structures.

The use of interactive multimedia appears to offer a promising mechanism to develop computer software which may serve this education goal and which also may become a part of an integrated computer aided design system.

Acknowledgments

The work reported here is supported by the Information Technology Laboratory, Waterways Experiment Station, US Army Corps of Engineers, Vicksburg, Mississippi. The expert system program, FRMECH, that much of this work is based on was written by Y.C. Kan under the direction of the first writer and was supported by the Center for Research in Computer Controlled Automation at Kansas State University . The bibliography data-base program BIBLI and associated data files were written by R.P Bernhardt as an M.S. thesis project under the direction of the first writer and was supported by the Naval Civil Engineering Laboratory Contract N62583/85 MT239. This support is gratefully acknowledged.

Figure 1. Multimedia Personal Computer

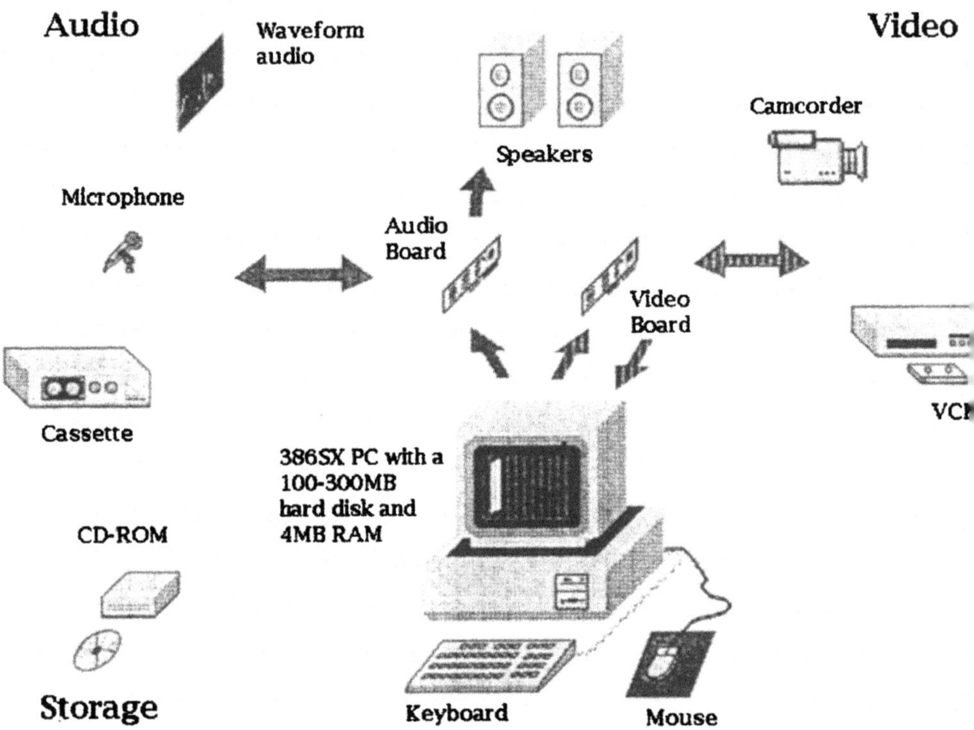

Figure 2. Multimedia Applications [14]

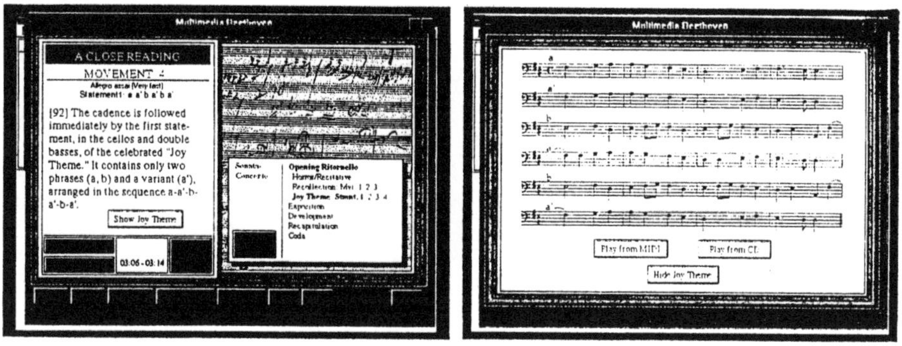

Sometimes sound is the whole point of an application, as with *Multimedia Beethoven: The Ninth Symphony.* This program, created by Robert Winter and the Voyager Company, features an interactive learning session where you can read about Beethoven's world, explore the structure of the symphony, and see and hear the symphony in detail.

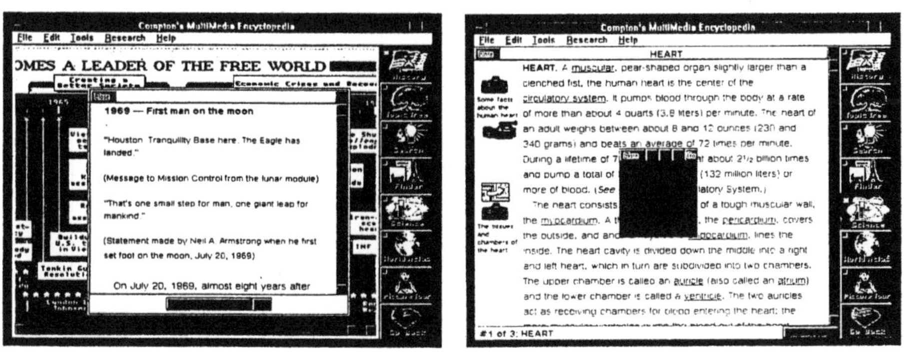

Both sound and animation make *Compton's Multimedia Encyclopedia for Windows*, from Compton's New Media, more than just a disk version of a print encyclopedia. You can hear famous phrases, see animations of how different things work, and jump easily from topic to topic.

Figure 3. Main Table of Contents

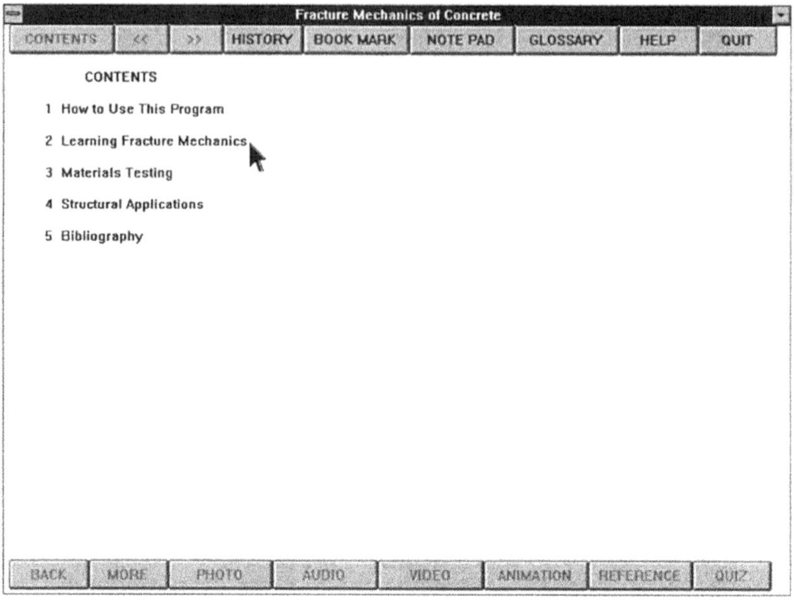

Figure 4. Learning Table of Contents

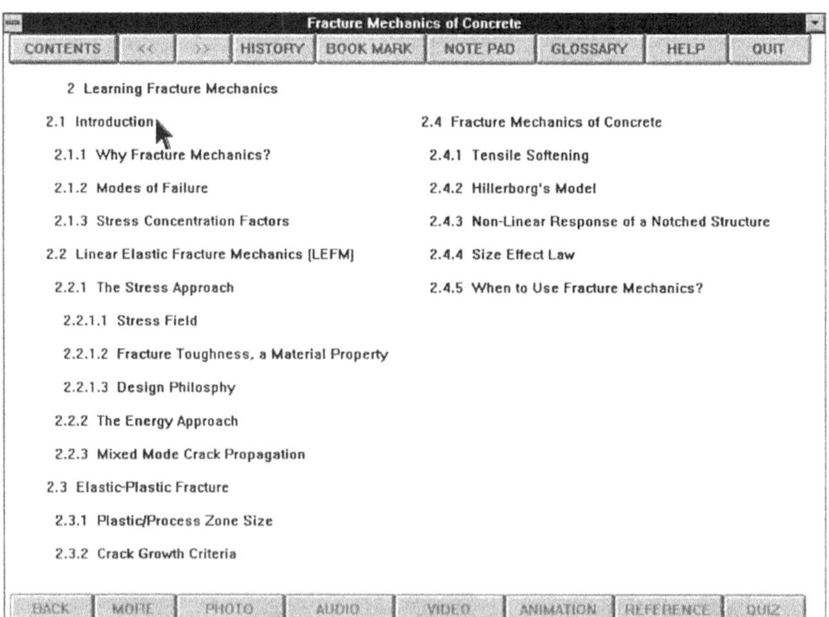

Figure 5. Introduction, Learning

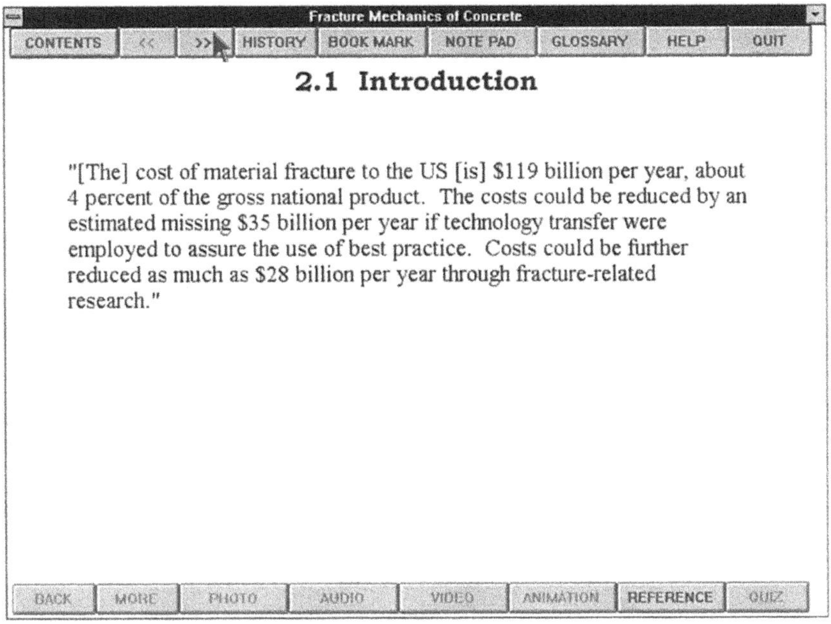

Figure 6. Why Fracture Mechanics, Learning

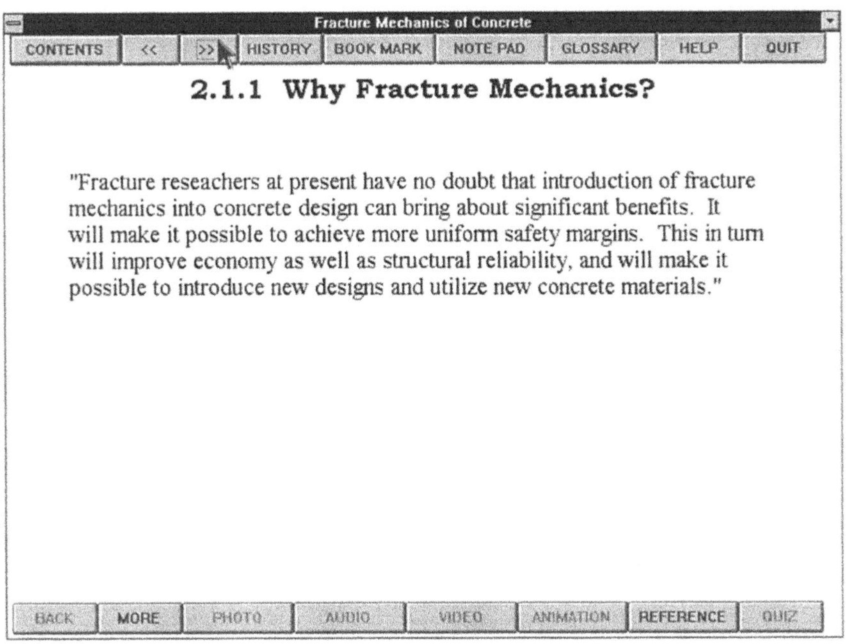

Figure 7. Modes of Failure, Learning

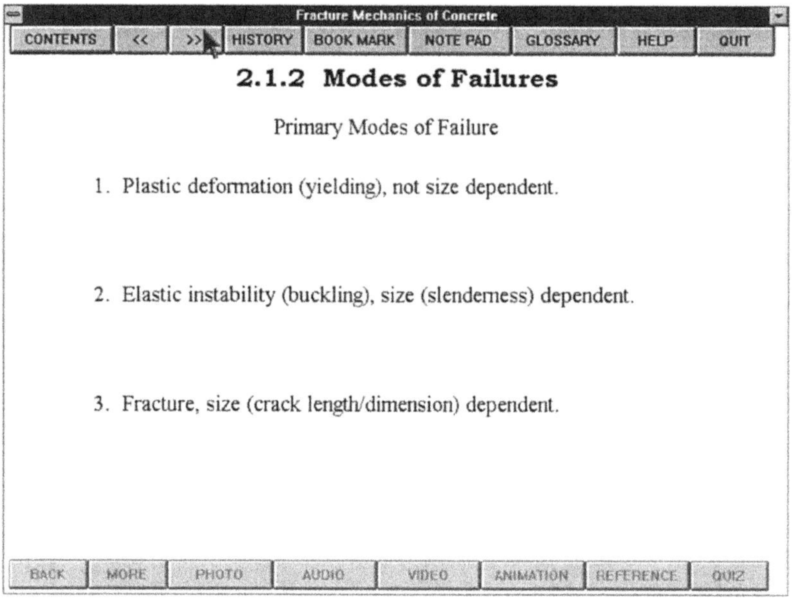

Figure 8. Stress Concentration Factors, Learning

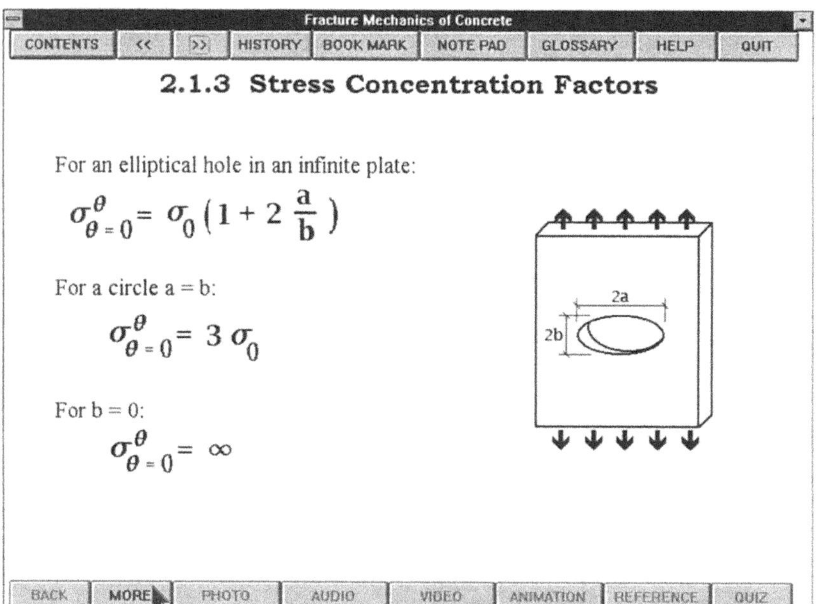

Figure 9. Mode I, Stress Concentration Factors, Learning

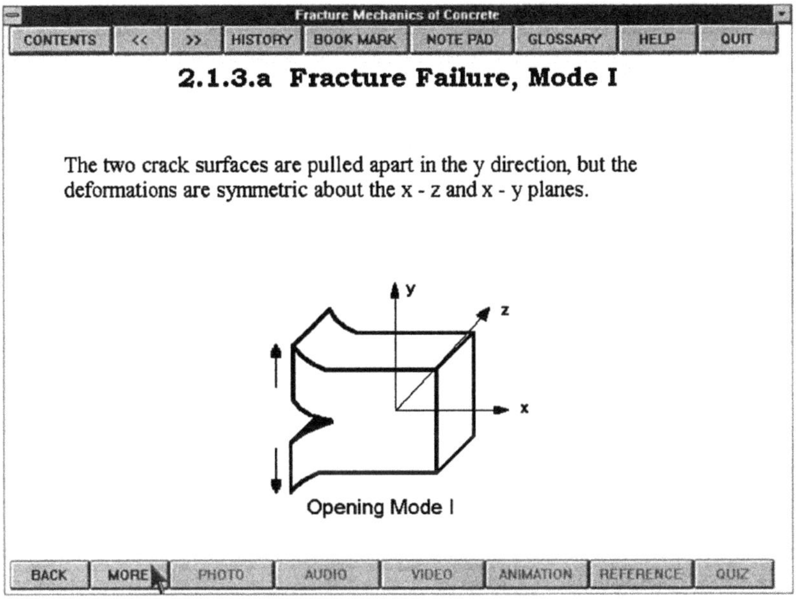

Figure 10. Mode II, Stress Concentration Factors, Learning

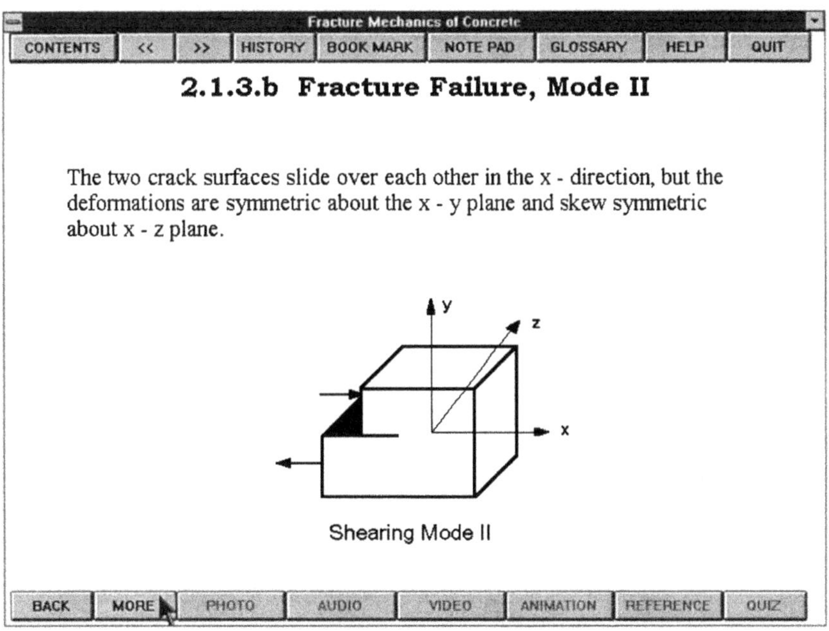

Figure 11. Mode III, Stress Concentration Factors, Learning

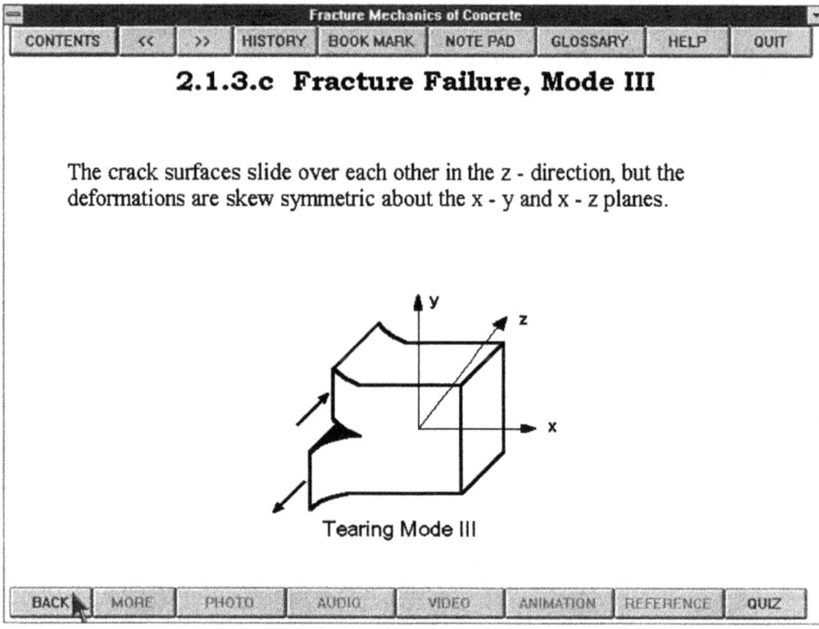

Figure 12. Testing Table of Contents

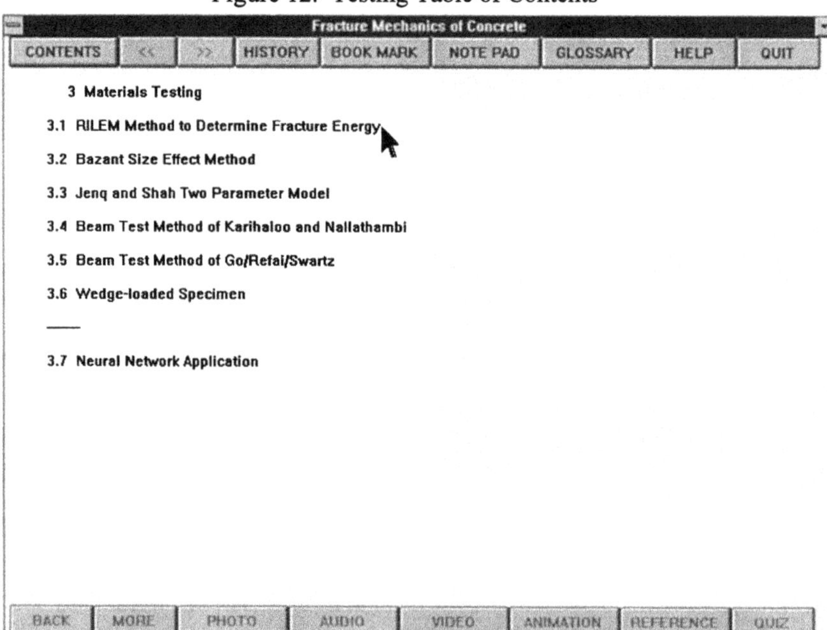

Figure 13. Background, RILEM, Testing

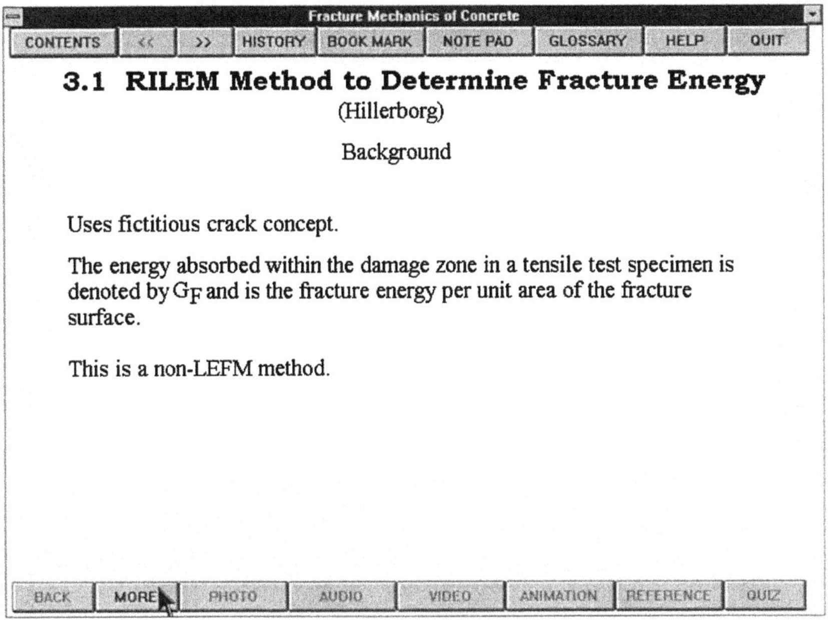

Figure 14. Specimens, RILEM, Testing

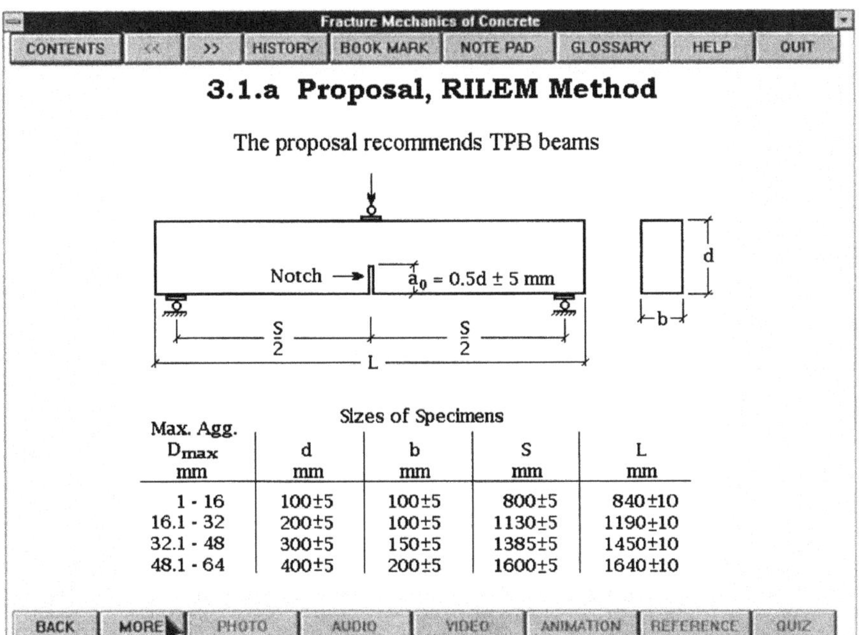

Max. Agg. D_{max} mm	d mm	b mm	S mm	L mm
1 - 16	100±5	100±5	800±5	840±10
16.1 - 32	200±5	100±5	1130±5	1190±10
32.1 - 48	300±5	150±5	1385±5	1450±10
48.1 - 64	400±5	200±5	1600±5	1640±10

Figure 15. Testing Procedures, RILEM , Testing

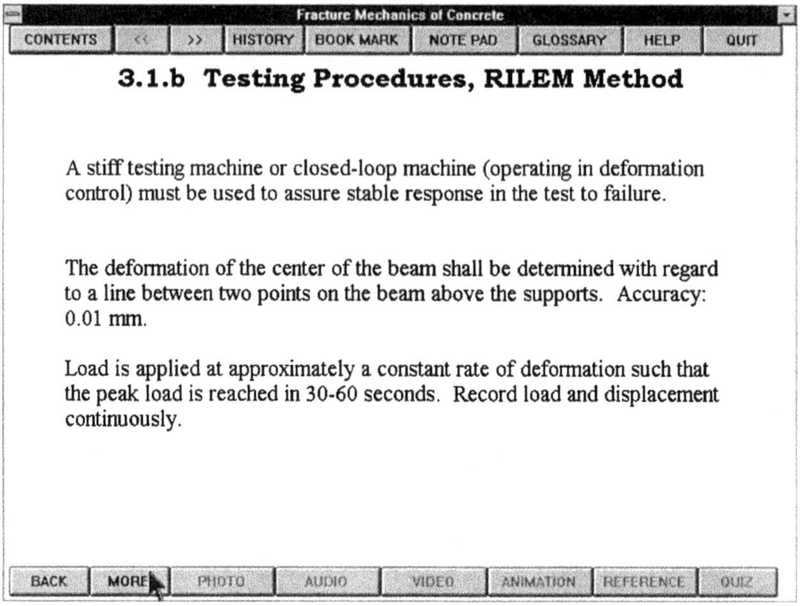

Figure 16. Note on Testing Procedures, RILEM, Testing

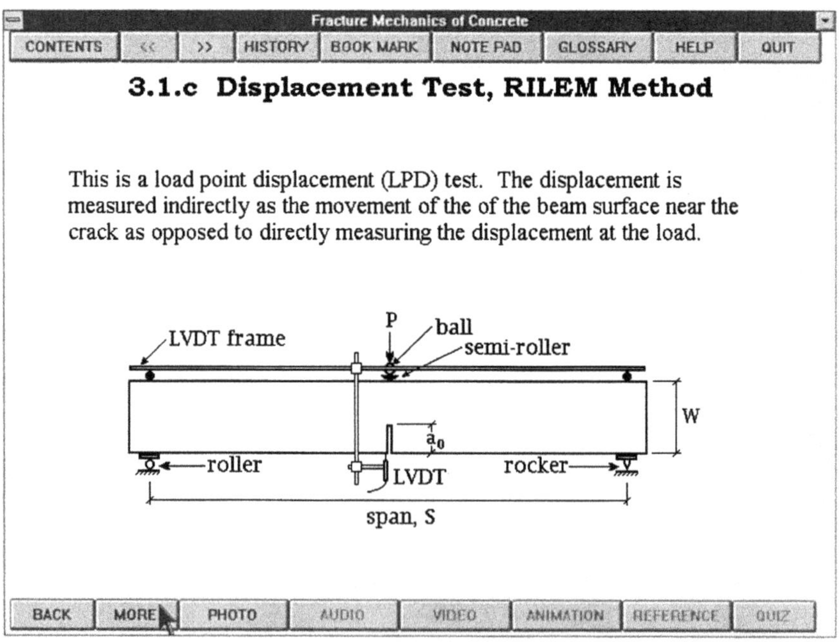

Figure 17. Data Plot, RILEM, Testing

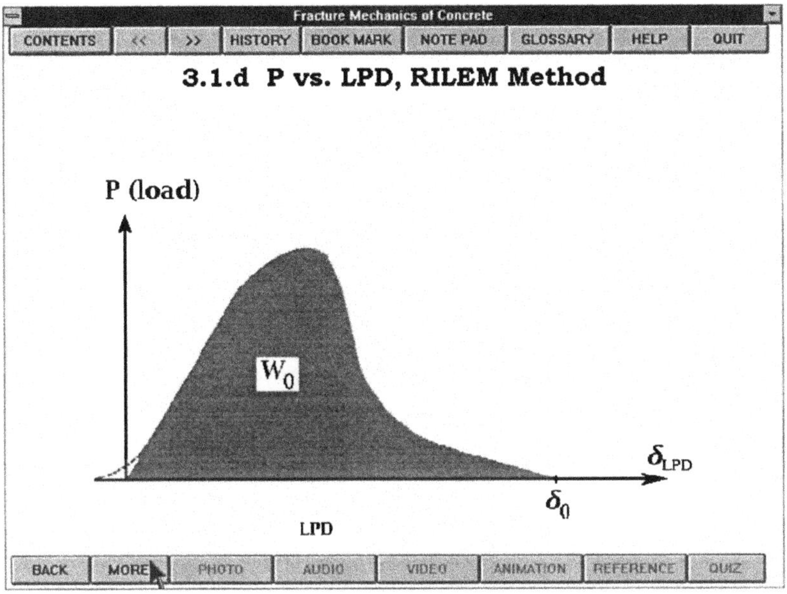

Figure 18. Data Evaluation, RILEM, Testing

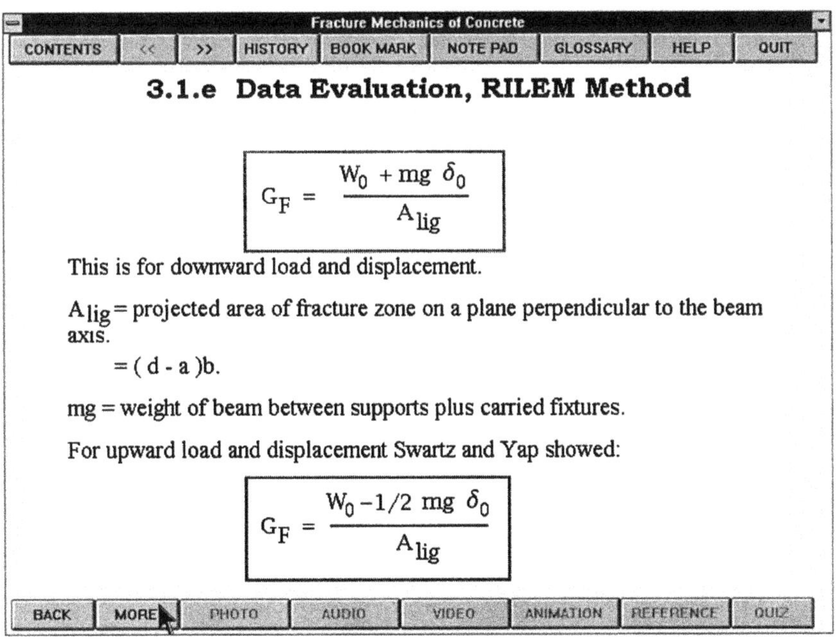

Figure 19. Discussion, RILEM, Testing

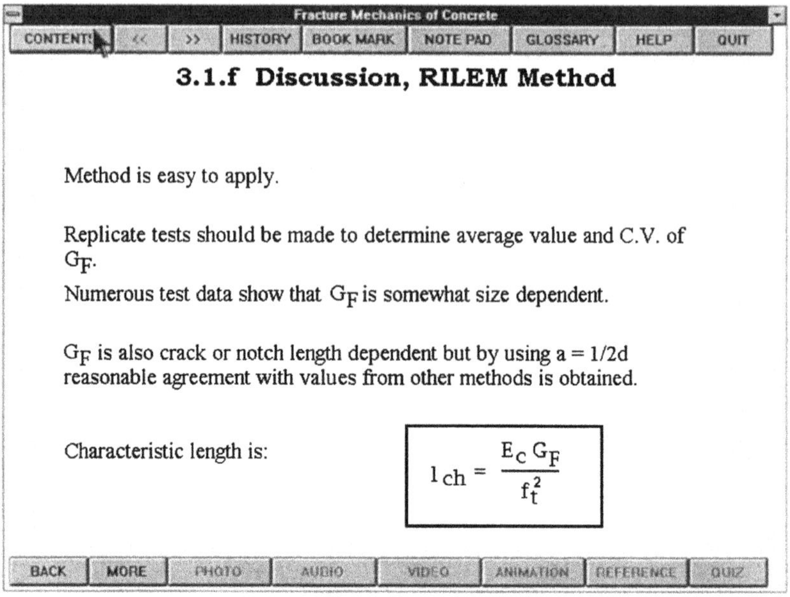

Figure 20. Background, Two Parameter Testing Model, Testing

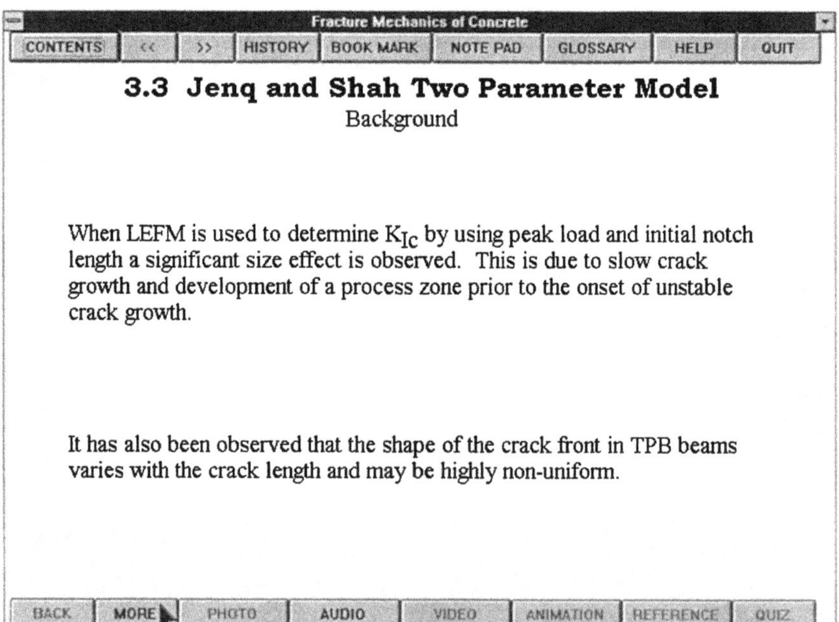

Figure 21. Background cont., Two Parameter Testing Model, Testing

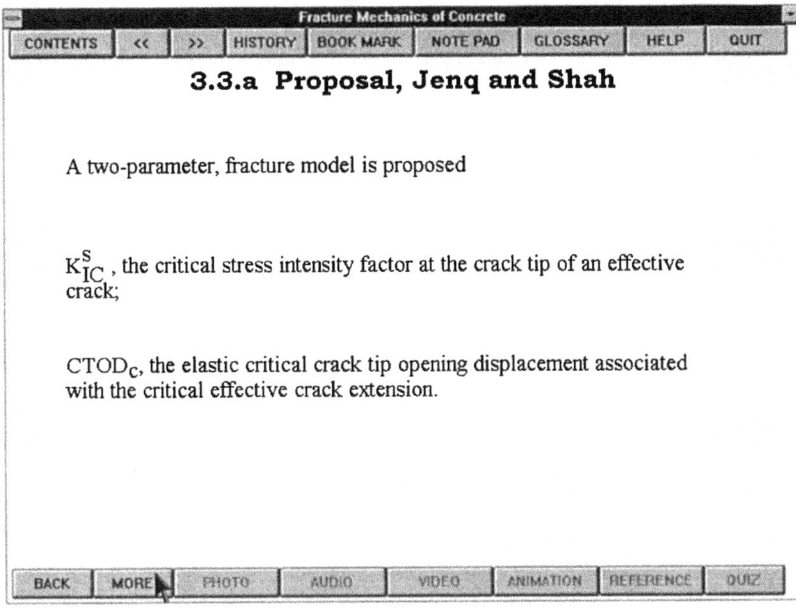

Figure 22. Specimen Geometry, Two Parameter Testing Model, Testing

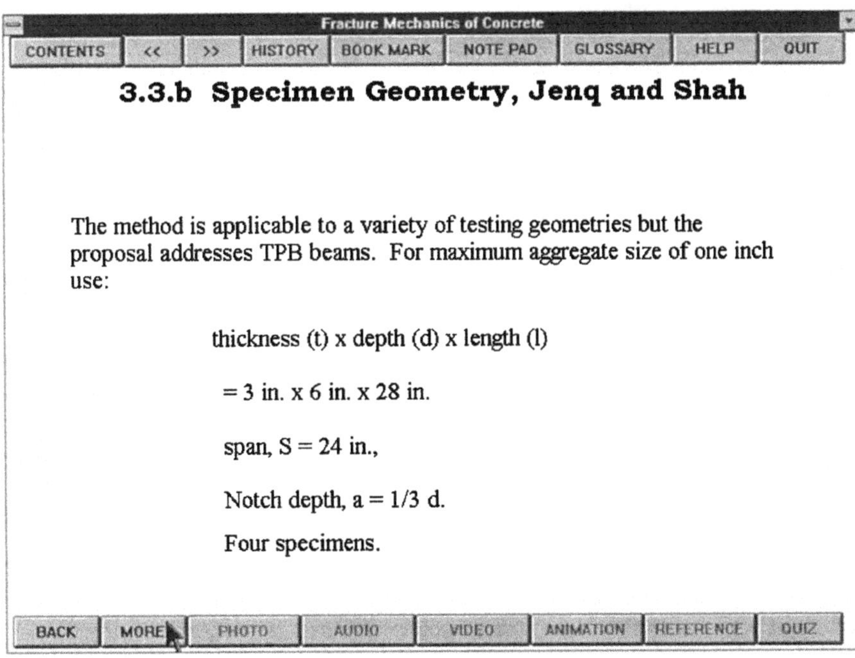

Figure 23. Test Setup, Two Parameter Model, Testing

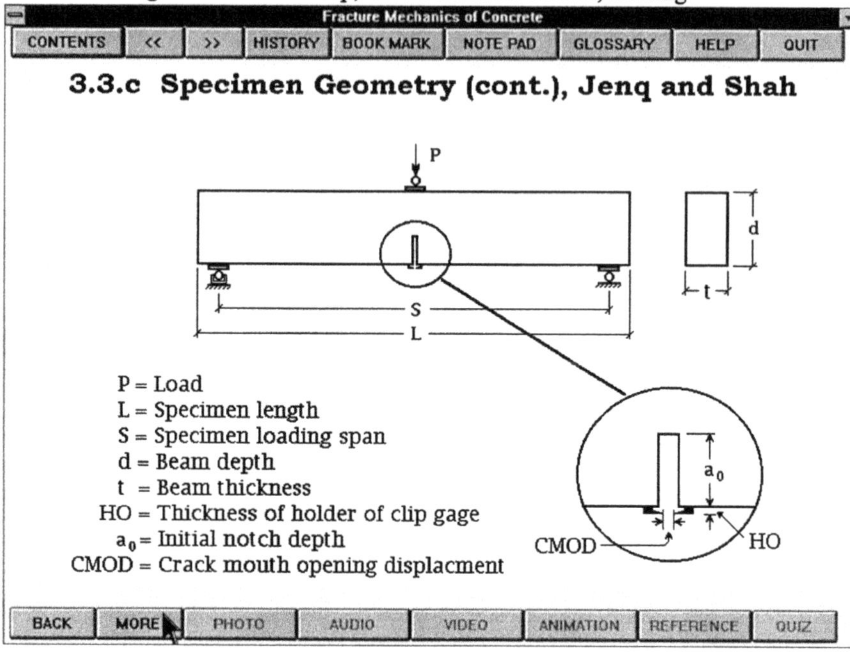

Figure 24, Testing Procedures, Two Parameter Model, Testing

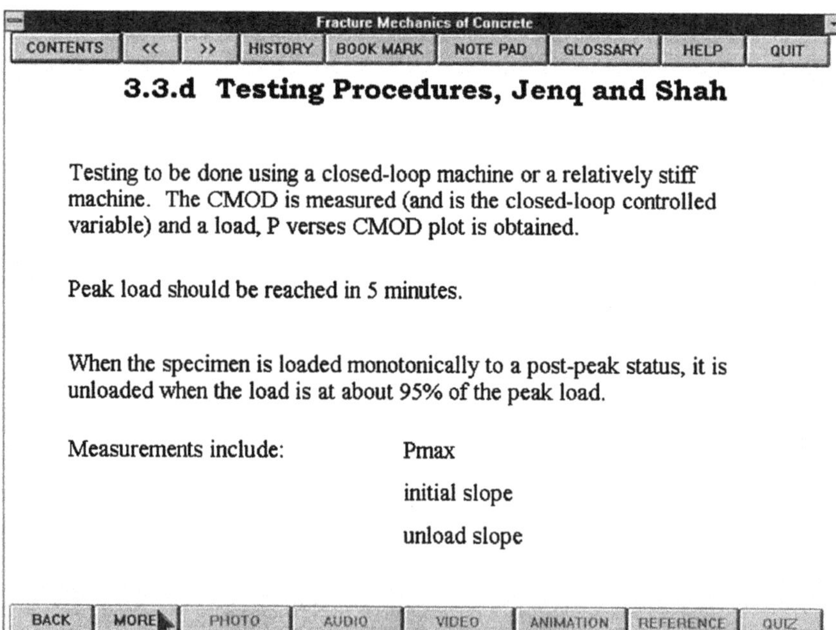

Figure 25. Data Evaluation, Two Parameter Testing Model, Testing

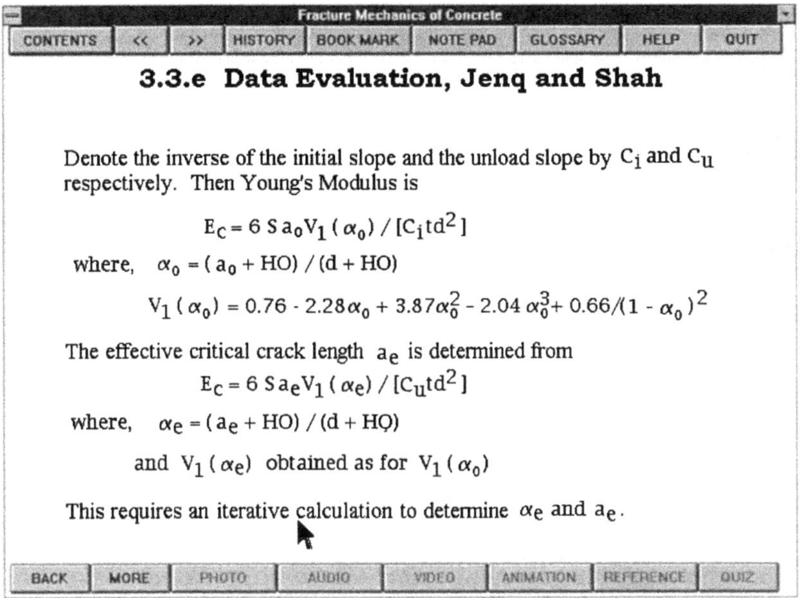

Figure 26. Iterative Algorithm, Two Parameter Testing Model, Testing

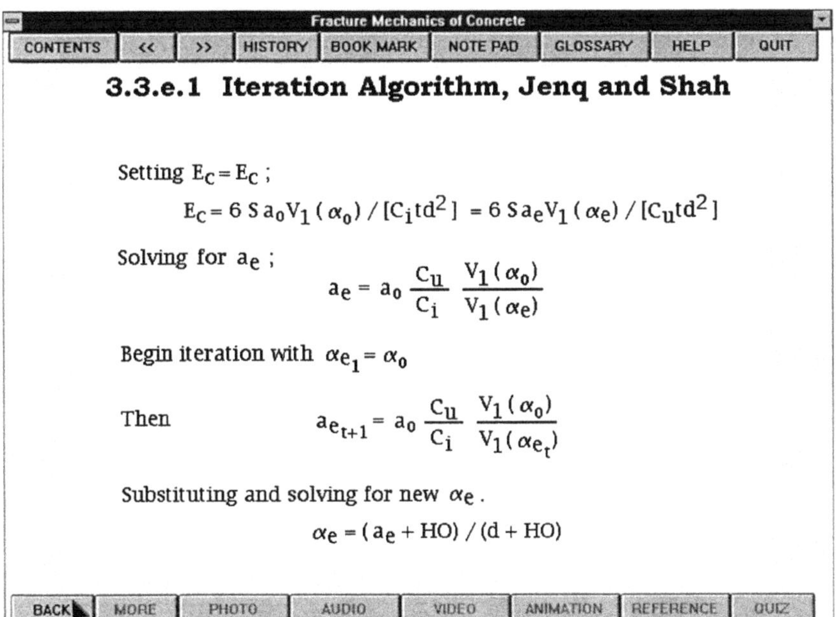

Figure 27. Data Evaluation cont., Two Parameter Model, Testing

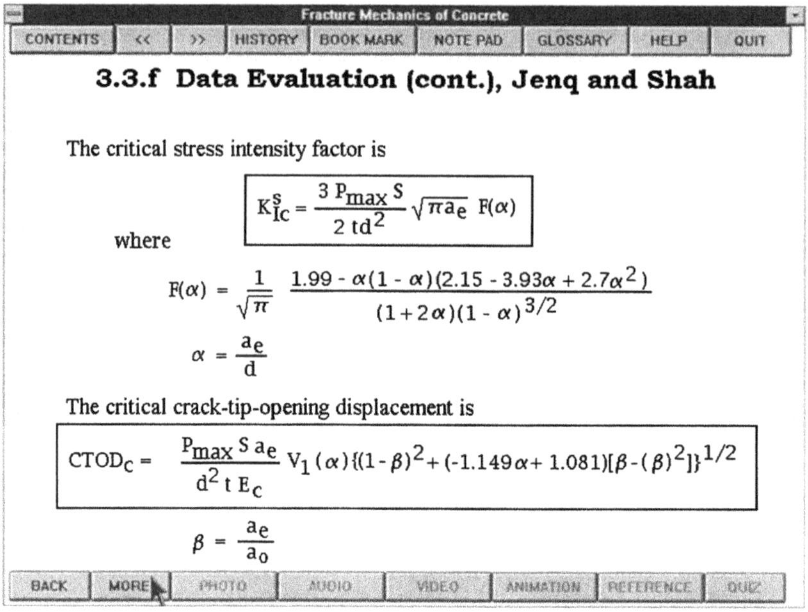

Figure 28. Discussion, Two Parameter Model, Testing

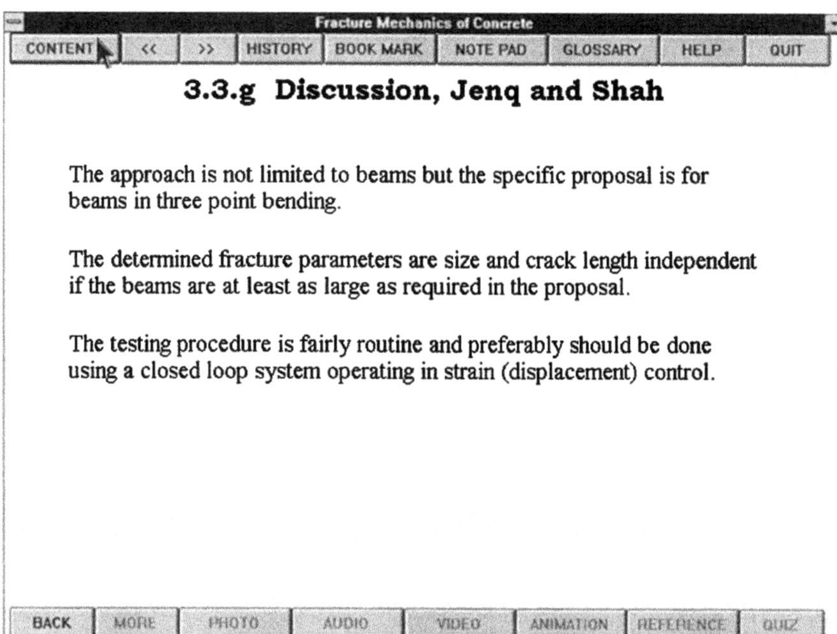

Figure 29. Structural Applications Table of Contents

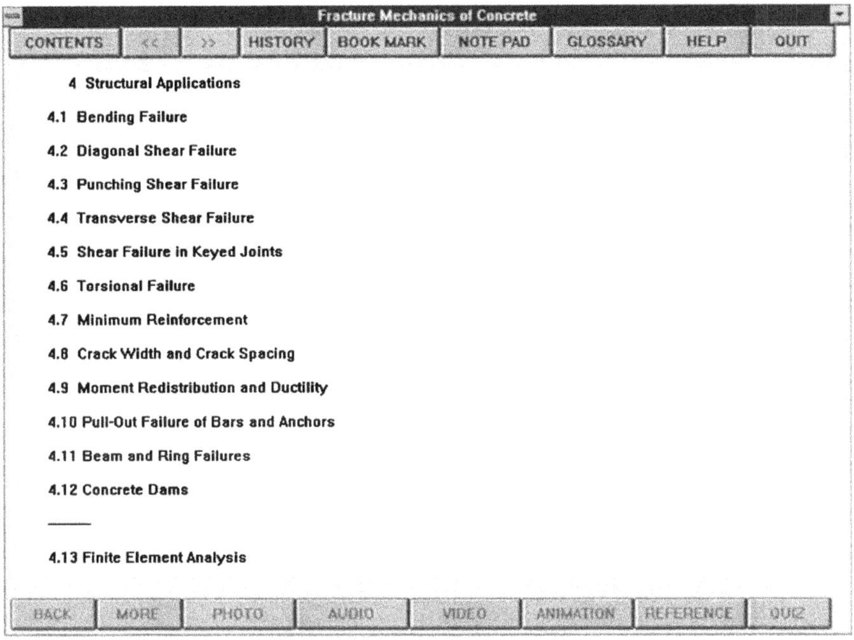

Figure 30. Sample File that Creates Figure 16

3.1.c Displacement Test, RILEM Method`
This is a load point displacement (LPD) test. The displacement is measured indirectly as the movement of the of the beam surface near the crack as opposed to directly measuring the displacement at the load.
\
{Bitmap31c}100x250y
|
{cont_arw}13105-
{back_arw}13102-
{camera1}13104-
*

References

1 Broek, D., *Elementary Engineering Fracture Mechanics*, Martinus - Nijhoff, The Hague, 1984.

2 Barsom, J.M. and Rolfe, S.T., *Fracture and Fatigue Control in Structures*, Prentice-Hall Inc., Englewood Cliffs, NJ, 1987.

3 "Standard Method of Test for Plane-Strain Fracture Toughness of Metallic Materials", ASTM Designation E-399-83, Vol. 03.01, ASTM Annual Standards, American Society for Testing and Materials, Philadelphia, PA, 1985.

4 Oucherterlony, F., "Fracture Toughness Testing of Rock with Core Based Specimens". *Engineering Fracture Mechanics*, Vol. 35, 1990, pp. 351-366.

5 RILEM TC50-FMC, "Determination of the Fracture Energy of Mortar and Concrete by Means of the Three-Point Bend Tests on Notched Beams", *Materials and Structures*, Vol. 18, 1985, pp. 285-290.

6 RILEM TC89-FMT, "Determination of Fracture Parameters (K^{S}_{Ic} and $CTOD_c$) of Plain Concrete Using Three-Point Bend Test", *Materials and Structures*, Vol. 23, 1990, pp. 457-460.

7 RILEM TC89-FMT, "Size-Effect Method for Determining Fracture Energy and Process Zone Size of Concrete", *Materials and Structures*, Vol. 23, 1990, pp. 461-465.

8 RILEM TC90-FMA, *Fracture Mechanics of Concrete Structures: From Theory to Applications*, ed. L. Elfgren, Chapman and Hall, London, 1989.

9 RILEM TC89-FMT, *Fracture Mechanics Test Methods for Concrete*, eds. S.P. Shah and A. Carpinteri, Chapman and Hall, London, 1991.

10 ACI Committee 446, *Fracture Mechanics of Concrete: Concepts, Models and Determination of Material Properties*, ACI 446, 1R-91, American Concrete Inst., Detroit, 1991, 146 pages.

11 Swartz, S.E., Kan, Y.C. and Hu, K.K., "An Expert System Approach to Applying Fracture Mechanics to Reinforced Concrete", *Applications of Fracture Mechanics to Reinforced Concrete*, ed. A. Carpinteri, Elsevier Applied Science, London, 1992, pp.579-605.

12 Fersko-Weiss, H., "3-D Reading with the HYPERTEXT EDGE", PC Magazine, May 28, 1991, pp 241-259.

13 Shneiderman, B. and Kearsley, G., *HyperText Hands-On!*, Addison-Wesley, MA, 1989.

14 Miller, M.J., "Multimedia", PC Magazine, March 31, 1992, pp. 112-123.

15 Creative Labs Inc., 2050 Duane Ave., Santa Clara, CA.

16 Media Vision Inc., 47221 Fremont Blvd., Fremont, CA.

17 Multimedia Technology: How it Can Work for You, A Taped Satellite Videoconference, Kansas State University, Manhattan, KS, April 30, 1992.

18 *Multi-media Computer Assisted Learning*, ed. P. Barker, Nichols Publishing, Inc, NY, 1989.

19 Miller, R.L., "Ten Good Reasons: Learning Benefits of Interactive Technologies", The Videodisc Monitor, February 1990, pp. 15-16.

20 Microsoft Windows 3.0, Microsoft Corporation, Redmond, Washington.
21 Borland C++ 3.0, Borland Company, Scott's Valley, California.
22 Petzold, Charles, *Programming Windows: The Microsoft Guide to Writing Applications for Windows 3*, Second Edition, Microsoft Press, Redmond, Washington, 1990.
23 *Microsoft Windows Multimedia Authoring and Tools Guide*, Microsoft Corporation, MicroSoft Press, Redmond, WA, 1991.

The following references were used in the preparation of the structure of the software package:

24 *Cognition, Education, and Multimedia: Exploring Ideas in High Technology*, ed. D. Nix and R. Spiro, Lawrence Erlbaum Associates, Inc., NJ, 1990.
25 Cook, P., "Multimedia Technology", *Interactive Multimedia*, ed. Ambron, S. and Hooper, K., Microsoft Press, WA, 1988.
26 El-Bibany, H., Paulson, B.C. JR., "Microcomputer/Videodisc System for Construction Education", *Microcomputers in Civil Engineering*, Vol. 6, 1991, pp. 149-160.
27 Hashim, S., *Exploring Hypertext Programming: Writing Knowledge Representation and Problem-Solving Programs*, Windcrest Books, PA, 1990.
28 Lemmen, K.A.M., Gillissen and W.J., Boon, K.L., "The Direct Use of Already Existing User Manuals for Computer Aided Instruction and Information Retrieval with the Aid of Hypermedia", *Computer Assisted Learning*, 3rd International Conference, ICCAL '90, Hagen, FRG, June 11-13, 1990, pp. 277-300.
29 Malasri, S., Olabe-Basogain, J.C., "Online Documentation for Engineering Software: A Hypertext Application", *Microcomputers in Civil Engineering*, Vol. 6, 1991, pp. 77-83.
30 Mühlhäuser, M., "Hyperinformation in Instructional Tool Environments", *Computer Assisted Learning*, 3rd International Conference, ICCAL '90, Hagen, FRG, June 11-13, 1990, pp. 245-264.
31 Plank, R.J., Burgess, I.W. and Brown, C.J., "Teaching Software for Structural Engineering", *Computer Assisted Learning*, 3rd International Conference, ICCAL '90, Hagen, FRG, June 11-13, 1990, pp. 301-308.
32 Woodhead, N., *Hypertext and Hypermedia*, Theory and Applications, Sigma Press, 1991.

The following references were used in the preparation of the substance of the software package:

33 ACI Committee 446, Fracture Mechanics, Aspects of Structural Behavior and Potential Code Improvements, ACI 446. IR-XX, American Concrete Inst., Detroit (in press).
34 *Applications of Fracture Mechanics to Reinforced Concrete*, ed. Carpinteri, A., Elsevier Applied Science, London, 1992.

35 Duga, J., W.H. Fisher, R.W. Buxbam, A.R. Rosenfield, A.R. Buhr, E.J. Honton, and S.C. McMillan. The Economic Effects of Fracture in the United States. Technical Report SP647-2, National Bureau of Standards, 1983.

36 Engineer Technical Letter No. 1110-8-16(FR), "Engineering and Design, Fracture Mechanics of Concrete Hydraulic Structures", US Army Corps of Engineers, 1991.

37 Saouma, V.E., "Fracture Mechanics of Concrete Dams", A Short Course, September 9-10, 1991.

38 Swartz, S.E., "Applicability of Fracture Mechanics Methodology to Cracking and Fracture of Concrete", Final Report to the Naval Civil Engineering Laboratory, 1985.

39 Swartz, S.E., Kan, Y.C., "Effect of Support Conditions on Fracture Energy Measurements for Concrete Beams", ECF 8 Fracture Behavior and Design of Materials and Structures.

40 Swartz, S.E., Kan, Y.C., "On the Validity of Indirect Measurement of the LPD for SEN Concrete Beams", *Fracture Processes in Concrete, Rock and Ceramics*, eds. van Mier, J.G.M., Rots, J.G., Bakker, A., E & F.N. Spon, London, 1991, pp. 771-778.

41 Guinea, G.V., Planas, J. and Elices, M., "Measurement of the Fracture Energy Using Three-Point Bend Tests: Part 1 - Influence of Experimental Procedures", *Materials and Structures*, Vol. 25, 1992, pp. 212-218.

42 Hillerborg, A., " The Theoretical Basis of a Method to Determine the Fracture Energy G_F of Concrete", *Materials and Structures*, Vol. 18, 1985, pp. 25-30.

43 Swartz, S.E. and Refai, T.M.E., "Influence of Size Effects on Opening Mode Fracture Parameters for Precracked Concrete Beams in Bending", *Fracture of Concrete and Rock*, eds. S.P. Shah and S.E.Swartz, Springer-Verlag, New York, 1989, pp. 242-254.

44 Jenq. Y.S. and Shah, S.P., "Two-Parameter Fracture Model for Concrete", *Journal of Engineering Mechanics*, ASCE, Vol. III, No. 10, Oct. 1985, pp. 1237-1241.

45 Karihaloo, B.L. and Nallathambi, P., "Notched Beam Tests: Mode I Fracture Toughness", Ch. 1 of *Fracture Mechanics Test Methods for Concrete*, eds. S.P. Shah and A. Carpenteri, Chapman and Hall, London, 1991, pp. 22-30.

46 dBase IV, Aston-Tate, Culver City, California.

24 TEACHING FRACTURE MECHANICS TO GRADUATE STUDENTS WITH WORKSTATION-BASED SIMULATION

A.R. INGRAFFEA and P.A. WAWRZYNEK
Cornell Fracture Group, Cornell University, New York, USA

Abstract

The use of an advanced, workstation-based fracture simulator for education in a graduate course in fracture mechanics is described. History of development, pedagogical usefulness, capabilities, and future developments for the simulator are described.

Keywords: Workstation, fracture simulator, fracture mechanics.

1 Introduction

This paper describes the use of a fracture simulation system which is the heart of a graduate course in engineering fracture mechanics, and a basis for graduate research in a variety of areas of linear and nonlinear fracture mechanics at Cornell University. The simulation system, the **FR**acture **AN**alysis **C**ode (FRANC), runs on an engineering workstation platform. It is highly interactive, and performs geometrical modeling, stress analysis, fracture mechanics, and visualization functions in an integrated manner. FRANC is used:

- To illuminate, in a self-paced manner, many of the fundamental concepts of linear elastic fracture mechanics (LEFM) and elastoplastic fracture mechanics (EPFM);
- As a stress intensity factor and crack growth rate calculator for problems in critical and subcritical crack growth;
- To illustrate the capabilities and shortcomings of modern numerical methods for fracture analyses; and

Teaching and Education in Fracture and Fatigue. Edited by H.P. Rossmanith. Published in 1996 by Taylor & Francis, 2 Park Square, Milton Park, Abingdon, Oxon, OX14 4RN. ISBN 0 419 20700 7.

- To perform practical case studies of actual fracture problems as part of a course-long project.

The objectives of the course and its particular pedagogy are first presented. The history of development, capabilities, and examples of use of FRANC within the course are next described. The future of the system, its relationship to experimental methods, and its wide distribution outside Cornell are also discussed.

2 Course Pedagogy

2.1 Course Objectives

The principal objective of the course, taught in the School of Civil Engineering, is to prepare students for work in fracture and fatigue during their thesis research and as practicing engineers. The course is complementary to another, taught in the Department of Theoretical and Applied Mechanics, which emphasizes fundamental and advanced theories of LEFM and EPFM. A student ideally leaves the course with practical knowledge and experience in solving many types of crack propagation problems in LEFM, a limited number in nonlinear fracture mechanics, and familiarity with ASTM standard fracture toughness testing procedures.

2.2 Course Content

The course is divided into three main areas with approximately equal coverage in time: theory and exact analytical procedures, numerical methods, and experimental methods. However, these topics are integrated as often as possible.

They are also augmented with a pleasant dose of historical perspective. For example, a set of about a dozen of the "classical" papers in fracture mechanics is distributed, read and discussed early in the course. Some significant development in a number of these is then revisited with the more modern analytical capabilities of FRANC.

Examples include:

- Finding what it takes using the finite element method to compute stress concentration factors as accurately as the analytical solutions of Inglis.
- Computing the energy release rate for the Griffith problem with a variety of finite element approaches such as crack incrementation, stiffness derivative, and modified crack closure.

They are also linked to societal issues such as economic and political influences. An example here is the socio-economic influences on the decision making process in repair/replacement of fatigue/fracture prone infrastructural elements such as tanks, bridges, and dams. Such recent serious failures in the US and abroad, such as the Schoharie Creek bridge [1], the Kölnbrein arch dam [2,3], and the Ashland oil storage tank [4] have been fertile ground for discussion and quantitative evaluation in this important area of education.

2.3 Course Timing

The course is offered to students who have completed at least one year of graduate study. At Cornell this implies that students have taken graduate courses in structural mechanics, elasticity, and finite element methods. Students enrolled in the course are frequently taking graduate courses in boundary element methods, plasticity, biomechanics, or rock mechanics simultaneously. This has led to project work in these courses in which FRANC has been the principal analysis/design tool.

2.4 Course Delivery Style/System

The course is delivered in a highly interactive style. Frequent discussions, and team solution of homework and projects is encouraged. Since the students taking the course usually come from three different departments, theoretical and applied mechanics, mechanical engineering, and civil engineering, multidisciplinary approaches involving a range of mathematical rigor often occur.

Access to FRANC occurs in two ways. Lectures are given in an electronic classroom equipped with an engineering workstation and overhead video projection. This means that the professor can perform real-time simulations in class, and the student can instantaneously react to ask "what if?" questions, critique the model and results, and/or suggest alternative approaches.

Outside the lecture, students have access to the numerous workstations in Cornell's Computer-Aided Design Instructional Facility. Here they have their own accounts, networking and filing systems, and document preparation access. This is also the site of most of the team play during homework solution and course project development.

3 FRANC: The *FR*acture *AN*alysis *C*ode

The development of FRANC began in 1985. Many of its capabilities were inspired by the programs FEFAP, FEFAP-G, and CRACKER, developed by Saouma [5], Gerstle [6], and Swenson [7], respectively. FRANC, however, is a completely new program based on radically different techniques for storing ad manipulating simulation data, which allows it to be more robust, maintainable, and extensible than its predecessors.

The initial development effort was two years, and resulted in version one of the so-called "research version" of the program. This version has served as the primary tool of investigation for a number of graduate theses at Cornell [8-10] as well as the basis of derivative programs for others [11]. The original version of the program was also used in the engineering fracture mechanics course offered at Cornell.

The research version of the program is continually being updated as new capabilities are added and existing capabilities enhanced. Such dynamic software, however, proved to be difficult to manage in an educational context, where an unstable software environment may disorient students. In 1986 a version of the program was frozen and became the basis of the first purely "educational version" of program, CU-FRANC. CU-FRANC was enhanced to incorporate a better use of color, to give better and more complete error and diagnostic messages, and to simplify portions of the interface that were conceptually difficult for new users. This version was also modified to include features of interest for pedagogical reasons but of little or no interest for research [12]. The educational version of the program has been used in conjunction with the fracture mechanics and finite element courses at Cornell, as well as being widely distributed through project SOCRATES (Study of Complementary Research and Teaching in Engineering Science) [13].

In 1991 work began on a completely new program, tentatively called FRED, FRANC-Educational, which will incorporate all the important features of FRANC in a completely new graphical user interface. A significant additional feature of FRED will include integrated capabilities for creating geometrical models and generating meshes, which must currently be done with an independent preprocessor.

3.1 Capabilities

The foundation for the crack propagation simulation capabilities of FRANC is a stress analysis capability using the finite element method. FRANC thus has all the basic features typical of moderately sized finite element programs. Material models include isotropic and orthotropic elastic materials, with some simple elastoplastic capabilities. The element library includes quadratic, quadrilateral and triangular continuum elements, as well as gap elements (with friction). For boundary conditions, nodal loads, surface tractions, body forces, thermal loads, initial stresses, and nodal displacements can be specified. Postprocessing and visualization capabilities include deformed meshes, line plots of stress, vector plots of principal stresses, and color contours of stress components. This is all done within the context of a highly interactive graphical interface, which is menue driven and employs direct manipulation to activate many of the features.

Along with basic finite element capabilities, FRANC incorporated many features specifically for fracture analysis and propagation. Most significant among these is the ability to remesh automatically to simulate crack nucleation and propagation. This is a four-step process illustrated in Figures 1 and 2. A new crack or an increment of an existing crack is either specified by the student or predicted by program. New cracks can begin from a boundary of the structure, or may be completely internal to the structure. The steps in the remeshing process are as follows:

1. Elements in the region of crack nucleation or growth are deleted. The size of this region is determined automatically by the program. Care is taken to retain information that describes structural boundaries and bi-material interfaces.
2. A crack is inserted or an existing crack is extended. All crack geometries are approximated by piecewise linear segments.
3. Quarter-point crack-tip elements are placed about the new crack tip. These elements are capable of modeling the proper stress and strain singularities dictated by LEFM. The placement of these elements is constrained to honor bi-material interfaces, if present.

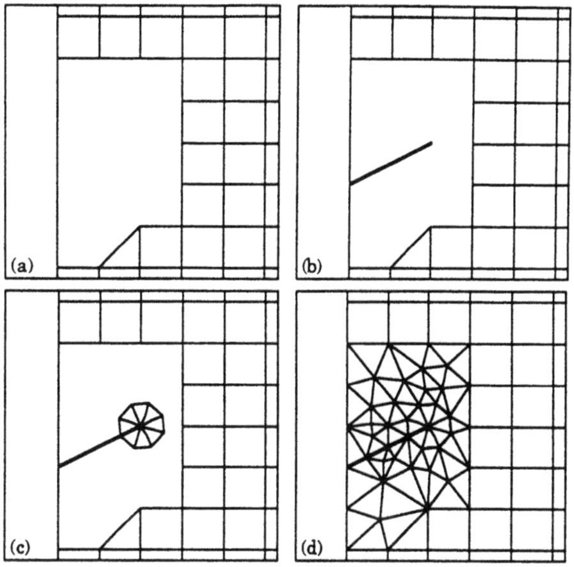

Figure 1. A typical crack nucleation sequence: a) elements are deleted, but the structural boundary is retained, b) the crack is inserted, c) singular elements are placed around the crack tip, and d) a transition mesh is generated.

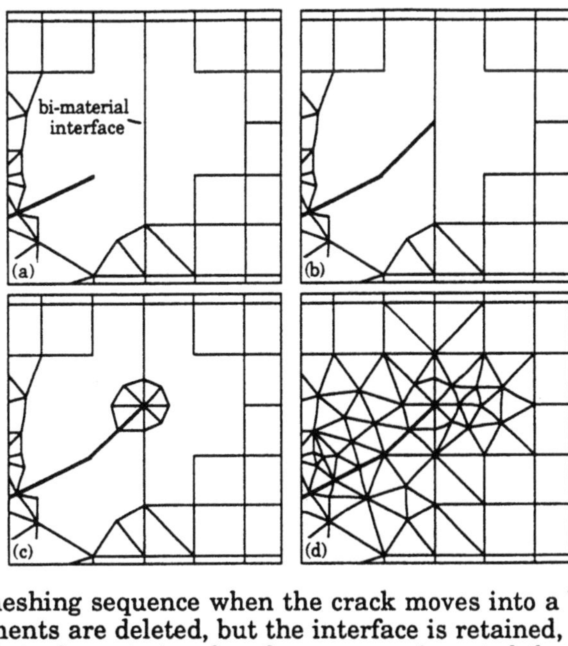

Figure 2. A remeshing sequence when the crack moves into a bi-material interface: a) elements are deleted, but the interface is retained, b) the crack is extended to the interface, c) singular elements are inserted that conform to the interface, and d) a transition mesh is generated.

4. A transition mesh is generated, which joins the crack-tip elements to the remainder of the elements and structural boundaries. The transition meshing algorithm automatically inserts new nodes where needed to enforce good element aspect ratios, and to provide a smooth transition from small crack-tip elements to relatively larger elements in the remainder of the mesh.

There are three independent techniques available within FRANC for determining stress intensity factors. The least accurate approach is the displacement correlation technique [14], which evaluates the stress intensity factors by substituting numerically determined displacements into the theoretical expressions for the crack-tip displacement fields. Stress intensity factors can also be determined by an equivalent domain evaluation of a J-integral [15]. This approach is very accurate when mode-1 loading only is present, but less so in mixed-mode configurations. The third technique is evaluating the modified crack closure integral [16]. In this approach, the product of the stresses in front of the crack, and the displacements of the crack faces, are integrated for the crack tip element. The technique allows a decoupling of modes, and this gives accurate results for mixed-mode configurations.

FRANC incorporates three different mixed-mode crack stability criteria, maximum circumferential stress [17], the maximum energy release rate [18], and the minimum strain energy density [19]. This allows the student to assess the stability of a crack subjected to mixed-mode loading. The information is displayed in an interaction diagram, which shows a failure envelope in a normalized stress intensity factor space. All cracks in a body are plotted in this space; an example is shown in Figure 3a. The same three stability criteria are used to predict the direction of crack propagation automatically. This is shown in Figure 3b.

Crack growth simulations are performed by way of the following incremental algorithm:

1. A crack is nucleated of a size and location specified by the student.
2. A stress analysis is performed.
3. Stress intensity factors are computed as is the direction of crack propagation.
4. The crack is extended and the body is remeshed.
5. Steps 2,3, and 4 are repeated until a student-selected termination condition is reached.

These steps can be performed manually, or in a completely automatic mode, with the program maintaining stress intensity factor versus crack length histories in either case.

FRANC has capabilities for multiple cracks, which makes it an ideal tool for investigating crack interaction, as well as the interaction of cracks with other geometrical features, such as holes. Figure 4 shows the interaction of two cracks emanating from holes and represents a typical application for the program.

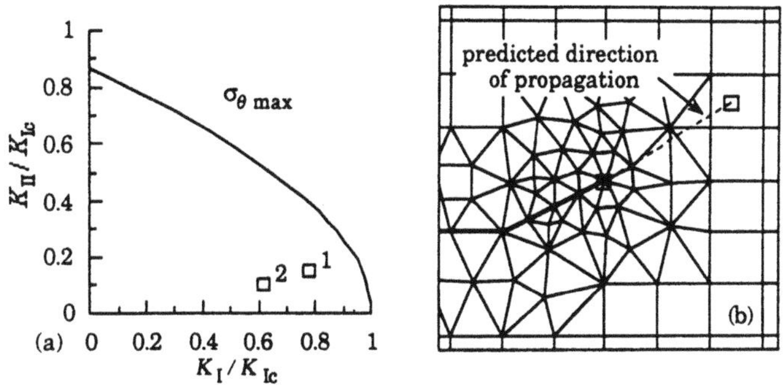

Figure 3 a) A mixed-mode failure interaction curve in the normalized stress intensity space, with the squares indicating two different crack tips, b) The direction of crack propagation automatically predicted by the program

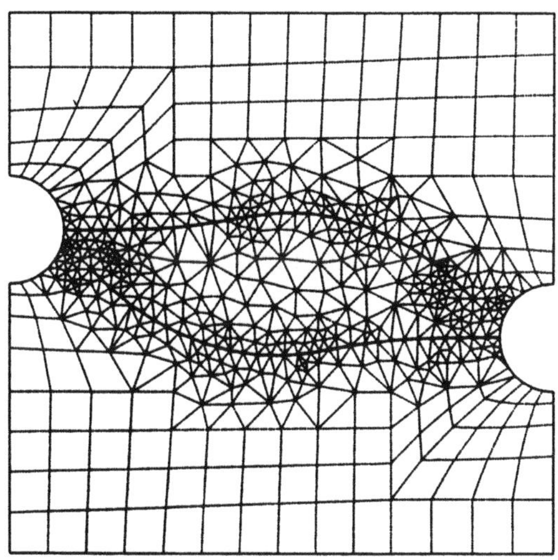

Figure 4 An example of automatic propagation showing crack interaction

3.2 Distribution

The first educational version of FRANC, CU-FRANC, mentioned above, was developed as part of a suite of advanced educational simulation systems in Project SOCRATES. This project was funded by the US Department of Education for three years, and has since been subsumed as part of SYNTHESIS National Engineering Education Coalition [20], a five year project supported by the US National Science

Foundation, industry, and eight engineering colleges. As part of the SOCRATES group, CU-FRANC has been distributed to over 50 schools worldwide. The various versions of the research version of FRANC has been distributed to another 20 schools, including many sites in Asia and Europe.

The new educational version, FRED, will eventually be available as public domain software to all educational users over the INTERNET through the National Engineering Education Delivery System, the major curriculum development project of SYNTHESIS.

4 Conclusions

FRANC has served an exciting, illuminating, and unifying role in fracture mechanics education at Cornell, and is being employed at an increasing number of institutions worldwide. It will continue to evolve in response to new workstation capabilities, and the demands from new developments in fracture mechanics theory and practice.

5 Acknowledgments

The authors would like to thank Ms Maya Srinivasan, Mr Glenn Shen, and Mr Michael Jackmin for their contributions to the continued development of FRANC. Financial support from the US Department of Education and the National Science Foundation is appreciated. Equipment support from the Digital Equipment Corporation and from Sun Microsystems is gratefully acknowledged.

6 References

[1] Swenson, D.V. and A.R.Ingraffea: Int.J.Fract., 51, 73, 1991.

[2] Linsbauer, H.N., A.R.Ingraffea, H.P.Rossmanith and P.A.Wawrzynek: J.Struct.Eng., 115,7,1599, 1989.

[3] Linsbauer, H.N., A.R.Ingraffea, H.P.Rossmanith and P.A.Wawrzynek: J.Struct.Eng., 115,7,616,1989.

[4] Gross, J.L., F.Y.Jokel, R.N.Wright and A.H.Fanny: Investigation into the Ashland Oil Storage Tank Collapse on January 2, 1988, NBSIR 88-3792, U.S. Department of Commerce, 1988.

[5] Saouma, V.E.: Interactive Finite Element Analysis of Reinforced Concrete: A Fracture Mechanics Approach, Ph.D. Thesis, Cornell University, 1981.

[6] Gerstle, W.H.: Finite and Boundary Element Modelling of Crack Propagation in Two-and Three-Dimensions Using Interactive Computer Graphics, Ph.D.Thesis, Cornell University, 1986.

[7] Swenson, D.V.: Modeling Mixed-Mode Dynamic Crack Propagation Using Finite Elements, Ph.D.Thesis, Cornell University, 1985.

[8] Wawrzynek, P.A.: Interactive Finite Element Analysis of Fracture Processes: an Integrated Approach, M.S.Thesis, Cornell University, 1987.

[9] Lin, S.W.S.: Case Studies of Cracking of Concrete Dams - A Linear Elastic Approach, M.S.Thesis, Cornell University, 1988.

[10] Lampkin, S.: Tow-Dimensional Numerical Evaluation of Near Wellbore Phenomena: Perforation Performance and Interacting Hydraulic Fractures, M.S.Thesis, Cornell University, 1990.

[11] Boone, T.J.: Simulation and Visualization of Hydraulic Fracture Propagation in Poroelastic Rock, Ph.D. Thesis, Cornell University, 1989.

[12] Srinivasan, M.: Fracture Analysis Code: A Computer-Aided Teaching Tool, M.S. Thesis, Cornell University, 1988.

[13] Ingraffea, A.R. and K.Mink: Academic Computing, 3,3,20, 1988.

[14] Shih, C.F., and H.G.Delorenzi: Int.J.Frat., 12, 647, 1976.

[15] Dodds, R.H.Jr. and R.P.Vargas: Numerical Evaluation of Domain and Contour Integrals for Nonlinear Fracture Mechanics. Report, Department of Civil Engineering, University of Illinois, Urbana-Champaign, 1988.

[16] Rybicki, E.R. and M.Kanninen: AIAA J., 2, 1977.

[17] Erdogan, F. and G.C.Sih: J.Basic Engng, 85,4,519, 1963.

[18] Hussain, M.A., S.U.Pu and J.Underwood: ASTM SPT 560,2,1974.

[19] Sih, G.C. and B.MacDonald: Engng Fract.Mech., 6,361,1974.

[20] Thomas, R.J.: PRISM, Amer.Soc.Engng Edu., 14,1991.

25 PC SOFTWARE ASSISTED TEACHING AND LEARNING OF DYNAMIC FRACTURE AND WAVE PROPAGATION PHENOMENA

H.P. ROSSMANITH and K. UENISHI
Institute of Mechanics, Technical University Vienna, Austria

Abstract

A teaching and learning oriented 2D Finite Difference Simulator has been developed to visualize the interaction of elastic waves with user-defined inhomogeneities such as material inhomogeneities (cracks, notches, faults) and dissimilarities (interface cracks, delaminations, etc.), damageable solids, solids under initial stress or strain, impact due to various obstacles and blasting effects. The program runs swiftly on a personal computer and is extremely user-friendly. In a course on dynamic phenomena this software, in conjunction with lab experiments and assisted by an adequate theoretical-analytical basis will provide students with a clear insight into dynamic fracture-wave interaction phenomena.

Key Words: numerical simulation, dynamic fracture, wave propagation, PC-assisted education, SWIFD.

1 Introduction

The interaction of elastic stress waves with material and structural inhomogeneities is of great importance in many fields of engineering such as for instance, in structural and materials testing in mechanical engineering, and safety assessment associated with mining induced seismicity in mining engineering.. Dynamic wave interaction problems, however, may become so complex that even if an analytical solution is available it is difficult for a student to grasp the complete picture of the interaction of cracks and waves. For many decades, photomechanics, in particular dynamic photo-elasticity has served as a powerful method and experimental tool which allows the

Teaching and Education in Fracture and Fatigue. Edited by H.P. Rossmanith. Published in 1996 by Taylor & Francis, 2 Park Square, Milton Park, Abingdon, Oxon, OX14 4RN. ISBN 0 419 20700 7.

visualization of the whole stress fields. Time limitation and non-availability of the rather sophisticated equipment, in general, prevents the teacher to live-demonstrate these experiments during a course. Therefore, there is a need for more effective tools for teaching and education of dynamic phenomena particularly those associated with fractures and waves.

A computer is the very tool that satisfies this demand. Even a medium-priced PC enables student and teacher alike to visualize and analyze such complex phenomena in an effective and pedagogical way. This contribution features the objectives of a dynamic numerical simulation course and its pedagogy. The development history, the wide range of features, and examples of the use of the SWIFD software in teaching and learning are described. Finally, future developments of the software will also be discussed.

2 Didactic Issues

This section brings to light some of the advantages of using student-oriented software in teaching and learning in engineering.

2.1 Objectives

The objectives of a course on dynamic fracture and wave phenomena, offered at the Institute of Mechanics at the Technical University Vienna either as part of the course on fundamental and advanced theories of continuum mechanics and fracture mechanics, or on its own, is to enable students to gain an improved understanding of the interaction between structural inhomogeneities and elastic wave disturbances. This course on dynamic fracture and waves is complementary to another course taught at the Institute of Materials Testing which emphasizes the more practical aspects of materials testing and lifetime assessment of structures as well as evaluation of failures and the analysis of failure cases. All these courses are taught at the Mechanical Engineering Department. Therefore, a mechanical engineering student ideally leaves the course with practical knowledge and experience in solving many types of wave propagation problems.

2.2 Structure of the Course

The course consists of three main areas: analytical procedures (70%, time), numerical simulations (15%), and laboratory experiments (15%). First, the classical theoretical background in fracture mechanics and wave propagation is presented with occasional reference to the historical development and the classical papers. The significant aspects of linear elastic fracture mechanics, nonlinear fracture mechanics, fatigue crack propagation, environmental effects and the application of the theory of fracture to various materials including metals, polymers, ceramics, rocks etc. as well as dynamic fracture theories are then highlighted and reviewed.

A new two-dimensional Finite Difference Simulator, SWIFD (Solids·Wave·Impact Fracture·Damage), has been developed at the Fracture - and Photomechanics Laboratory of the Technical University in Vienna. SWIFD is a highly user-interactive software and has been designed to run on a personal computer, for PC's are more accessible to students than workstations. SWIFD supports the dynamic part of the course. The course is offered to graduate and undergraduate students. However, a rigorous base in mathematics and physics is most helpful for full appreciation of the course if taught at graduate level.

3 Photomechanics via Numerical Simulations

Dynamic photoelasticity provides the experimenter with a sequence of "snap-shots" of the space-time continuum of a dynamic physical process. These photographs are normally obtained by the application of high-speed photographic cameras which are capable of producing from a few to several hundred images at framing rates between 20,000 up several million frames per second. For animation purposes it has been customary to either link these photographs to produce a "jerky" kind of animation or to analyze the recordings and then interpolate and perhaps even extrapolate the resulting data. The tremendous increase of computing power during the last ten years has forced a change of paradigm in dynamic photomechanics. Now, dynamic photomechanic recordings (isochromatic fringe patterns, holographic recordings or other pattern) serve as a means to control computer simulations, assist in the development and assessment of appropriate software for numerical simulations.

4 SWIFD: The Finite Difference Simulator

The first version of the finite difference code was developed in 1992 [1] under the name SPRO (Stress Wave **PRO**pagation in Solids), and was intended for researchers and students who are already familiar with wave propagation in elastic media. SPRO has been a completely new finite difference simulator. The use of new techniques for performing calculations and storing and manipulating data enables the new software to run swiftly under DOS on a PC.

Responding to the ever-increasing requests for more user-friendliness the existing versions of the program are continually being updated and new functions are being added. In 1995, the program has obtained its own "window" and all parts hitherto separated, have been incorporated into one single code. This led to renaming of the software, SWIFD.

SWIFD can advantageously be used:

- To get **accustomed** with the fundamental concepts of wave propagation in elastic solids;
- To **visualize** the interaction of elastic waves with user-defined structural inhomogeneities such as e.g. cracks of arbitrary shape;
- To **analyze** the various time- and space-dependent physical quantities including displacements, velocities, stress components, energy, and stress intensity factors; and

- In a more advanced course, to **acquire** a fundamental knowledge of damage mechanics and of fault stick-slip mechanisms.

Although SWIFD comes in two versions, a research oriented one and a teaching and learning oriented one, a number of basic characteristics apply to both versions.

The teaching and training version of SWIFD allows a student to investigate problems of the following kind:

- Computation of the reflection and transmission coefficients of plane waves of arbitrary time history that impinge on a material discontinuous or graded interface;
- Visualization of the formation and behavior of Rayleigh surface waves when they propagate along stress-free boundaries such as open cracks;
- Simulation of the incidence, reflection, transmission, refraction and diffr-action of waves about a crack tip; and
- Evaluation of dynamic stress intensity factors for mixed-mode fracture problems and stationary cracks.

4.1 Features

SWIFD (Fig.1) enables the user to visualize the interaction of elastic waves with interfaces, cracks, cut-outs, inclusions, boundaries, faults in monolithic as well as stratified media. The material can be either damageable or non-damageable, and static pre-loading may be applied. All parts of the program package are menu-driven, which accepts input from the keyboard and the mouse. Default values for most parameters are predefined.

The build-in graphics editor assists the user in the construction of very complex two-dimensional problems (Fig.2). The functions defined interactively by the user include the:

- selection of *material parameters* (menu in Fig.2);
- definition of *discontinuous* or *graded interfaces* with infinitesimal or finite width for any domain partition of the model (see Fig.2);
- selection as a function of time and space (in the submenu Boundary Conditions in Fig.3) of a wide variety of *load configurations* (normal and shear loads) such as dynamic point forces at any point of the model, dynamic stress distributions along selected sections along inner and outer boundaries, as well as sequential and spatially distributed blasting events;
- selection of almost arbitrary *mixed-mode boundary* conditions (menu driven in the submenus Geometry in Fig.2 and Boundary Conditions in Fig.3), in terms of displacements or stresses prescribed along sections or at full on outer and/or inner boundaries of a solid in order to treat e.g. the problem of a frictionless rigid punch indenting a 2D solid; and
- definition of normal and inclined *faults* with a wide variety of fault conditions.

Figure 1. The *Initial Display*

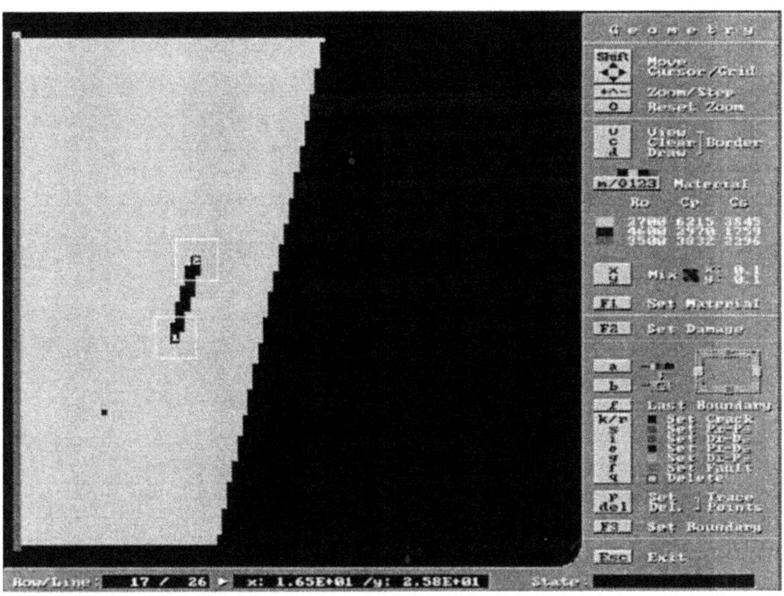

Figure 2. The submenu *Geometry*

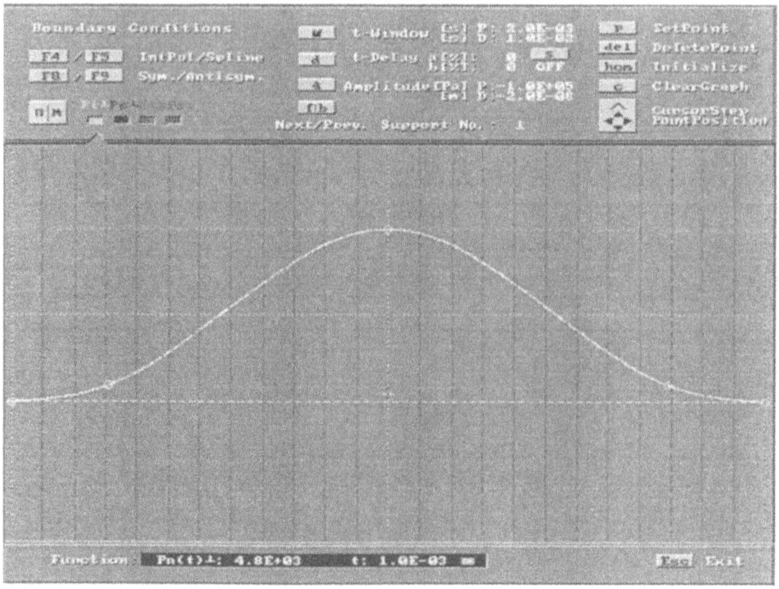

Figure 3. The submenu *Boundary Conditions*

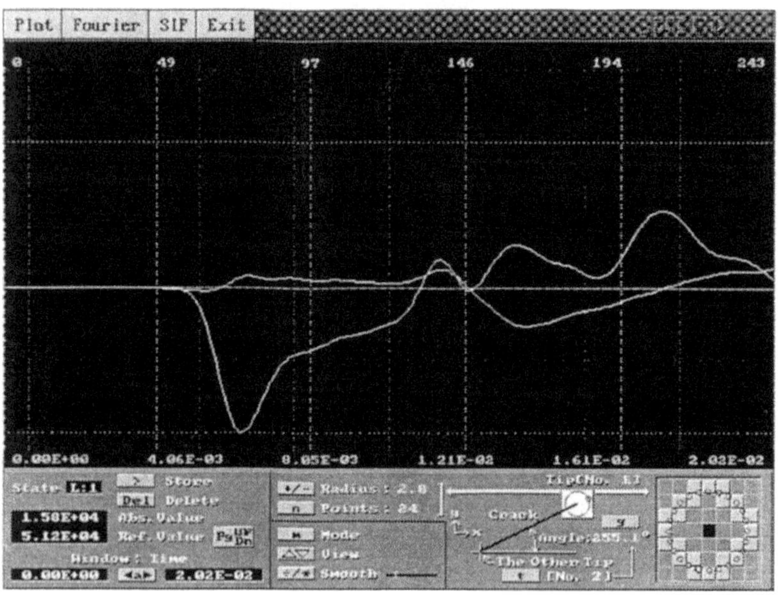

Figure 4. The submenu *Analysis*

Guided by dynamic photomechanics work, in the numerical simulations whole field displacement patterns pertaining to a sequence of time instances are stored for data processing. The post-processor allows the analysis of the data of this (even dense) sequence of numerical snapshots and a wide spectrum of graphic modes assists the presentation of the results for output on the screen, on files or else. SWIFD also enables the user to analyze the various time and spatial distributions of physical quantities, for example, displacements, velocities, stress components, and stress intensity factors (see the submenu Analysis in Fig.4). Of particular interest in teaching and training in dynamic fracture mechanics is the time history of the stress intensity factor for a dynamically loaded structure or structural component. In SWIFD this is accomplished by using a polar-coordinate type of presentation of the stress intensity factors where for a given fracture criterion the student can easily identify the locus of fracture, the fracture time, and the values of the stress intensity factors.

5 Example Problem

The capabilities of SWIFD will be demonstrated by investigating the diffraction of a plane longitudinal stress wave about the tips of a crack of finite length which is located close to a free boundary. The wave impinges at a shallow angle as seen in Figure 5. The sequence of four numerically generated isochromatic fringe patterns (lines of constant maximum shear stress) shown in Figure 5 illustrate the wave-crack interaction process. In Figure 5a, the longitudinal compressive wave impinges on crack tip A, it is diffracted in Figure 5b and is reflected from the crack face as well as from the free surface to the right in Figure 5c. The compressive stress at the crack tip gives rise to a negative stress intensity factor which is physically irrelevant. The final snap-shot, shown in Figure 5d pertains to the situation where, due to the boundary-reflected PP-tensile wave which now hits the crack. The crack is subjected to a state of mixed-mode loading where the resulting normal opening and shear stresses are reflected in the two stress intensity factor traces given in Fig.6. A polar graph representation (shown in the submenu SIF in Fig.7) allows one to study the conditions for fracture. In fact, in this problem the combined action of crack normal opening and crack in-plane sliding does yield stress wave induced fracture initiation as the combined stress intensity factor does exceed the fracture toughness of the material (indicated by the white solid line ellipse in Fig.7). One can observe from Figure 7 that the initial phase of wave diffraction is governed by nearly pure crack compression up to the time when the reflected wave hits the crack. Then the mode-1 contribution is accompanied by a considerable mode-2 contribution which finally leads to fracture at the time and conditions indicated in the polar diagram.

6 Future developments

Currently, a SWIFD version is developed which will run under MS-Windows. Being fully aware of the major drawback of the present program - lack of 3D-capability of a

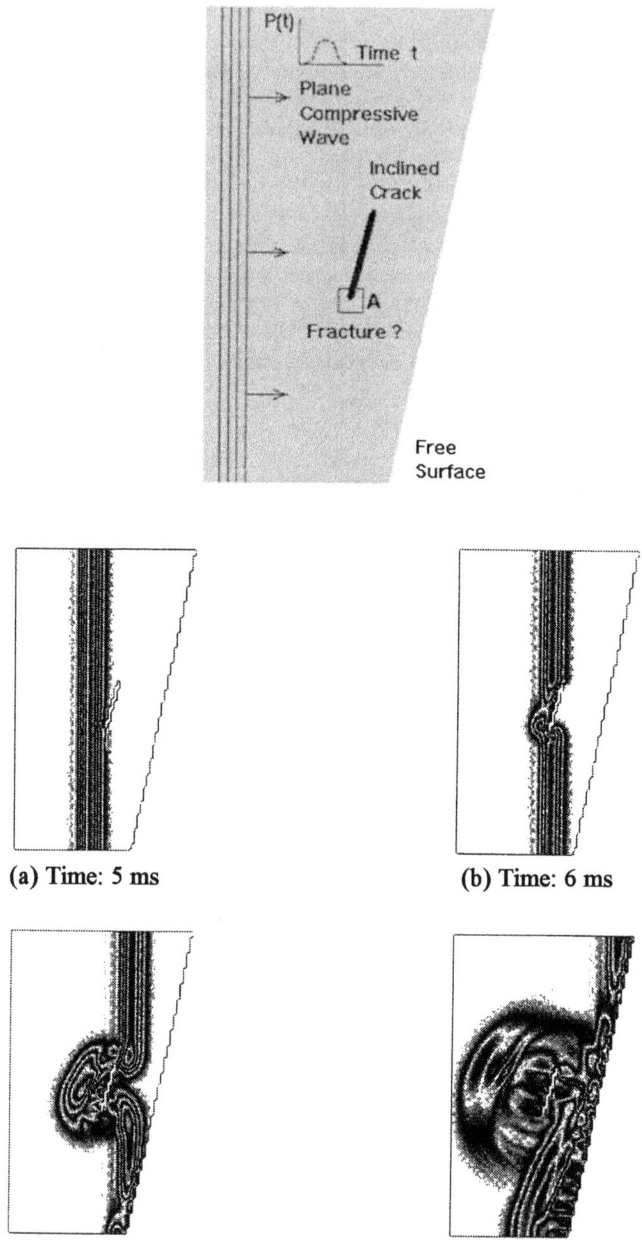

(a) Time: 5 ms

(b) Time: 6 ms

(c) Time: 7.5 ms

(d) Time: 10 ms

Figure 5. Wave diffraction about a finite length crack close to a free surface

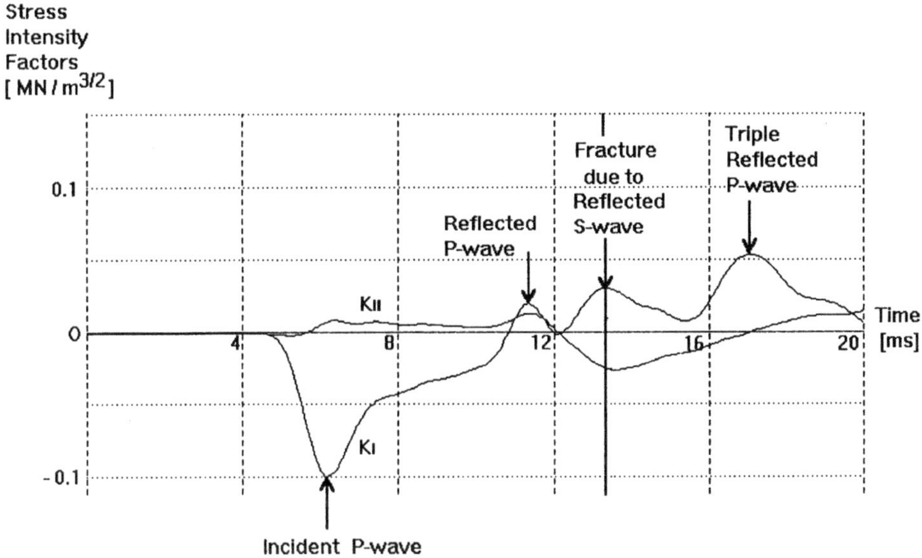

Figure 6. The two stress intensity factor traces

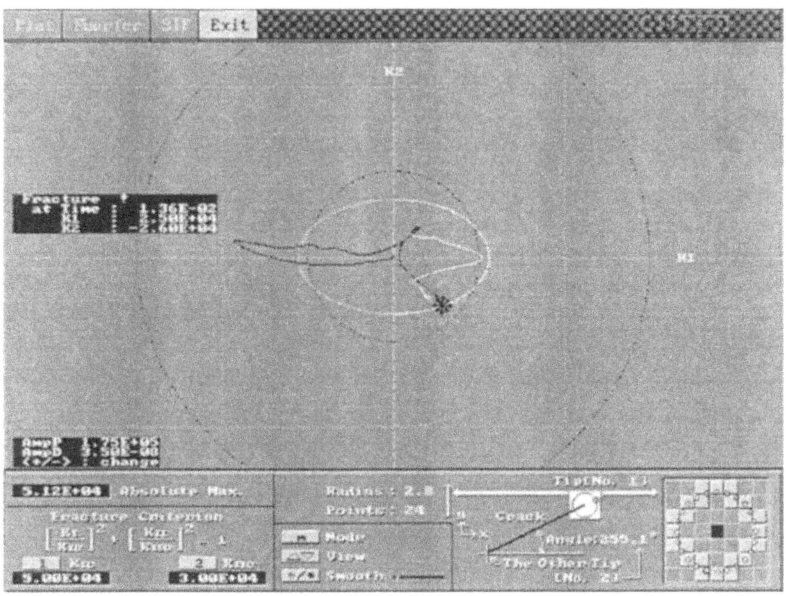

Figure 7. The submenu *SIF*

2D version - the program will be expanded to include a full 3D-capability. Obviously, computer time and storage space limitations dictate the pace of development and availability of a low-cost PC version within easy reach of students and teachers alike.

7 Conclusions

The software SWIFD, an educational tool, has been designed for research as well as teaching and training for students on various levels of education. The menu-driven software provides students and the teacher with an extremely versatile and useful graphical tool for visualization of dynamic processes via numerical simulation and ensuing animation. This offers a clear insight into dynamic fracture and wave phenomena. The program package is continually being updated in response to and to meet demands from students and educators.

8 Acknowledgments

This work is part of an initial study on numerical simulation of fault dynamics and blasting sponsored by the Austrian National Science Foundation Project No.P10326-GEO. The Finite Difference Simulator SWIFD has been developed at the Fracture and Photomechanics Laboratory of the Institute of Mechanics at the Technical University Vienna. The authors would like to thank Prof.S.Selberherr of the Institute of Microelectronics at the TU Vienna for his kind cooperation and graciousness in terms of CPU time free of cost.

References

[1] H.P.Rossmanith and K.Groschupf: SPRO User's Manual, Version 1.1, Fracture and Photomechanics Laboratory, Institute of Mechanics, Technical University, 1992.

[2] H.P.Rossmanith and K.Uenishi: SWIFD User's Manual, Version 1995, Fracture and Photomechanics Laboratory, Institute of Mechanics, Technical University,1995.

Proposals and Improvements

Teaching and Education in Fracture and Fatigue. Edited by H.P. Rossmanith. Published in 1996
by Taylor & Francis, 2 Park Square, Milton Park, Abingdon, Oxon, OX14 4RN. ISBN 0 419 20700 7.

26 ON INCONSISTENCY OF TERMINOLOGY RELATING TO TEACHING AND EDUCATION ON FATIGUE AND FRACTURE

V.P. NAUMENKO
Institute for Problems of Strength, National Academy of Sciences of the Ukraine, Kiev, Ukraine

Abstract

The adoption of the Unified Fracture Mechanics Terminology (UFMT) is one possibility in the improvement of the existing methodology for teaching and education in fatigue and fracture. In this paper the definitions of some basic and related terms included in different standards on fracture mechanics terminology are confronted with one another. Alternative definitions of the same terms are offered. By this strategy the author tries to show that the definitions of basic and related terms appropriate for the UFMT should emerge as a consequence of harmonizing a crack model with an actual crack and then both taken together with a fracture model, laboratory test methods and failure assessment codes.
Key Words: Fracture mechanics, basic and related terms, definitions, standard terminology.

1 Introduction

Consistency of terms and definitions is a key element of any method and methodology for teaching and education. In the ever growing amount of publications used at present for teaching and education in fatigue and fracture one can often encounter confusing interpretations of terms and definitions relating to the same physical phenomenon. A need for the UFMT had long been felt and was continuously becoming more urgent. It was pointed out when the ESIS Technical Committee on Fracture Terminology was being organized [1].

Inherent separation and some inconsistency of terms and definitions in different fields of fatigue and fracture research are the reality which cannot be ignored. Under the influence of objective and subjective factors such

Teaching and Education in Fracture and Fatigue. Edited by H.P. Rossmanith. Published in 1996 by Taylor & Francis, 2 Park Square, Milton Park, Abingdon, Oxon, OX14 4RN. ISBN 0 419 20700 7.

distinctions are varying unpredictably all the time. Therefore, the adoption of the UFMT should be considered as a provisional trade-off based on the widest consensus of different points of view. At the same time, it is vital for the UFMT to use thoroughly justified definitions of such basic terms as: **crack, crack parameter, crack-end region, crack-end-region parameter, crack extension, crack-extension resistance, distinctive crack-state, fracture, fracture parameter, fracture resistance** and **fracture toughness**. Nowadays, the number of essentially different interpretations of these terms is close to or may be even coincides with that of their interpretators. The question is - which of them should be given preference to when elaborating the UFMT? The apparently evident answer is - these should be those definitions which have been included into the available standards on fracture mechanics terminology.

By drawing the attention of specialists to a controversial character of some standard definitions, the author has in mind to initiate a discussion on the UFMT. The lacking confusion and even discrepancy of basic and related definitions inhibit communication, teaching and education in the field of fatigue and fracture as well as the insight into information published. Fundamental importance of a consistent definition of the fracture mechanics terms was brought to light by the so-called "energy" discussion started in 1988, that is briefly reflected in Turner's remarks and the Editor's note [2].

2 Previous terminology elaborations

The first standards relating to fracture mechanics terminology was the ASTM Standard E616-77 [3]. It contains definitions, descriptions of terms, symbols and abbreviations approved for use in standards on fracture testing. The area of its application turned out to be wider than the recommended one. Thus, using the current edition of E616-89, the Proceedings of the 8th European Conference on Fracture have been prepared. Yet, the level of the terminology consistency attained up to date is far from that needed in teaching and education on failure analysis and safety engineering.

The necessity of the UFMT development was discussed by the authors of refs [4, 5]. It was proposed in ref. [4] to start this work with establishing a list of terms and definitions relating to mode I fracture testing. Further revisions and widening of this list were expected to result in covering all aspects of fracture by the UFMT. In ref. [5], along with the importance of a global fracture mechanics terminology, the author pointed out the importance of compiling a multilingual dictionary of fracture research, failure analysis and its applications in various fields of engineering.

Many of the terms, definitions and abbreviations given in ref. [4] have been borrowed from the previous edition E616-82 of standard [3]. Thus, designation codes for specimen configuration, applied loading and crack orientation adopted in this standard were reproduced without any changes. At the same time, ref. [4] includes the definitions lacking in E616-82, as well as the new definitions of such terms as **real crack** and **ideal crack**. Note that a modified version of the above mentioned designation codes was developed later by the authors of ref. [6].

Some results from E616-82 and ref. [4] were used in the course of developing the State Standard of Ukraine SSTU 2442-94 [7]. The SSTU 2442-94 differs from the available prototypes by the extra-wide scope of the subject area and by its objective to solve precisely fundamental problems of fracture mechanics terminology. In the opinion of the authors, standard [7] - "establishes the terms and definitions of the basic fracture mechanics notions". And finally, preliminary version [8] of "A compendium for the integrity of structures" designated as ESIS TC7-1-94D should be mentioned. This document aims at providing a unified terminology and nomenclature in the field of structural integrity. It deals only with the most common concepts.

3 Standard definitions of some basic and related terms

As an illustration, standard definitions of some basic and related terms derived from the generic term **crack** are confronted with one another. The following format of itemization is adopted:

A - basic term and related term;
B - definition of the basic term from E616-89;
C - the same from SSTU 2442-94;
D - the same from ESIS TC7-1-94D;
E - definition of the related term from E616-89;
F - the same from SSTU 2442-94;
G - the same from ESIS TC7-1-94D;
H - a brief comment.

A - **crack** and **ideal crack**:
B - is not defined;
C - defect of the continuity in a solid, that is bounded by two opposite non-interacting surfaces, the distance between which is negligibly small as compared to other characteristic dimensions of the defect, and by a narrow end region with not completely disturbed interatomic bonds;
D - a sharp defect such that in the stress free state the two faces touch each other along contact areas which occupy most of the crack surface;
E - a simplified model of a crack used in elastic stress analysis. In a stress-free body, the crack has two smooth surfaces that are coincident and join within the body along a smooth curve called the crack front; in two-dimensional representations the crack front is called the crack tip;
F - a crack in a non-loaded body taken in the form of an infinitly narrow slit formed by parallel non-interacting flanks and an ideally sharp tip;
G - a crack which has two smooth surfaces joining within the body along a smooth curve called the crack front and which are coincident in the stress free body;
H - each reference has its distinctive definition, if any, of the same term. The lack of the definition *B* implies that the term **crack** is useless in standards on fracture testing. The definitions *C*, *D* and *F*, *G* are misnomers for several reasons. Firstly, the term **crack** is defined via the term **defect**. The latter is uncertain, as well as are the terms **opposite non-interacting surfaces**, **narrow end region**, **contact areas** and **ideally sharp tip**.

Secondly, in accordance with items *F* and *G*, an ideal crack, as such, is nothing else but a crack itself. Thirdly, an actual crack, for example, a branched one may be bounded not only by two (as in standards on fracture testing), but also by four, six, etc. "opposite non-interacting surfaces", as well as by the same number of "opposite interacting surfaces" or "opposite partly interacting surfaces". In many cases, for example, in ductile tearing of a thin plate with a through-the-thickness crack the distance between the "opposite non-interacting surfaces" may be of the same order as the crack length. According to item *C* any crack has a narrow end region, i.e. "opposite interacting surfaces". Along with it in the definition of the related term (item *F*) this attribute of the crack dies out. And finally, the terms **stress-free body** (see item *E*) and **non-loaded body** (see item *F*) reflect different phenomena, as do the terms **coincident surfaces** (items *E* and *G*) and **infinitly narrow slit** (item *F*).

A - **crack parameter** and **crack size**:

B, C, D - are not defined;

E - a lineal measure of a principal planar dimension of a crack. This measure is commonly used in the calculation of quantities descriptive of the stress and displacement fields and is often also termed **crack length or depth**;

F - a linear measurement (*author*: this is word-to-word translation) of the crack extent in a certain direction;

G - is defined via the term **crack length** that is the shortest distance between the crack front and the location of the largest crack displacement for an embedded crack or for a through crack;

H - in all instances any definition of the basic term **crack parameter** is lacking. Moreover, there are no indications that the term **crack size** may be defined differently from the term **ideal crack size**. These facts together with the contents of items *E*, *F* and G as a whole lead to the disputable conclusion that any crack in any event may be described adequately by its planar dimensions solely. It is worth noting that the definition *G* is not adaptable to kinked or curved cracks.

A - **crack-end region** and **ideal crack-end region**:

B - is not defined;

C - is not defined, but nevertheless the term **narrow end region** is inserted into the definition of the term **crack**;

E, F, G - are not defined.

To summarize: The terms **crack-end region** and **ideal crack-end region** are either not compatible with other terms or useless at all.

A - **crack-end-region parameter** and **crack-tip opening displacement**:

B, C, D - are not defined;

E - the crack displacement due to elastic and plastic deformation at variously defined locations near the original (prior to load application) crack tip. The crack displacement is the separation vector between two points (on the surfaces of a deformed crack) that were coincident on the surfaces of an ideal crack in the undeformed condition. The crack tip is the crack front in a two-dimensional stress-free body;

F - relative displacement of the relevant near the crack-tip points on the flanks that are chosen beforehand for a mode I crack. The crack tip is an arbitrary point of the crack front;

G - the crack displacement at a distance close to the tip of the crack, for instance at the location of the tip before loading, or at the intersection between the crack surface and lines through the tip inclined at 45° degree to the x_1 axis. The crack displacement is the separation vector between two points (on the surfaces of a deformed crack) that were coincident on the surface of an ideal crack in the undeformed condition. The crack tip is the crack front in two dimensional representations;

H - in all instances the definition of the generic term **crack-end-region parameter** is lacking. And vice versa, one of the crack-end-region parameters, namely, **crack-tip opening displacement** is defined in all instances. There are no indications that the term **crack-tip opening displacement** may be defined differently from the term **ideal-crack-tip opening displacement**. It is pertinent to note that in relation to an actual crack the terms **undeformed condition** and **stress-free body** (items *E* and *G*) have different meanings. From simple comparison of items *E*, *G*, on the one hand, and item *F* on the other hand it follows that the latter is a misnomer phrased in a confused manner. Such phraseology is typical for a great deal of the definitions presented in [7]. In addition, the definition *F* shows inherent discrepancy between the actual and the declared subject areas of the SSTU 2442-94.

A - **crack extension** and **crack-extension resistance**:

B - an increase in crack size;

C - is defined via the term **crack growth** that is the process of increasing the area of free crack surfaces formed by fracture;

D - is not defined;

E - a measure of the resistance of a material to crack extension expressed in terms of the stress-intensity factor, *K*, crack-extension force, *G*, or values of *J* derived using the *J*-integral concept;

F - a measure of the resistance of a material to stable crack growth expressed by the value of one of the fracture mechanics parameters;

G - is defined via the term **stable crack propagation fracture resistance** that is the evolution of the stress intensity factor (of the strain energy release rate) during slow stable crack propagation;

H - it is of interest to recall the subject areas of the terminologies declared in [3, 7, 8]. The definition *B* should be appropriate for use in certain standards on fracture testing. Such is indeed the case. The definition *C* should be appropriate for use as one of the basic notions of fracture mechanics. Such is not indeed the case. Fracture as the separation process has its exact antithesis, that is, the restoration process. When both processes are caused, for example, by some combination of uniaxial tension, high hydrostatic pressure and high temperature, the area of "free crack surfaces" may not increase in the course of crack growth. The lack of definition *D* along with the availability of the definition *G* implies that the term **crack-extension resistance** may be defined without any use of the term **crack extension**. And finally, the actual subject areas of the definitions *F*, *G*, and *E* (narrow and wide) are exactly the opposite of that (wide and narrow) declared in [7, 8] and [3].

A - **distinctive crack state** and **start of crack-extension**:

B, *C*, *D*, *E* - are not defined;

F - is defined via the term **crack initiation** that is the beginning of crack growth;

G - is not defined;

H - in all instances any definition of the basic and related terms is lacking. Actually, the definition *F* by itself is nothing else but a synonym for the term **crack initiation** which on its own accord is a poor choice. Nonetheless, the related terms are extensively used in standards [3, 7, 8]. The vital importance of the generic term **distinctive crack state** is underlined by the following list of deriviative terms: start of crack-extension, onset of stable crack-extension, onset of slow stable crack-extension, onset of unstable brittle crack-extension, pop-in, initiation of a fast-running crack and shortly after crack-arrest [3]; critical value of fracture mechanics parameter, onset of fracture process, crack initiation, crack arrest, crack branching, critical crack size, critical value of the stress intensity factor, *K*, at the instant of the crack initiation and onset of unstable crack-extension [7]; critical value of *K*, critical value of the strain energy release rate, stop of a propagating crack; various critical values of the *J*-integral, various critical values of the crack tip opening displacement, stable crack propagation initiation, crack propagation initiation, unstable crack propagation, at, or greater than, 0.2 mm crack growth, and pop-in [8].

4 Some alternative definitions

The following format for itemizing the alternative definitions is adopted:

A - basic term and related term;

B - definition of the basic term;

C - definition of the related term.

A - **separation process** and **restoration process**:

B - a process that gives rise to the formation and/or extension of stress-free surfaces within a body;

C - a process that gives rise to the elimination and/or contraction of the stress-free surfaces within a body.

A - **fracture** and **fracture resistance**:

B - the separation process causing the newly formed surfaces to move away from one another without removing a material out of the body or transferring it within the body;

C - a measure of the resistance of a material to fracture expressed in terms of fracture parameters.

A - **crack** and **ideal crack**:

B - a cavity in a body enclosed by mating surfaces that have been formed in the course of the separation process on its own or in combination with the restoration process. Both processes are going on without removing a material out of the body, transferring it within the body or adding it into the body. A curve joining a pair of the mating surfaces is called the **edge line**; in two-dimensional representations the edge line is called the **crack edge.** In the undeformed condition, the ratio of the greatest values of transverse and planar dimensions of the crack cavity should not exceed some characteristic points which distinguish a crack from a pore and other discontinuities in the given material;

C - a simplified model of a crack used in stress-strain analysis. The ideal crack has smooth mating surfaces that are joined in pairs within the body along smooth curve (curves) called the **ideal edge line (lines)**; in two-dimensional representations the edge line is called the **ideal crack edge.** In a stress-free body the distance between the mating surfaces of an ideal crack depends on the procedure chosen for the crack modelling and may be a function of the material properties, as well as of the geometry and sizes of the cracked body.

A - **front line** and **ideal front line**:

B - a line on a surface of a crack that coincides with the observed crack front prior to load application; in two-dimensional representations a pair of the mating front lines are called the **crack tip;**

C - variously defined line on a surface of an ideal crack; in two-dimensional representations a pair of the mating ideal front lines are called the **ideal crack tip.**

A - **crack parameter** and **crack size**:

B - a characteristic of a crack in a stress-free body used for the determination of ideal crack parameters.

C - a crack parameter characterizing the extent of the crack cavity in a certain direction.

A - **ideal crack parameter** and **ideal crack size**:

B - a characteristic of an ideal crack in a stress-free body used for calculation of quantities descriptive of the stress and displacement fields;

C - an ideal crack parameter characterizing the extent of the ideal crack cavity in a certain direction.

A - **crack displacement** and **ideal crack displacement**:

B - the separation vector between two points on mating surfaces of a deformed crack that had chosen locations on these surfaces in the undeformed condition;

C - the separation vector between two points on mating surfaces of a deformed ideal crack that had chosen locations on these surfaces in a stress-free body.

A - **crack-front opening displacement** and **ideal crack-front opening displacement**:

B - a crack displacement due to elastic and plastic deformation between two points located on the front lines of mating surfaces; in two-dimensional representations the crack-front opening displacement is called the **crack-tip opening displacement;**

C - an ideal crack displacement between two points located on the front lines of mating surfaces; in two-dimensional representations the ideal crack-front opening displacement is called the **ideal crack-tip opening displacement.**

A - **crack-end region** and **ideal crack-end region**:

B - are those portions of the mating surfaces of the crack within which the cohesive forces for a material are varying between their original value in undeformed condition and zero at the instant of splitting the material bonds;

C - are those portions of the mating surfaces of the ideal crack which are bounded by the ideal edge line on one side and by the ideal front line on the other.

A - **crack-end-region parameter** and **ideal crack-end-region parameter**:

B - the crack parameter characterizing its crack-end region;

C - the ideal crack parameter characterizing its crack-end region.

A - **crack extension and ideal crack extension**:

B - a relocation of any point of the edge line and/or the front line caused by the fracture process;

C - a relocation of any point of the ideal edge line and/or the ideal front line in the course of fracture modelling.

A - **fracture parameter and crack-extension resistance**:

B - a quantitative measure that is commonly used to characterize the local stress-strain field around the ideal crack-end region;

C - a measure of the resistance to the extension of a crack expressed in terms of the ideal crack parameters;

A - **distinctive crack state** and **start of crack-extension**:

B - a demarcation point between the sequential stages of single-type behavior of a material within the crack-end region;

C - the distinctive crack state corresponding to the transition from the apparent crack extension in the course of front line blunting with no stable crack extension to the formation of new stress-free surfaces within the crack-end region.

5 Discussion

Any definition approved for use in one standard or another is bound to fall into one, and only one, terminological system that in its turn originates from a single concept of fracture. The commonly accepted (conventional) theories and models of fracture exemplify such concept and serve as an origin of standard definitions in the above list. The main confusion and inconsistency between these definitions may be attributed to standard [7] developed under the supervision of A.Ya. Krasowsky. In contrast to the conventional theories of brittle fracture, designated in [9, 10] as a "decohesion" model, he introduced the concept of a "coalescence" model. Within the framework of the latter, the so-called K_μ-model was proposed in 1972 [11]. In the subsequent discussion both concepts will be represented by the K_μ-model as the most appropriate for comparison analysis.

The alternative definitions in the above list originate from the so-called ρ-theory of brittle fracture [12, 13, 14]. Basically, this theory and the K_μ-model are distinguished by their approaches to crack and fracture modelling. Let us outline both of these approaches through the use of an infinite plate containing a centered crack of length $2c_p << 2W$ and $2H$ (Fig. 1a). A uniform loads, σ and $q = k\sigma$, may be either tensile or compressive.

In any fracture analysis an actual crack is represented by its simplified model, i.e. by an ideal crack. Within the K_μ-model approach to our two-dimensional problem this is a planar slit whose principal dimension $2a$ is equal to visually identified crack length $2x_{pu}$. It appears as if we forget about the availability of a considerable amount of useful information on actual crack in the undeformed condition. The case in point is its transverse dimensions, $\delta_u(x)$; residual stresses, $\sigma_{pu}(x)$, induced by the material transformations at the preceding stages of the fracture-process zone evolution; a physical crack-end region of length $r_{pu} = c_{pu} - x_{pu}$ where c_{pu} is the physical half-length of the crack, and cohesive forces, $p_u(x)$, acting within the crack-end region. Here index "u" denotes the undeformed condition of the plate in Fig. 1a when $\sigma = q = 0$. Other distinctive states of a stationary crack are shown in Fig.1b.

Thus, the K_μ-model approach to crack modelling is based on a unrestricted use of the following assumptions:

$$a = x_p = x_{pu},\ \delta(x) = 0,\ r_p = 0,\ \sigma_p(x) = 0 \text{ and } p(x) = 0. \qquad (1)$$

It must be emphasized that this representation of an ideal crack is entirely related to a stress-free body. Associated (undeformed) condition of the actual crack is characterized by the following equalities:

$$x_p = x_{pu},\ \delta(x) = \delta_u(x),\ r_p = r_{pu},\ \sigma_p(x) = \sigma_{pu}(x) \text{ and } p(x) = p_u(x). \qquad (2)$$

As a consequence, the fracture modelling involves only unidirectional extension of the slit with the zeroth end region. Such oversimplification or, more properly, depriving an actual crack of its salient features has no admissibility limits.

A different approach to modelling a crack and fracture was proposed within the framework of the ρ-theory [12, 13, 14]. Parameters of an actual crack refer to a stress-free body when $\sigma = q = 0$, $\sigma_p(x) = 0$ and $p(x) = 0$. In addition to c_p and r_p parameters, it is advantageous to represent an actual crack by the crack-tip opening displacement, δ_p, or crack-tip opening angle, α_p. The c_p, r_p, δ_p and α_p values are determined by experimentation prior to crack and fracture modelling. All of them reflect the elastic response of the crack profile and thus each by itself is a single-valued function of the load parameters σ and k (see Fig. 1a). The relevant test method is outlined in [14].

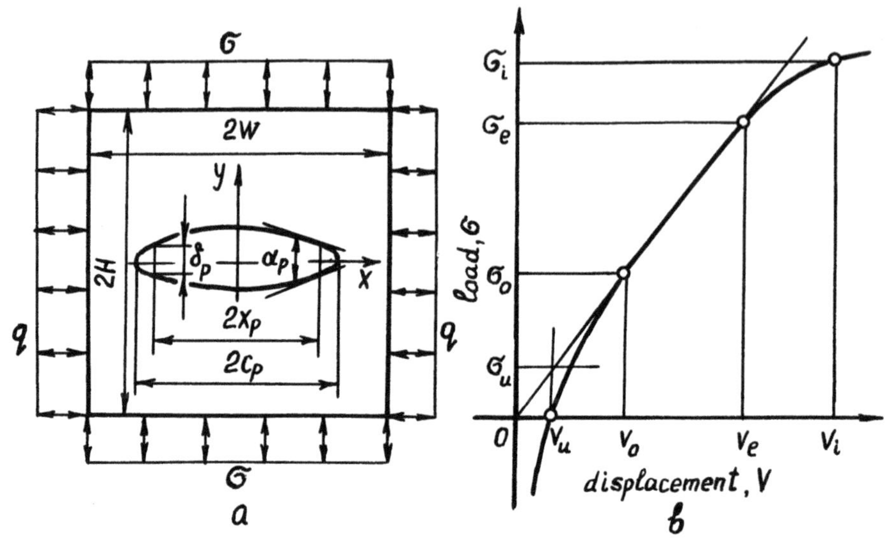

Fig. 1 Interior crack geometry (a) and tensile load, σ, vs transverse displacement, v, diagram (b).

The procedures for modelling the crack and fracture are mutually related. Each involves three sequential steps. The crack is modelled by: 1) an equivalent elliptical cavity with stress-free surfaces; 2) an equivalent slit with stress-free surfaces; 3) an equivalent slit with Dugdale strip end region. Accordingly, fracture is modelled by: 1) omnidirectional extension of the elliptical cavity; 2) unidirectional extension of the slit with the zeroth end regions; 3) unidirectional extension of the slit with Dugdale strip end regions. In this fashion three sets of ideal crack parameters are established: 1) parameters of the elliptical cavity; 2) parameters of the slit with the zeroth end-regions; 3) parameters of the slit with Dugdale strip end-regions. The cavity parameters are of a key importance because they serve as links between the actual and the ideal cracks. The relationships between the parameters of the ideal and actual cracks for our problem (Fig. 1a) are presented in [14].

To form a justified basis for the UFMT the most appropriate concept of fracture is to be chosen. As guidelines in choosing one concept or another the following assertions may be used:

- a crack model is harmonized with an actual crack and they both taken together with a fracture model, laboratory test methods and failure assessment codes;
- a wide range of the practicality is preferred to a narrow one.

The first step towards harmonizing the main constituents of any concept of fracture implies that the incompatibility of assumptions (1) related to the stress-free body and condition (2) related to the undeformed condition is taken into account or eliminated.

This step is absent in the framework of the K_μ-model, and conversely, the procedure of the harmonization mentioned above is outlined in the ρ-theory. As a consequence, the effects of in-plane constraint and load biaxiality are not described by the K_μ-model, and conversely, they are inherent in the ρ-theory approach [12, 13, 14]. Furthermore, a unified analysis of the mode I crack behaviour, for example, under uniaxial and biaxial loading in tension and/or compression (Fig.1a) is impossible to perform within the K_μ-model approach. Along with it different details of such analysis in the context of the ρ-theory are presented elsewhere [12, 13, 14].

It is pertinent to mention some peculiarities of the generalized K_μ-model [15], cited by the authors of ref. [9] as "A generalized fracture model and a criterion which take into account the stress state at the crack tip. The authors of ref. [15] obtained generalized relationships in a closed form ...". These relationships predict an unbelievable (hundreds times) drop in the resistance to brittle fracture when passing from the crack-tip plane stress conditions (K_C) to the plane strain conditions (K_{IC}). But no less paradoxical is the prediction related to ductile fracture. In this instance the value of K_{IC} may be dozens of times as large as the value of K_C. It is self-evident that such predictions might not originate from the conventional theories of fracture. They reflect the key features of the K_μ-model and follow immediately from the heart of this model.

6 Conclusions

The currently available standard definitions of the basic terms considered above are for the most part either lacking or confused and inconsistent. They may be radically changed with changing the basic models of crack and fracture. In this manner the alternative definitions of the same terms are generated. Due to fatal subjectivity of the author they may be well apart from sophistication. Nevertheless, the emergence of the alternative definitions by themselves will stimulate the development of the UFMT and the refinement of the standards on fracture mechanics terminology.

Acknowledgements

I am grateful to Dr. Ian Milne, Prof. H.P.Rossmanith and Prof. K.J.Miller for their support in the emergence of this paper.

References

1 The Editor (1989) EGF Committee on Terminology. *Official Newsletter of the European Group on Fracture*, No. 10, Summer/Autumn, p. 7.
2 Turner, C.E. (1992) Dissipation Rate for Describing Tearing Resistance. *Official Newsletter of the European Structural Integrity Society*, No. 19, Summer, pp. 10-11.
3 ASTM Committee E-24 (1989) Standard Terminology Relating to Fracture Testing. E616-89, Originally Published as E616-77. *Annual Book of ASTM Standards*, Vol. 03.01, pp. 614-625.
4 Pisarenko, G.S., Naumenko, V.P. and Rakovsky, V.A. (1986) On the Fracture Mechanics Terminology, *Problemy Prochnosti*, N.7, pp. 113-121.
5 Kausch, H.H. (1990) Is Now the Time for Global Terminology in Fracture Mechanics? In: *Fracture Behaviour and Design of Materials and Structures,* **Proc. of** ECF8, (ed. D.Firrao), EMAS, Vol. 1, pp. 17-22.
6 Panasyuk, V.V. and Yarema, S.Ya. (1987) Designation System for Fracture Toughness Test Specimens *(in Russian), Fiziko-Khimicheskaya Mekhanika Materialov*, No. 1, pp. 122-124.
7 Krasowsky, A.Ya., Orynyak, I.V. and Torop, V.M. (1994) Strength Calculation and Testing, FRACTURE MECHANICS, Terms and Definitions. *The State Standard of the Ukraine*, Published in Ukrainian under the Designation SSTU 2442-94, 23 p.
8 ESIS Technical Committee 7 (1994) A Compendium for the Integrity of Structures. *Preliminary Draft*, ESIS TC7-1-94D, 23 p.
9 Krasowsky, A.Ya. and Pluvinage, G. (1993) Structure Parameters Governing Fracture Toughness of Engineering Materials. *Materials Science*, Vol. 29, No. 3, May-July, pp. 113-123.
10 Krasowsky, A.Ya. (1994) Models for Fracture of Materials, in *Resistance of Materials to Deforming and Fracture*, the Reference Source, (ed. V.T.Troshchenko), Naukova Dumka, Kiev, pp. 373-384.
11 Pisarenko, G.S. and Krasowsky, A.Ya. (1972) Analysis of Kinetics of Quasi-Brittle Fracture of Crystalline Materials, in *Mechanical Behaviour of Materials*, The Society of Materials, Kyoto, Vol. 1, pp. 421-432.
12 Naumenko, V.P. (1987) Brittle Fracture and Strength of Materials in Compression and Tension. *Preprint of the Institute for Problems of Strength of Academy of Sciences of the Ukr SSR* (in Russian), Kiev, 38 p.
13 Naumenko, V.P. (1991) Modelling of Brittle Fracture in Tension and Compression, in *Fracture Processes in Concrete, Rock and Ceramics*, Proc. RILEM/ESIS Conference, (ed. J.G.M. van Mier, J.G.Rots and A.Bakker), E & F.N. Spon, London, Vol. 1, pp. 183-192.
14 Naumenko, V.P. (1995) A Framework for a Unified Methodology of Crack Assessment in Components, to be Published in the *Proceedings of Symposium "Risk and Economic Evaluation on Failure and Malfunction of Systems"*, E & FN Spon and Chapman & Hall.
15 Krasowsky, A.Ya. and Vainshtok, V.A. (1978) A Failure Criterion of Materials Taking into Consideration the Form of the Stressed State at the Crack Apex. *Strength of Materials*, Vol. 10, No. 5, pp. 559-565.

27 CLASSIFICATION OF ELASTO-PLASTIC FRACTURE MECHANICS CRITERIA

G. PLUVINAGE
Laboratory of Mechanical Reliability , University of Metz, France

Abstract

A classification scheme for elasto-plastic fracture mechanics criteria is proposed. This classification includes seven classes of criteria based on local and global stress, local and global strain, local and global energy, and a criteria for interpolation between two limiting states is proposed as an additional class.

Key words: EPFM, fracture criteria, stress, strain, energy, classification.

1 Introduction

Although lecturing in fracture mechanics is mainly concerned with linear-elastic fracture mechanics (LEFM) some aspects of elasto-plastic fracture mechanics (EPFM) such as a descriptive explanation of the J-integral and the crack opening displacement concepts are often added. However, with regard to fracture criteria, many more have been proposed in EPFM and there are only very few books which present a general overview on this topic [1-2].

The major difficulty associated with this important part of fracture mechanics is presented by the existence of a very large number of these criteria.

A thorough presentation of fracture criteria requires a classification of the domains of validity and a comparison between these criteria. A rough classification may be based on those classes of criteria related to stress, strain and energy. An interpolation method between the two well-defined limiting states of brittle fracture and plastic instability is added.

Teaching and Education in Fracture and Fatigue. Edited by H.P. Rossmanith. Published in 1996
by Taylor & Francis, 2 Park Square, Milton Park, Abingdon, Oxon, OX14 4RN. ISBN 0 419 20700 7.

A more adequate description of the major aspects of EPFM criteria can be achieved by utilizing the classes listed in Table 1. These seven classes contain 30 major elasto-plastic fracture criteria.

Table 1 Basic classification of EPFM fracture criteria

local stress criteria	local strain criteria	local energetic criteria
global stress criteria	global strain criteria	global energetic criteria
	interpolation criteria	

Two important aspects shall be mentioned:

- EPFM is applicable for defects which are either not very short or not very long; otherwise an ultimate strength criterion is required for very short defects and a plastic instability criterion is needed for very long defects;

- EPFM can be considered as a modification of the lower bound of what is termed the noxiousness of a defect; this lower bound is related to LEFM with respect to the most severe defect (crack) and the worst situation (brittle fracture).

These two major aspects can be discussed on the basis of a Feddersen diagram (see Fig. 1) where the global critical stress is plotted versus the ratio between defect size and width of the specimen.

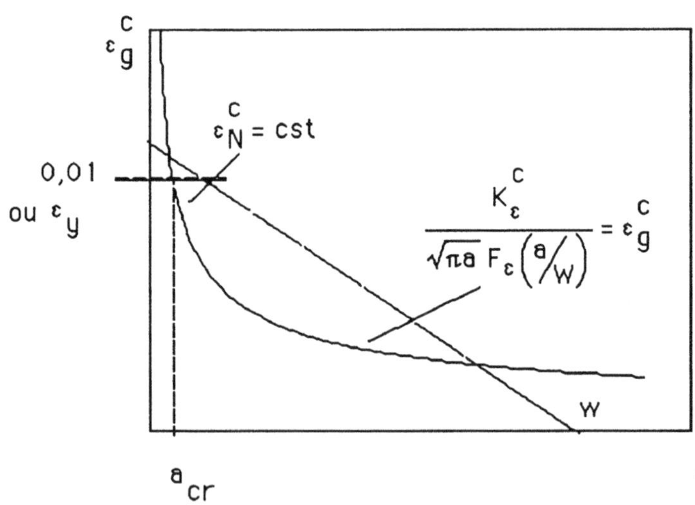

Figure 1 Fedderson Diagram showing the various Areas I,II, and III explained in the text

In this diagram three distinctive areas can be identified:

Area I : for short defects where the ultimate strength criterion may be applied,
Area II : for medium size cracks; the field of fracture mechanics,
Area III: for long defects where a limit analysis may be applied.

The particular influence of the stress gradient at the crack tip on the critical global stress σ_g^c is unveiled by the fact that in Area II the LEFM curve is below the line (σ_N^c=R_m), where σ_N^c is the nominal critical stress and R_m is the ultimate strength. The line defined by (σ_N^c=R_m), is a hypothetical line governed by the assumption that at fracture initiation there is no stress gradient at the tip of the defect.

The influence of stress relaxation due to plastic deformation and the reduction of the stress gradient at the tip of the defect in EPFM is expressed by the fact that in Areas I and III the EPFM line is above the LEFM line in the Feddersen diagram.

An objective presentation of all elasto-plastic fracture criteria must be based on the fact that there is no preference for any particular criterion. Of course, some of the EPFM fracture criteria are very popular, widely known and being utilized in the standards in many countries, and much research has been performed with regard to their applicability and their limitations.

However, all of these EPFM criteria share a common deficit: they rely on the assumptions used in continuum mechanics. This means that they are not intrinsic with respect to the material structure and they are dependent on the geometry and the load path. It is there necessary to complement this classification with the limitations and the applicability of these criteria. Some criteria are applicable in a fracture situation which comes close to LEFM, others, however, work better at conditions close to plastic instability, and some of them can be used for all situations. At present, no unique general criterion is known which is indiscriminately applicable in EPFM.

In fact, one has a wide choice from about 30 criteria and this wide selection does present some difficulties. Even worse, these criteria do not share the same domains of validity. These difficulties need to be illustrated and assessed in a general and clear overview of all the possibilities in this field.

2 Global stress criteria

Global stress criteria derive from LEFM and are used when the non-linearity in the load versus displacement diagram is not too pronounced. There criteria are, in fact, and extension of LEFM. The basic idea lies in calculating an equivalent critical stress intensity factor K_{Ie}^{c} via LEFM without any restrictions and assumptions with respect to small scale yielding. The relationship is as follows:

$$K_{Ie}^{c} = \sigma_g^{c} \sqrt{\pi . a}\ F_\sigma(a/w) \tag{1}$$

where σ_g^c is the critical global stress, 'a' is the crack length and $F_\sigma(a/w)$ is a geometrical correction factor.

The apparent fracture toughness, K_x^*, is related to the equivalent critical stress intensity factor K_{Ie}^c by using a non-dimensionless correction factor φ. The subscript 'x' defines the apparent fracture toughness:

$$K_x^* = K_{Ie}^c \sqrt{\varphi} \tag{2}$$

It is important to note that the correction factor, φ, appears in a square root function. This is because the correction factor is sometimes defined via the critical strain energy release rate, which is proportional to the square power of the critical stress intensity factor.

There are numerous possibilities for the introduction of a correction factor all of which can be classified into three groups:

- correction with respect to crack length,
- correction with respect to critical global stress, and
- correction with respect to fracture toughness.

The correction with respect to the crack length is based on the concept of the effective crack length which consists of the sum of the actual crack length and the plastic zone size.

The effective crack length can be obtained either

- directly by adding to the crack length the length of the plastic zone calculated according to Irwin [3] or Dugdale [4], or,
- indirectly from the load displacement diagram through the initial tangent modulus E* or the secant modulus E_S [5] or the value of the compliance C (a_{eff}) [6] equal to the ratio of critical displacement and critical load.

The definition of the equivalent critical global stress, $\sigma^c_{g,q}$ is obtained from the load-displacement diagram using a 5% offset procedure [7] or the equivalent energy method where the actual non-linear curve is replaced by a straight line which, as seen in Fig.2, yields the same area under the curve [8].

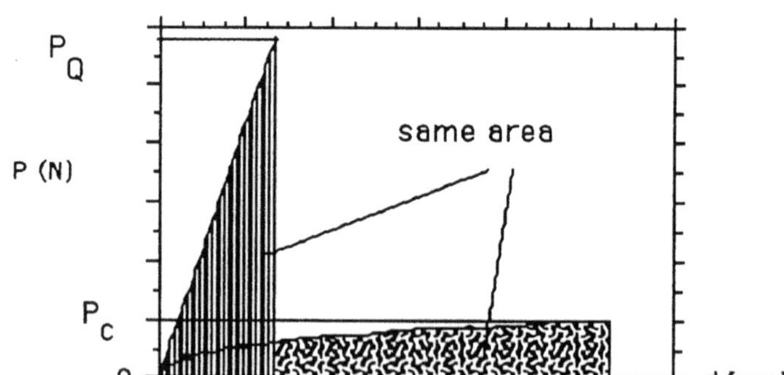

Figure 2 Method of equivalent energy: the critical load P_c is replaced by the equivalent load P_Q

Another possible definition of the apparent fracture toughness is based on the relationship between stress intensity factor and critical crack opening displacement. This apparent fracture toughness has been compared with the equivalent critical stress intensity factor by Newman [9] and Heald, et al [10]. The correction factor as derived by Witt [8] is defined as the square root of the ratio between the work done in the non-linear situation, U_B, and the work done in brittle fracture, U_A.

The various definitions of the criteria in LEFM are summarized in Table 2.

Table 2 Comparison of LEFM fracture criteria

IRWIN	$\rho_0 = \left[1 + \frac{1}{2}\left(\frac{\sigma_g^c}{Re}\right)\right]^2$
DUGDALE	$\rho_1 = \sec\left(\frac{\pi\,\sigma_g^c}{2\,Re}\right)$
ALLEN	$\frac{d_c}{P_c} = C(a\,eff)$
JONES	$\varphi_j = \sqrt{\frac{E^*}{E_s}}$
ASTM E 399	$\sigma_{g,q}^c = \frac{P_q}{B.\,W}$
EQUIVALENT ENERGY	$\sigma_{g,eq}^c = \sqrt{\frac{U}{2.C}}$
WITT METHOD	$\varphi_w = \sqrt{\frac{U^B}{U^A}}$
HEALD, SPINK and WORTHINGTON	$\varphi_2 = \mathrm{Ln}\,\sec\left(\frac{\pi\,\sigma_g^c}{2\,Re}\right)$

3 Criteria based on local stress

Fracture criteria based on the concept of a global stress are like a black box as no information about the microscopic fracture mechanisms is needed for the application of these criteria. For elasto-plastic fracture, these criteria are not intrinsic with respect to the material and they are geometry and load path dependent. If the correct stress distribution at the crack tip is known and associated with a local stress fracture criterion it is expected that failure can be correctly predicted by using a unique parameter which is material intrinsic.

Fracture criteria based on local stress are characterized by two parameters: a critical failure stress and a characteristic distance. These two parameters define a particular state which is associated with an a priori unknown critical stress distribution.

Adopting a simple spring atomic model at a sub-microscopic level Barenblatt [11] derived that the value of the maximum cohesive stress is equal to $E/2\pi$ where E is Young's modulus. Failure will occur when two neighboring atoms are displaced by a distance equal to $b_o/4$ where b_o is the atomic distance (Fig.3a).

At a microscopic level the criterion proposed by Heald et al [10] assumes that fracture occurs when crack surfaces are displaced by a distance equal to the critical crack opening displacement, δ_c. The cohesive stress is assumed to be constant and equal to the ultimate strength R_m (Fig.3b).

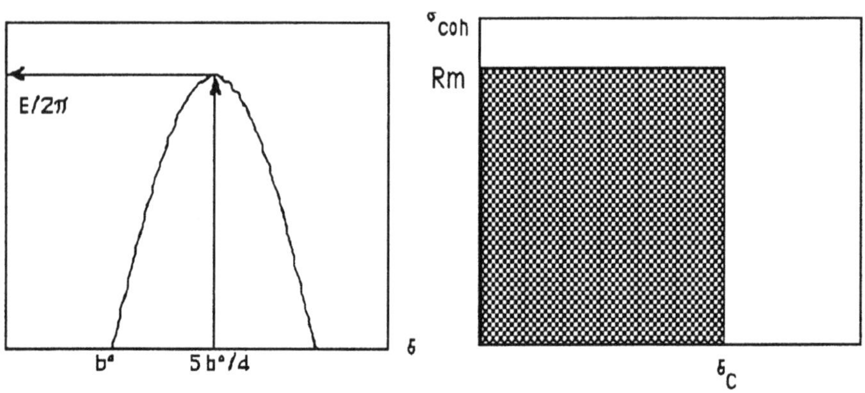

a) b)

Figure 3 Local stress fracture criteria according to the approaches by a) Barenblatt and b) Heald et al

The most popular local stress criterion is the one by Ritchie et al [12] which postulates that fracture will occur when the stress distribution reaches the critical cleavage stress, σ^*_c, over a characteristic distance, X_c. This criterion predicts that when the yield stress increases, the fracture toughness will also decrease. Alternative forms of this criterion are due to Krasowsky and Pissarenko [13] and Holzmann et al [14].

In the theory of stress concentrations by Irvine and Quirk [15], the critical stress distribution is assumed to be constant and equal to the ultimate strength prevailing over the characteristic distance.

Probabilistic fracture criteria based on local stress can be applied using the well-known Weibull distribution [16] in order to predict the probability of fracture under a constant Weibull stress, σ_w, acting in the entire process volume, V_{el}. This model is useful to explain the so-called scale effect and in a statistical way to define the safety factor.

The two characteristic parameters, the critical failure stress and the characteristic distance, are compared in Table 3 for the various failure criteria based on the concept of local stress.

Table 3 The 8 criteria in LEFM

Criterion	critical failure stress	caracteristic distance
Barenblatt (11)	$E / 2\pi$	$b_0 / 4$
Heald, Spink and Worthington (10)	Rm	d_C
R K R, Krasowsky and al Hollzman and al (12) (13) (14)	σ^*_C	X_C
Irvine and Quirk (15)	Rm	X_C
Weibull (16)	σ_W	V_{el}

4 Global energy criteria

The knowledge of the work done during fracture, U, provides one of the most practical ways to estimate the fracture strength. The work done for fracture is complex and consists of one part associated with the creation of two new fracture surfaces and another part for the plastic work. The latter can eventually be divided into a part localized in the ligament area and another part near the force contact point. The work for creating damage as part of the plastic work spent in the material can be taken into account. By definition, the true fracture strength work is the only work to create two new fracture surfaces. There are two ways to separate this work from the entire work done for fracture:

- the first method is based on an incremental approach where the work done during fracture is evaluated in two situations: first for a crack of length 'a' and then for a crack of length 'a+da' where 'da' is an incremental crack advance (Fig.4a); and

- the second method postulates that the fracture strength work is proportional to the work done during fracture (Fig.4b).

These two methods lead to two categories of global energy criteria.

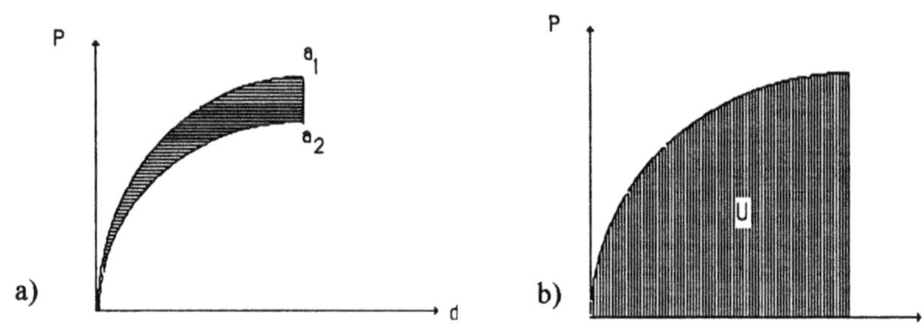

Figure 4 The two global energy based fracture criteria:
a) incremental approach, and b) global approach

The first category of fracture criteria is derived from the critical linear strain energy release rate, G_c, defined in the form:

$$G_C = -\frac{dU}{da} \tag{3}$$

and this definition has been extended by Liebowitz [17] to include the critical non-linear work performed during fracture, U_{nl}:

$$\hat{G}_C = -\frac{dU_{nl}}{da} \tag{4}$$

It has been proved that for elasto-plastic fracture the well-known J-integral [18] is precisely equal to the derivative of the work done during fracture.

In the Generalized Theory of Fracture [19] the fracture parameter I_c is also equal to this quantity, but has also the property of being proportional to the uniform strain energy density.

The fracture criteria of the second category postulate that the fracture strength is proportional to the work done during fracture if the entire work is dissipated on the ligament area. If this condition is not satisfied, it is more difficult to determine the fracture resistance. All these global energy based fracture criteria are summarized and compared in Table 4.

Table 4 Global energy based fracture criteria

Critical strain energy release rate (Irwin)	G_C	$G_c = -\dfrac{dU}{da}$
Non Linear critical strain energy release rate (17)	$\tilde{G}_c$	$\tilde{G}_c = -\dfrac{dU_{U_n}}{da}$
J integral (18)	$J_{Ic} = \int w^*_C dy - Ti \dfrac{du_i}{d_{oc}} ds$	$J_{IC} = -\dfrac{dU_C}{da}$
Generalized Theory of Fracture (19)	$I_c = K, w^*_{0.c}\, a$	$I_c = -\dfrac{dU}{da}$
critical Energetic parameter J_{IC} (20)	$J_{IC} = \dfrac{u\,\eta}{Bb}$	
Essential Work of Fracture (21)	Γe	$\Gamma_e = \dfrac{U_c}{Bb} - Bb\, W^*_{PL}$

Turner [20] has postulated that the fracture strength J_{Ic} is proportional to the critical work done during fracture per unit of ligament area. The coefficient of proportionality is termed η. It is necessary to mention that the energy parameter J_c is only equal to the J-integral if some conditions related to geometry and thickness of the specimen are satisfied.

In the fracture criterion based on the concept of Essential Work [21] the fracture strength Γ_e is equal to the work done during fracture per unit of ligament area minus the surface plastic work. The J_{Ic}-criteria based on the critical energy parameter J_{Ic} is one of the most popular fracture criteria. It can be easily measured with a single specimen and has been proposed as a standard. Some restrictions with respect to ligament geometry and thickness lead to a limitation of the validity of this fracture criterion to situations of quasi-brittle fracture conditions. The J-integral is a useful computational quantity, however, for highly non-linear fracture the criteria based on the concept of essential work for fracture or the generalized theory of fracture is preferable.

5 Local energy-based fracture criteria

Similarly to the local stress criteria, local energy criteria are described by two parameters, the local strain energy density and a characteristic distance. There are only two fracture criteria of this type, one proposed in Hungary by Guillemot [22] and the other one in the USA by G.Sih [23]. These criteria are listed in Table 5.

The critical strain energy criterion is based on the notion of a micro-specimen cut out from the vicinity of the crack tip which is loaded in tension. The micro-sample has a diameter X_C equal to the characteristic distance. This criterion incorporates the effect of triaxiality by the way of a stress concentration factor.

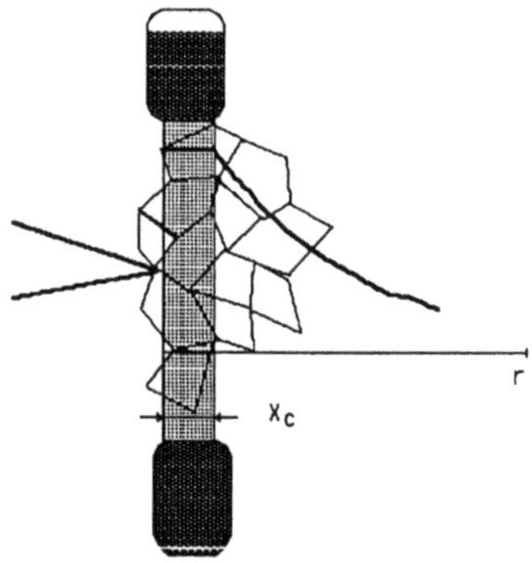

Figure 5 Concept of a micro-specimen taken from the vicinity of the crack tip

The criterion proposed by Sih [23] gives the local strain energy density through the stress intensity factor. This parameter has a r^{-1} dependence and it is necessary to define the point where this strain energy density is to be calculated. The critical strain energy density, S_C, is defined as the ration between the strain energy density and the 'core' distance, r_C, is independent of the distance r. This criterion is helpful to predict crack bifurcation under conditions of mixed-mode fracture.

Table 5 Strain energy criteria

critical strain energy density	W_C^x	X_C
critical strain energy density factor	$S_C = \frac{W_C^*}{r_C}$	r_C

6 Local strain criteria

This kind of criterion is similar to the other local criteria using stress and strain energy density parameters. Like the other criteria, they are characterized by two parameters: a local critical strain and a characteristic distance over which the critical local strain acts.

There are three criteria of this type of which one is very popular and has even led to a standard: the critical crack opening displacement. The other two have been proposed by Randall [24] and Osborne and Embury [25].

The critical crack opening displacement criterion for fracture was proposed by Wells in 1961 [26]. This criterion postulates that fracture occurs when the crack opening displacement reaches a critical value δ_C. This criterion is a local strain criterion by way of the concept of the micro-specimen proposed by Cottrell [27].

It is assumed that at the crack tip a micro-specimen of diameter equal to the size of a micro-structural unit (such as the grain size, d_g, the spacing of the inclusions, l_i, etc.) is stretched in tension with a resulting displacement equal to the critical crack opening displacement. The critical local strain built up in this micro-specimen is assumed to be equal to the local failure strain, ε_f, in tension.

For cases where fracture initiates form a notch, Randall [24] has proposed to take the characteristic distance equal to the notch radius, ρ, and he defines the fracture toughness as the product $\varepsilon^c_1 \rho^{1/2}$ (Fig.6).

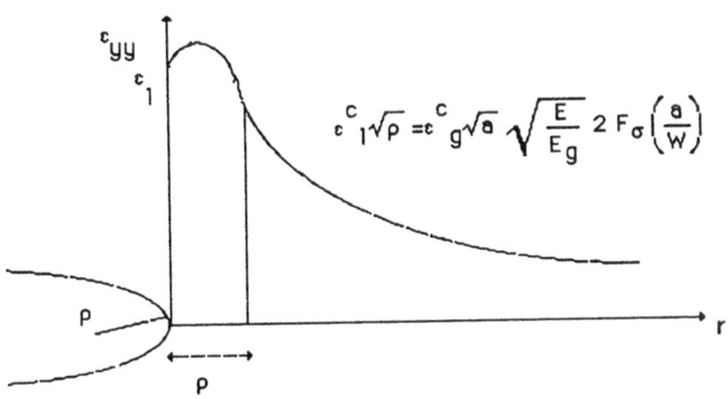

Figure 6 Distribution of strain along the ligament used for the definition of fracture toughness

The two characteristic parameters of each of these local strain fracture criteria are listed in Table 6.

Table 6 Fracture criteria based on the local strain concept

	Parameter 1	Parameter 2
Randall (24)	$\varepsilon^c_l = \varepsilon_f$	r
Osborne and Embury (25)	$\varepsilon^c_l = \frac{\varepsilon_f}{3}$	X_C
critical crack opening displacement (26)	$\varepsilon^c_l = \varepsilon_f$	l_i, d_g

7 Global strain criteria

This kind of fracture criteria has not been well developed. A strain intensity factor k_ε can be defined in LEFM according to the following formula:

$$K_\varepsilon = \varepsilon_g \sqrt{\pi a}\; F_\varepsilon(a/w) \quad (5)$$

where $F_\varepsilon(a/w)$ is a geometrical correction function.

The only global strain criterion which has been proposed in the literature is due to Soete [28] from Ghent University in Belgium. Soete postulates that in an engineering structure no material part will show global ductility of the order of less than 1%. In addition, the allowable defect is defined via a Feddersen diagram (shown in Fig. 1) by:

$$\varepsilon^C_g = f(a/w) \quad (6)$$

The allowable defect is defined as the defect which corresponds to the transition between the regimes of applicability of LEFM and the ultimate strength criterion. Soete suggests that the critical global strain at this transition is approximately 1%. This criterion has not been very well developed for several reasons. One of the major difficulties is the necessity of using large-scale equipment for the characterization of the critical defect.

8 Criteria of interpolation between two limiting states

In the field of fracture, two limiting states are particularly well defined:

- brittle fracture governed by LEFM, and
- plastic collapse governed by a limit analysis.

Brittle fracture is characterized by a unique characteristic material parameter, the fracture toughness, K_{Ic}, whereas plastic collapse is governed by a different characteristic material parameter, the ultimate strength, R_m.

EPFM can be considered to cover the intervening states limited on the one axis by brittle fracture and on the other axis by plastic fracture collapse, characterized by the two parameters, the fracture toughness K^*_C and the critical global stress σ^C_g., respectively. The apparent fracture toughness is defined by using the well-known LEFM formula given in eq.(1).

Two non-dimensional parameters define the non-dimensional apparent fracture toughness $k_R = K^*_c/K_{Ic}$ and the non-dimensionalized critical global stress ratio $S_R = \sigma^c_g/R_m$.

In the (k_R, S_R)-plane brittle fracture is defined by a point with coordinates (0,1) and plastic collapse is associated with the point with coordinates (1,0). Any elasto-plastic states of fracture are characterized by points with coordinates (S_R,K_R) which form the so-called interpolation curve which interpolates between the states of pure brittle fracture and plastic collapse.

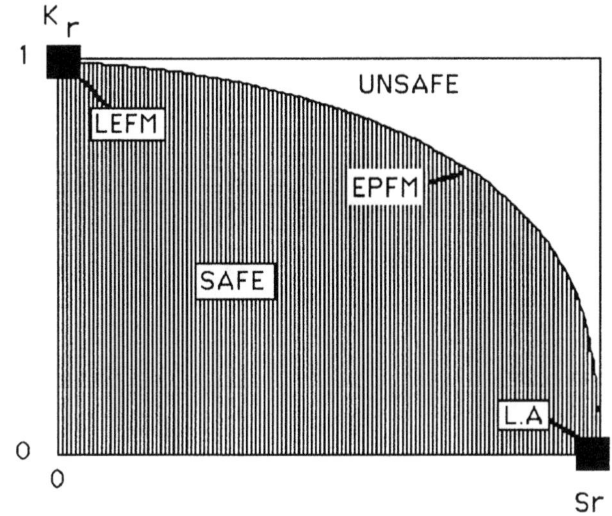

Figure 7 Principle of the Failure Assessment Diagram:
EPFM interpolates between the states of brittle fracture (LEFM)
and plastic collapse (LA).

Figure 7 shows that the interpolation curve separates the area in which structural integrity is preserved from the regime of failure (unsafe). The interpolation curve defines states of critical conditions. Such a diagram is termed Failure Assessment Diagram.

The failure curve or interpolation curve can be obtained from various consideration. Several differing theories define the interpolation curve which connects the two limiting states:

- the R6 procedure as proposed by the CEGB is based on the Dugdale plastic zone model [29],
- an extension of the R6 procedure based on the J-integral has been proposed by EPRI and takes into account the effect of strain hardening [30],
- Newman [31] has proposed an empirical formula using two parameters: the fracture toughness K^*_N and a parameter m_N,
- a method based on the J-integral (Annex A16 of RCC-MR) [32].

These various interpolation approaches and their associated mathematical expressions are given in Table 7.

Table 7 Methods to interpolate in the Failure Assessment Diagram between the states of brittle fracture and plastic collapse.

R6	$K_r = S_r^c \left\{ \left(\frac{8}{\pi} \right)^2 \ln \sec \left(\frac{\pi S_r}{2} \right) \right\}^{-1/2}$
EPRI	$K_r^2 = \frac{Sr^2}{M_e Sr^2 + M_N S_r^{n+1}}$; $M_e = \frac{J(aeff)}{J(a)}$; $M_N = \frac{J(a, n)}{J(a)}$
RANDALL	$K_r = 1 - m_N(S_r)$ $K_r = \left[1 - m_N S_r\right]^2 \cdot \frac{Re}{Rm}$
RCC . MR	$K_r = \frac{1}{\sqrt{\left(\frac{E \varepsilon_{ref}}{S_2^c \cdot Re} \right) + \left(\frac{(S_{ec})^3 \cdot Re}{2 E \cdot \varepsilon_{ref}} \right)}}$

These criteria are now widely used for structures characterized by high technological standard and high level of reliability such as nuclear power plants. In many countries, these criteria are now proposed to be incorporated into the standards.

9 Conclusions

For a better understanding of the concepts and methods of elasto-plastic fracture mechanics a classification of all fracture criteria is required. Here, a seven category classification is presented which is based on strain, stress, and energy using local, global and interpolation methods.

The advantage of this classification is that it shows all the different existing methods in EPFM and extends the students knowledge beyond the most popular criteria such as the ones based on the J-integral and the crack opening displacement. One must though remember that all these criteria describe the fracture toughness with by a unique parameter but it is indeed known that there is an influence of the geometry and the loading path on the characteristic material parameters.

References

[1] Pluvinage,G.: Mecanique elastoplastique de la rupture. Editions CEPADUES, 1989.

[2] Toth, L.: Private communication, 1995.

[3] Irwin, G.R.: Analysis of stresses and strains near the end of a crack traversing a plate. J.Appl.Mech 24, 361-364, 1948.

[4] Dugdale, D.S.: Yielding of steel sheets containing slits. J.Mech.Phys.Solids 8, 100-104, 1960.

[5] Jones, G.T.: Post yield fracture mechanics analysis and its application to turbo-generator design. 3rd Int. Conference on Fracture, Munich, 1973.

[6] Allen, R.J.: The calculation and use of the plasticity correction in toughness. British Rail Research Department Report ME 70/74, 1971.

[7] ASTM Standard E 813-81: J_{Ic} measure of fracture toughness, 810-828, 1981.

[8] Witt, F.J.: The application of the equivalent energy procedure for predicting fracture in thick pressure vessels. Practical application of Fracture Mechanics to Pressure Vessel Technology, 163-167, 1970.

[9] Newman, J.C.Jr.: Fracture analysis of surface and through cracked sheets and plates. Engg Fract Mech 5, 667-685, 1973.

[10] Heald, P.T., Spink, G.M. and Worthington, P.J.: Post yield fracture mechanics. Materials Science and Engineering, 10, 129-136, 1972.

[11] Barenblatt, G.I.: The Mathematical Theory of Equilibrium Cracks. *Brittle Fracture,* 1959.

[12] Ritchie, R.O., Knott, J.F., and Rice, J.R.: On the relationship between critical tensile stress and fracture toughness in mild steel. J.Mech Phys Solids 21, 359-410, 1973.

[13] Pisarenko, G.S. and Krasowsky, A.: Proc 1st Conf Mech Behavior of Materials. JSMS Kyoto, 421-432, 1971.

[14] Holzmann, M., Vlach, B. and Bilek, Z: The effect of microstructure on the fracture toughness on structural steels. Int.J. Pressure Vessels and Piping, 1-9, 1981.

[15] Irvine, W.H. and Quirk, A.: An elastic plastic theory of fracture mechanics and its applications to the assessment of the integrity of pressure vessels. Practical Application of Fracture Mechanics to Pressure Vessel Technology, London, 1977.

[16] Weibull, W.: A statistical theory of strength of materials. Royal Swedish Inst. of Engineering Report No 151, 1939.

[17] Liebowitz, H. and Eftis, J.: On non-linear effects in fracture mechanics. Engg. Fracture Mechanics 3, 267-281, 1971.

[18] Rice, J.: A path independent integral and the approximate analysis of strain concentration by notches and cracks. J.Appl.Mechanics 44, 35, 1968.

[19] Andrews, E.H.: Recent development in the fracture mechanics of polymers. Makromol.Chem.Suppl.2, 189-206, 1979.

[20] Turner, C.E.: Methods for post-yield fracture safety assessment. Post-Yield Fracture Mechanics (Ed. D.G.H.Latzko), 23-210, 1979.

[21] Cotterell, B. and Mai, W.: Plane stress ductile fracture. Advance in Fracture Research, Vol.4, 1683-1695, 1981.

[22] Guillemot, L.F.: Brittle fracture of welded materials. Proc. 2nd Commonwealth Welding Conference, London, C.7, 353-382, 1965.
[23] Sih, G.C.: Strain energy density factor applied to mixed-mode crack problems. Int.J.Fracture, 10, 3, 1974.
[24] Randall, P.N.: Effects of crack on the gross strain crack tolerance of A 533 B steel. J.Engg. Industry ASME B 95 (1), 935-944, 1972.
[25] Osborne, D.E. and Embury, J.D.: The influence of warm rolling on the fracture toughness of bainitic steels. Metallurgical Trans Vol.4, 2051-2061, 1973.
[26] Wells, A.A.: Application of fracture mechanics at and beyond general yield. British Welding Journal 10-11, 563-570, 1963.
[27] Cottrell, A.H.: Trans.Amer.Inst.Min.Metall.Petro.Engineering 212, 192, 1958.
[28] Soete, W.: An experimental approach to fracture initiation in structural steels. Fracture ICF 4, 775-804, 1977.
[29] Shih, C.F:, Kumar, V.and German, M.D.: Studies on the failure assessment diagram using the estimation method and J-controlled crack growth approach. ASTM SPT 803, II, 239-261, 1983.
[30] Kumar, V., German, M.D. and Shih, C.F.: An engineering approach for elastic-plastic fracture analyses. EPRI Report NP 1931, USA, 1981.
[31] Newman, J.C.Jr.,: Fracture analysis of surface and through cracked sheet and plates. Engg.Fract.Mechanics 5, 667-683, 1973.

28 FUNDAMENTALS OF PHYSICO-CHEMICAL MECHANICS OF FRACTURE: PURPOSES AND CONTENTS OF THE NEW EDUCATIONAL COURSE

S.A. SHIPILOV
Institute of Physical Chemistry, Russian Academy of Sciences, Moscow, Russia

Abstract

A new educational course "Physico-Chemical Mechanics of Fracture" the theme of which is environment-assisted cracking (EAC) of engineering materials is described and discussed. The course has been created by the author for materials and engineering students. The main aim of this course is to impart to the students certain skills for complex studying of the synergistic action of a large body of different parameters on possibility, mode and velocity of the EAC of materials under in-service conditions.

1 Introduction

Since the early 1960s, environment-assisted cracking (EAC) of engineering materials is one of the topical problems of materials science. This problem is directly related to the problems for maintenance of the safety and reliability of costly engineering structures, namely nuclear power plants, fossil fired power plants, oil and gas pipelines, oil production platforms, large ships and oil tankers, equipment of chemical plants, aircraft and aerospace technologies, etc. EAC of materials is a major cause of the premature failure of various components of given structures. The losses resulting from the EAC of materials exceed dozens of billion dollars yearly and are escalating throughout the world. High degree of the degradation of materials used in components of these structures further increases the probability of catastrophic failures as well as failures with serious consequences for the environment.

Dr.A.R.C.Westwood of Martin-Marietta Laboratories was among the first to specially emphasize as early as 1977 that "any course on fracture for engineers must contain an adequate consideration of

Teaching and Education in Fracture and Fatigue. Edited by H.P. Rossmanith. Published in 1996 by Taylor & Francis, 2 Park Square, Milton Park, Abingdon, Oxon, OX14 4RN. ISBN 0 419 20700 7.

environmental factors" [1]. However, up to now there are only a few Universities where a course, the subject of which is EAC of materials, is read. It seems to me that there are three causes why similar course is not read at other Universities. First, there is a lack of specialists who know the EAC problem and understand its specificity. Second, most if not all of top-level specialists are working in Research Centres and Laboratories rather than they are teaching in Universities. Third, multi-interdisciplinary character of this subject requires that a specialist (teacher) know at least the individual divisions of such disciplines as materials science, fracture mechanics, and physical chemistry equally. Of course, the last one is difficult to achieve because each given discipline is extensive in itself, and training of the specialist, who could know the divisions required, is difficult to imagine.

In this paper the prerequisites for creature of a new educational course "Physico-Chemical Mechanics of Fracture", the theme of which is EAC of engineering materials, and contents of this course are described and discussed.

2 Prerequisites for creature of the course

2.1 EAC of materials as a phenomenon

EAC is increasingly used as a collective term encompassing a whole spectrum of phenomena from idealized case of purely mechanical fracture, on the one hand, to purely corrosion processes (or dissolution), on the other hand, and includes practically all possible modes of the fracture of engineering materials under in-service conditions. These are stress corrosion cracking (SCC), corrosion fatigue, hydrogen induced cracking, hydrogen embrittlement, liquid metal embrittlement, irradiation assisted SCC, neutron irradiation embrittlement, etc.

SCC is a cracking process caused by the conjoint action of a corrosive environment and nominally static or slowly increasing tensile stress. Corrosion fatigue is a cracking process caused by the conjoint action of environment and alternating stress, while the environment does not have to be corrosive. Hydrogen induced cracking is a cracking process due to absorbed hydrogen atoms. Hydrogen embrittlement is a loss of ductility of certain materials containing hydrogen. Liquid metal embrittlement occurs in materials under nominally static or slowly increasing stress in the presence of particular liquids. Irradiation assisted SCC is environmental cracking of materials exposed to irradiation. Neutron irradiation embrittlement occurs in certain materials after exposure to neutron irradiation.

Nature of these fracture modes varies from one class of material to another. However, all given fracture modes are much similar to each other. In these processes the mechanical work of external forces (for example, applied tensile stresses), and physico-chemical interaction between environment and material are combined in a specific manner. These processes can produce macroscopically brittle fracture, even when they occur in materials which are ductile in a simple tensile test, and at stresses which may be far below those required for general yielding. There is a wide range of susceptibility among various combinations of materials and environments. The possibility, mode and velocity of the processes depend on a number of parameters which are traditionally divided into three major groups: the groups concern parameters related to material, environment and loading condition [2].

The parameters of material are determined by its composition, heat treatment, microstructure, which depends on the quantity and character of the defects (e.g., grain dimensions, presence and dimensions of incipient microcracks), and surface condition. The parameters of environment are phase condition, chemical composition, temperature, pressure, etc. The loading condition - nominally static, slowly increasing or alternating stress - is characterized by nature (applied or residual stress), type (tension, bend, torsion) and velocity of mechanical attack application. What is more, EAC is appreciably defined by nature of forces operating between material and environment in their interface.

2.2 Comments on the traditional courses

The multiformity of the parameters which define possibility, mode and velocity of the EAC of materials exclude the chance of their isolated consideration in "pure" state by the studying of the EAC phenomenon. The statement seems evident one. However, the traditional courses, which are read to materials and engineering students, are oriented namely to the isolated consideration of these parameters.

For example, in the course "Mechanics of Deformable Solid", which is read to engineering students, the idealized model is considered in the form of a medium with the properties of continuity, homogeneity and isotropy instead of real materials. This model allows us to obtain the right notions about the behaviour of a certain conventional solid under its deformation and fracture. But those "right" notions turn out substantially inexact and insufficient ones with an attempt to estimate the deformation and fracture of real engineering materials under in-service conditions. In the course "Corrosion", which is read to materials students, unfortunately very little attention is given to the approaches of fracture mechanics, and the corresponding section of the course provides for the study of "the effect of corrosion on the mechanical properties of materials" or "the effect of mechanical stress on velocity of corrosion". In other words, the one of the most important sections of "Corrosion" is taught on the level of 50-60-ies years. Of course, this level does not correspond to the modern insights about nature of the EAC of materials. It is also to be noted that the engineering students study no individual divisions of physical chemistry (in particular, the sections "corrosion" and "electrochemistry") at all, while the materials students study it not quite enough. The chemical students (physical chemists), as a rule, do not study the materials science, they have not any notions about the fracture mechanics, etc. Thus the training of specialists, who will be engaged in studying and preventing the causes of failures of engineering structures under in-service conditions (namely the failures due to EAC of materials), seems to be insufficiently considered.

It should be emphasized that no fundamental concepts of such scientific disciplines as fracture mechanics, materials science, physical chemistry, etc. (in itself) can explain nature of the EAC of materials and promote to choose the engineering materials with known resistance to EAC as well as the effective means of protection against this fracture mode.

2.2 Notion of the new course

2.2.1 Scientific aspects of the EAC problem

Let us briefly consider only certain of the scientific aspects of EAC problem. This problem is that an increase of the strength or corrosion resistance of metals is accompanied with the inversion of their consumption qualities under in-service conditions. The higher strength and/or corrosion resistance of metals the lower is their cracking resistance in aqueous solution, while the solution can be of corrosion-inactive in the ordinary sense. For example, the double rise of the yield strength of steel (from 800 to 1600 MPa) is accompanied with 10 times (from 70 to 7 MPa$\sqrt{m}$) decrease of the threshold stress intensity K_{ISCC}, which corresponds to the start of corrosion crack growth [3]. Moreover, the "plateau crack growth velocity" in steel of 1700 MPa yield strength in hot water can be 10 million times more than that in steel of 800 MPa yield strength [2]. Depending on the solution composition, velocity of corrosion fatigue crack growth can be either much higher (by dozens or more times) or somewhat lower than the fatigue crack growth rate in air. Hence the velocity of crack growth is determined not only by mechanical, metallurgical and environmental (in itself) parameters and their combination but also by electrode potential, the insignificant change of which can lead to the acceleration of crack growth by a hundred or more times [4].

What is more, the EAC problem is directly related to the problems for maintenance of the safety and reliability of costly engineering structures, namely nuclear power plants, fossil fired power plants, oil and gas pipelines, oil production platforms, large ships and oil tankers, equipment of chemical plants, aircraft and aerospace technologies, etc. At present, EAC of materials is a major cause of the premature failure of various components of these structures [5].

2.2.2 Starting idea of the new course

Three major groups of parameters influencing environment-assisted failure in the components of conventional structure are shown in the shape of circles in the Figure. Contrary to the similar pictures published elsewhere [2], Figure is intended to show the educational courses which are necessary to train specialists in the independent section of failure problem. It is seen, that no one of the shown courses (materials science, physical metallurgy, strength of materials, fracture mechanics, general chemistry, physical chemistry) covers all parameters influencing the failure. Therefore in the place where the abbreviation "EAC" is usually written [2], i.e. where the three circles overlap, it is necessary to write the name of a new educational course which can be denoted as "Physico-Chemical Mechanics of Fracture".

"Physico-Chemical Mechanics of Fracture" must be considered, in my opinion, as a new independent field of failure science which synthesizes mainly the knowledge and the approaches of such scientific disciplines as materials science, fracture mechanics, physical chemistry, etc. The subject of "Physico-Chemical Mechanics of Fracture" is to understand the role and mutual influence of physical, physico-chemical, metallurgical and mechanical variables on the kinetics and mechanism of nucleation, growth and retardation of

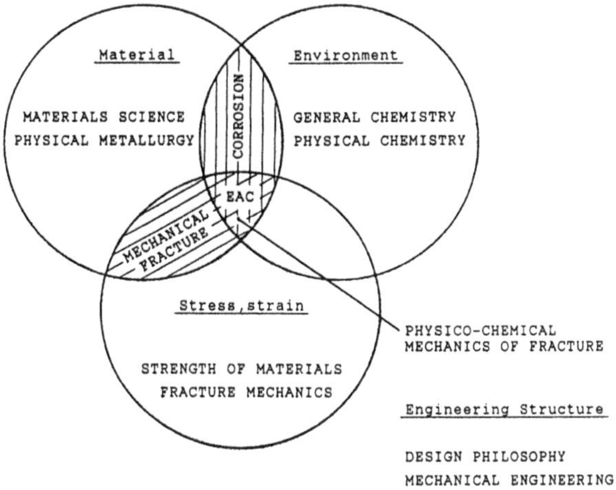

Figure. Three groups of parameters influencing environment-assisted failure, and educational courses required to train top-level specialist in this failure field.

environment-assisted cracks in engineering materials. The methodology consists in investigation of synergistic action of physico-chemical processes inside the crack and physical processes (e.g., transformation of microstructure) in the local fields of materials ahead of the crack tip, while these processes are subjected to the effect of mechanical stresses, on the kinetics and mechanism of cracking. The aim of this course is to impart to the materials students certain skills for complex studying of the synergistic action of a large body of variables on the possibility, mode and velocity of the EAC of materials. The course can be used for engineering students also.

3 Contents of the course

The course "Physico-Chemical Mechanics of Fracture" has been divided into three parts. Prerequisites and reference concepts of the course, while the concepts deal with electrochemical theory of corrosion, fracture mechanics, methods for stress corrosion testing, etc., are given in Part 1. The number of the hours depends on main specialization and background of the students. The special features of the EAC of engineering materials of various types are shown in Part 2. Among these materials are carbon steels, high strength steels, stainless steels, titanium alloys, aluminum alloys, magnesium alloys, nickel-base alloys, etc. The particular examples of the material cracking in various solutions simulating the technological and natural environments are considered also. The advice of the choice of materials is given at the end of this part. Engineering aspects of the EAC problem are shown in Part 3. These are shown on the examples of the failure of components of power plants, oil and gas pipelines, field equipment, equipment of chemical plants, the marine and coastal structures, the aircraft and aerospace technologies, etc. The main effective means of protection against this fracture mode are reviewed also.

The course was developed for students of the Higher School of Materials Science at the Lomonosov Moscow State University and Department of Physical Metallurgy and Metals Heat Treatment at the Bauman Moscow State Technical University. The course is dedicated to the School's students of the ninth semester and Department's students of the eleventh semester. It should be emphasized that there are some differences in the training of students in these Universities. The point is that the School is aimed at training materials scientists, whereas the Department is aimed at training materials engineers. Because of this, the attention of the School's students is usually focused on the studying of fundamental (physical and chemical) aspects of some materials problems, while the attention of the Department's students is usually focused on the studying of engineering aspects of the like problems. The difference in approach to the general problems is reflected in the different dimensional range of interest of these student groups. Broadly speaking, the School's students have been concerned with microscopic parameters such as the grain size, the spacing of dispersed particles, Burgers vector, etc. While the Department's students have been concerned with macroscopic parameters such as the plate width or thickness, the flaw depth and tip radius, the plastic zone size, and the crack opening displacement. Of course, the differences shown are appreciably conventional ones, and top-level training of students in both Universities permits me to read the course without essential changes.

4 Conclusions

A new educational course "Physico-Chemical Mechanics of Fracture", the theme of which is EAC of engineering materials, has been created by the author for materials and engineering students.

The subject of "Physico-Chemical Mechanics of Fracture" is to understand the role and mutual influence of physical, physico-chemical, metallurgical and mechanical variables on the kinetics and mechanism of nucleation, growth and retardation of environment-assisted cracks in the materials.

The methodology consists in investigation of synergistic action of physico-chemical processes inside the crack and physical processes in the local fields of materials ahead of the crack tip (while these processes are subjected to the effect of mechanical stresses) on the kinetics and mechanism of cracking.

The aim of this course is to impart to the students certain skills for complex studying of the synergistic action of a large body of variables on the possibility, mode and velocity of the EAC of materials.

5 Acknowledgements

The author wishes to acknowledge Profs Yu.D.Tretyakov and B.A. Prusakov for approval of the new course, and permission to offer this course at the Lomonosov Moscow State University and the Bauman Moscow State Technical University.

6 Appendix

Contents of the course:

Part one.	Prerequisites and reference concepts of the physico-chemical mechanics of fracture
Chapter 1.	Environment-assisted cracking of materials as a major cause of the failure of engineering structures
Chapter 2.	Electrochemical theory of corrosion*
Chapter 3.	Models of environment-assisted cracking
Chapter 4.	Review of fracture mechanics*
Chapter 5.	Methods for stress corrosion testing
Part two.	Environment-assisted cracking of engineering materials
Chapter 6.	Stress corrosion cracking
Chapter 7.	Corrosion fatigue
Chapter 8.	Hydrogen embrittlement
Chapter 9.	Liquid metal embrittlement
Part three.	Environment-assisted cracking of materials under in-service conditions
Chapter 10.	Materials of the components of nuclear power plants
Chapter 11.	Materials of the components of fossil fired and hydro-electric power plants
Chapter 12.	Materials of oil and gas pipelines, field equipment
Chapter 13.	Materials of the equipment of chemical plants
Chapter 14.	Materials of marine and coastal structures
Chapter 15.	Materials of the components of aircraft and aerospace technologies

Volume of the chapters denoted by (*) depends on main specialization and background of the students.

7 References

1. Westwood, A.R.C. (1978) Panels - Fracture, education and society, in *Advances in Research on the Strength and Fracture of Materials*, Vol.4 (ed.D.M.R.Taplin), Pergamon Press, New York, pp.243-5.
2. Speidel, M.O. (1984) Stress corrosion cracking and corrosion fatigue - fracture mechanics, in *Corrosion in Power Generating Equipment* (ed.M.O.Speidel and A.Atrens), Plenum Press, New York, pp.85-132.
3. Brown, B.F.(1971) Stress corrosion cracking of high strength steels, in *The Theory of Stress Corrosion Cracking in Alloys* (ed.J.C.Scully), NATO, Brussels. pp.186-203.
4. Shipilov, S.A. (1994) The role of hydrogen induced cracking in corrosion fatigue crack propagation, in *UK Corrosion and Eurocorr'94 Conference Papers*, Vol.4, The Chameleon Press Ltd, London, pp.40-8.
5. Shipilov, S.A. (1995) Environment-assisted cracking of materials as a significant cause of engineering systems malfunction, to be published at *the Symposium on 'Risk and Economic Evaluation on Failure and Malfunction of Systems'*, 30 October-1 November 1995, Lisbon, Portugal.

29 A PROPOSAL FOR A GRADUATE INTER-DISCIPLINARY COURSE ON PROBABILISTIC FRACTURE MECHANICS

P.F. FRUTUOSO E MELO
COPPE/UFRJ Eng. Nuclear, Brazil
F.L. BASTIAN
COPPE/UFRJ, Eng. Metalurgica e de Materiais, Brazil
P. KALEFF
COPPE/UFRJ, Eng. Oceanica, Rio de Janeiro, Brazil

Abstract
Based on the identification of the need to provide engineering graduates with training in the field of probabilistic structural analysis of structures subject to cracking, a taraining program is presented. The outline is that of a graduate course on what is called probabilistic fracture mechanics but understood as an interplay between fracture mechanics, probabilistic strength assessment and structural analysis. Due to the highly interdisciplinary content of such a course the state of the art in each one of the three fields is reviewed, aimed at the identification of interdisciplinary trends and/or openings. A critical analysis of the existing results and tools follows, regarding the definition of the scope of the course and the identification of possible flaws in a prospective interface between the specialist subjects. The outline of the course is presented and expected or possible limitations to an adequate understanding are discussed. Finally the practical implementation of the subjects is discussed, focused on the availability of a variety of operational tools capable of speeding up the training process.
Keywords: Interdisciplinary cooperation, probabilistic fracture mechanics, reliability engineering, statistical analysis, structural reliability, structural analysis.

1 Introduction

Fracture mechanics is a very important tool in structural design since cracking due to fatigue loads, creep or stress corrosion among others is a common failure mechanism in structures.
To date, the method to forecast crack propagation under given conditions of stressing and environment is well established and its use as an anlysis tool is staightforward.

Teaching and Education in Fracture and Fatigue. Edited by H.P. Rossmanith. Published in 1996 by Taylor & Francis, 2 Park Square, Milton Park, Abingdon, Oxon, OX14 4RN. ISBN 0 419 20700 7.

In practical structural design situations however loading and material properties may be of random nature. As a consequence, crack propagation may only be forecast in a probabilistic sense. This means that instead of establishing a direct relationship between crack length and elapsed time or number of cycles, a third parameter i.e. the associated probability, must be considered and the usual deterministic limit state criteria like the attainment of a critical stress intensity factor must be restated in terms of allowable probabilies. Moreover, the global stress field of complex structures may not be known directly but only as a result of one or many steps of structural analysis.

Although each one of the subjects mentioned, i.e. fracture mechanics, probabilistic strength assessment and structural analysis, is well developed in literature, experts in each one of these fields will hardly have any experience outside the scope of their specific realm. Thus, the design of complex structures subject to random loading and sensitive to crack propagation presents a challenge which will not in general be adequately dealt with, individually or as a team, by experts from each one of the fiels since there is even the lack of a common language.

Our scope in proposing a graduate interdisciplinary course in probabilistic fracture mechanics is to offer graduates in each one of the specialist subjects mentioned, the opportunity to gain insight into the other fields and to master the specific tools, aiming at the development of a unified approach to typical structural design situations.

In what follows, the state of the art in each one of the three fields will be reviewed, aimed at the identification of interdisciplinary trends and/or openings. A critical analysis of the existing results and tools will follow, regarding the definition of the scope of the course and the identification of possible flaws in a prospective interface between the specialist subjects. The outline of the course will be presented and expected or possible limitations to an adequate understanding will be discussed. Finally the practical implementation of the subjects will be discussed, focused on the availability of a variety of operational tools capable of speeding up the training process.

2 State of the art

Although fracture mechanics is a relatively recent subject a lot of experience is already available in the interpretation and evaluation of its basic aspects. Adequate crack propagation laws, be it in the linear elastic or in the elasto-plastic fatigue ranges [1], be it in stress corrosion processes both in a normal [2] and in a temperature affected environment [1,3] are already available and a lot of effort is being put into further developments in that direction. Crack nucleation and sub-microscopic growth mechanisms are also well documented [4] and analytical forecast models are becoming increasingly more frequent [5]. Finally, load range and loading pattern dependence in fatigue crack growth are being studied in increasing depth and a growing amount of information has been gathered in the recent past [6]. As a paralell development, the typical scatter usually found in experimental investigations like those conducted to evaluate the coefficients of the crack propagation laws or the values of critical stress intensity factors, is no longer being viewed as a nuisance but as an indication that due consideration must be given to the random character of the relevant parameters [7].
Reliability methods, originally devised to forecast and folow-up the behavior of

complex mechanical and electrical systems [8] [9], developed into a widely applicable tool adequate to deal with present day challenges like the determination of the expected life span of whole industrial systems or parts thereoff, the assessment of the safety levels of nuclear power plants [10] or the development of safety or economy oriented maintenace schemes for all types of facilities were failure of individual components happens at random [11] [12]. But beyond that, reliability methods also penetrated the relatively isolated environment of structural design and evolved into a specialized tool to evaluate the safety of stuctures [13], both at the level of the structural members seen as components [14], as well as for the whole structure seen as a system [15]. The basic feature of structural reliabilty is the need for an adequate description of the relevant structural failure mechanisms since, unlike to what happens with components of the sound-failed type, tipical of electrical or mechanical systems, sructural failure will occur when a given limit state, definable in terms of a given set of parameters, deterministic and random, is attained. When failure due to cracking is considered, the limit state will be attained when the stress intensity factor reaches its critical limit, beyond which unstable crack propagation will follow.

Structural analysis, originally an exercise in formulation and solution, numerical or analytical, of the partial differential equations representing the equilibrium of deformable bodies, evolved into the art of translating the geometrical information of a structure into the input code of some finite element analysis program. Finite element programs are widespread and avilable at almost any degree of sophistication be it in terms of the solution algoritms be it in terms of userfriendliness regarding the generation of input data or the availability of graphical output features and allow for the solution of practically any existing structural configuration in terms of displacements and stresses. However, it is still the responsibility of the user to feed the finite element program with the relevant data necessary to generate the correct solution to the desired problem. Although originally deterministic, finite element programs evolved to consider some of the intervening parameters, mainly loads, as random variables. The so called stochastic finite elements [16] allow for the discretization in time of a number of random processes and are a good substitute to classical spectral analysis of stochastic variables since nonlinearities leading to general non-Gaussian processes are better dealt with.

3 The available tools

In each one of the specialized subjects mentioned above, operational tools were developed aimed originally to face challenges specific to their own field of application. These tools are reviewed and commented below.

Classical fracture mechanics is concerned on one hand with the calculation of critical crack sizes or stresses in components with cracks or fabrication defects under monotonic loading in the linear elastic or elasto-plastic regimes, both in agressive or non-agressive environments and, on the other hand, with the forecast of the residual life of components under sub-critical crack growth due to fatigue loading, environment assisted cracking or creep conditions. Basic to both aspects, when linear elastic fracture is considered, is the necessity of the knowledge of the stress intensity factor, characteristic to the configuration under analysis. Stress intensity factors for several geometric configurations are presently available and may be found in handbooks [17,18,19] or may be obtained directly, using experimental, analytical or numerical methods [20,21]. Once the pertinent stress intensity factor is defined for a given geometry the sub critical crack growth may be forecast via empirical crack growth data available in several compilations like for example [22]. The definition of the critical stresses or crack sizes may be found from the critical stress intensity factors specific to the material dealt with. When elasto-plastic fracture is being considered, one of the stress intensity factor equivalents, i. e. the J-integral or the crack tip opening diplacement, are to be determined and from their critical values, the critical crack lengths or stresses [23,24]. Alternatively, commercial software packages are available to deal with specific situations. In either case however, the user must have a precise knowledge of the fracture mechanics concepts involved and also of the response of the material to the specific mechanical and environmental conditions.

Structural reliability algorithms assign failure probabilities to structural arrangements in which some of the defining parameters are random. Such parameters may be related to loads, material properties, geometry or failure mechanisms. Whereas the basic algorithms may vary, it is a common feature to all methods to alternate steps of structural analysis with steps of (re-)definition of the intervening random parameters. It is thus mandatory that some structural analysis procedure be available. Due to the relatively high demand of reanalyses necessary, it has been a common practice to simplify the structural definition or analysis procedures in order to reduce the necessary computer effort. Whatever the algorithm used or the structural analysis procedure chosen, it is still the responsibility of the user to, on one hand, correctly define the distribution functions of the random parameters and, on the other, establish the appropriate limit state function.

Structural analysis programs are purpose oriented, in general closed, software packages allowing the user to evaluate stress fields in structures or other, geometrically complex, material configurations given the applied loads, static or dynamic and the material characteristics, linear or non-linear. As already mentioned it is the user's responsibility to provide the adequate input to a finite element program and also to adequately interpret the output. Most Finite Element programs have more or less adequate user interfaces which allow for an almost purely geometric definition of the necessary mesh of elements and for visual interpretation of the output stress field, rendering them quite satisfactory for the immediate pupose they are developed for, i. e., the analysis of general structural systems. No matter how user friendly however, the decisions regarding the most adequate elements to use, the minimal (and sometimes the maximal) degree of subdivision and (in some cases) the necessary simplifications of the original geometry, combined with an adequate understanding and interpretation of the program output, demand special training and are known by the general designation of structural modelling.

Probabilistic fracture mechanics, as understood in this proposal, should be regarded as the interdisciplinary field unifying the above tools in order to treat problems where:

a) the stress field may not be uniform, demanding one or more steps of structural analysis to be determined and not necessarily deterministic since it may be the result of the response of a structure to non-deterministic loads or may depend on non deterministic material properties;

b) the behavior to fracture will depend on the possibly non deterministic character of the crack propagation laws due to the random nature of the measured coefficients as well as on the load sequencing which may be non deterministic when the loads acting on the structure are random; and

c) the expression of failure, although based on objective criteria (the attainment of a critical stress intensity factor), may depend on a number of random variables and thus demand a probabilistic description.

4 The training proposal

While developing the ideas for the training proposal, our aim was to highlight the basic and advanced aspects of each one of the individual fields involved but favoring the practical side of each subject in order to offer the trainees the opportunity to effectively manipulate the specific tools and gain insight into the fundamentals. Moreover, since the expected target trainees comprise a heterogeneous set, composed of graduates in metalurgical, mechanical (and the like) and systems engineering, we envisaged a possibly broad and self contained character for each one of the subjects.

What we present in the following is our view of the basic topics that should be covered and also of the manner in which they should be taught in order to improve the efficiency of the training process.

- Basic Fracture Mechanics

Should be viewed as the admission gate to the understanding of the physics of crack initiation and propagation and as an introduction to the basic concepts related to the evaluation of crack growth. Fundamental subjects like Griffith's theory and linear elastic and elasto-plastic fracture mechanics should be followed by applied subjects like the evaluation of environmental assisted cracking, applied elasto-plastic fracture mechanics, fatigue and corrosion fatigue cracking and creep crack growth.

- Advanced Fracture Mechanics

Should serve the purpose of providing the operating tools which will allow for the treatment of practical situations via advanced elasto-plastic fracture mechanics. These should comprise among others the crack tip opening displacement and the J-integral methods as well as the analysis based on the crack growth resistane curves (R curves).

- Basic Finite Element Stress Analysis

Here the finite element method should be viewed as a working tool to determine stresses and displacements in structural configurations of arbitrary shape. A review of the basic concepts of elasticity should be followed by a description of the basic types of elements and of the stress and displacement fields they are capable to emulate. Exercices on modelling simple structural configurations should follow in order to allow a practical interpretation of the typical output results. Advanced exercices should comprise special geometric configurations representing cracks or notches, complex structural configurations and an introduction to nonlinear structural response aimed at a prospective modelling of crack growth situations.

- Basic Probability and Statistics

Should comprise a review of statistics understood as a tool to represent random data. Distribution functions should be presented as convenient means to condense and give meaning to aparently scattered information. Exercises in ordering of and distribution fitting to data (one- and multi-dimensional, originating preferably from fracture mechanics and structural loading measurements) should be encouraged. Probability theory should follow basic statistics. Probability should be introduced as an effective means to forecast the behaviour of random processes. The assignment of probabilities to values or ranges of values of random variables should be practiced.

- Basic Reliability Engineering

Experience has shown us that this course should give an overview of failure modeling by probability distributions, followed by some reliability testing tools, including censored data analysis and Bayesian methods, as well as reliability growth models. Next, some discussion on loads and capacity of components and systems should be introduced to allow for an efficient discussion of structural reliability. If there is available time, a few concepts of structural reliability should be introduced here to motivate the trainees. Systems reliability should be introduced next, including structural systems. Then maintenance models should be approached, with examples taken from diverse fields of application and then stochastic models for failure and repair. A final discussion on available methods for the analysis of complex systems is due, including special topics, like common cause failures and human errors

- Structural Reliability Methods

Should be accepted as working tools to evaluate the failure probabilities of structural configurations. Wide use should be made of existing software modules sparing the trainees' mental resources to deal with the central issue of such methods which is the adequate description of the failure mechanisms of the structural configurations under analysis. Simple exercices in selection of probability distributions appropriate to describe the relevant random parameters of general structural configurations should be followed by the description of the random parameters relevant to crack propagation situations. Some simple failure mechanisms should be used to allow for the understanding of the use of the software modules selected and realistic crack propagation situations should follow to provide an adequate closure.

- Statistical Analysis of Reliability Data

The aim of this course is to give the trainees the tools for drawing conclusions from field data on component failures, defects, cracks, etc. Software packages should be introduced to allow for modeling and estimating parameters. Probability paper plotting, regression analysis and test of hypotheses are fundamental. Emphasis should be given to fracture mechanics data, of course.

- Probabilistic Fracture Mechanics

Should be understood as the closure of the training programme. It should be comprised by a number of application exercices and a course work in which all the concepts treated would be reviewed regarded as integral parts of a complete analysis procedure leading to the assignment of failure probability levels to given practical situations in applied fracture mechanics. Some suggestions for such problems are discussed in the following section.

5 Practical implementation

There are many ways in which a training proposal like the one presented above might be implemented. To our view the most productive manner is to set up a number of practice inspired exercices, prefferably adapted to the field of interest of the individual trainees. Such exercices may develop into master's and doctoral dissertations.

Typical situations taken from practice in which the use of probabilistic fracture mechanics is advised are, among others:

- The evaluation of the residual life of pressure vessels in hydrocarbon processing facilities
- The analysis of crack propagation in the bracings of offshore structures
- The evaluation of the integrity of piping subject to high pressures and temperatures typical to the heat exchanging devices of industrial facilities and nuclear reactors

6 Conclusions

In the above we have discussed the feasibility and proposed the outline of what might become a regular specialization program for graduates willing to become active in the field of structural design and analysis of structures sensitive to several types of cracking under actual operating conditions.

Our aim in discussing the basic implications related to the implementation of such a program was to show that it is quite feasible, even at the present stage where very few experts will have the allround knowledge necessary to face the pertinent design challenges themselves, without external aid. But teamwork and expert interplay is exactly the answer to such a state of affairs and we hope to have contributed positively to the development of an increasing technical cooperation between experts from the, yet, independent fields comprising the, inevitably, interdisciplinary field of probabilistic fracture mechanics.

7 References

1. Kanninen, M.F. and Popelar, C. H. (1985) *Advanced Fracture Mechanics*, Oxford University Press, Oxford.
2. Hertzberg, R.W. (1983) *Deformation and Fracture Mechanics of Engineering Materials*, John Wiley & Sons, New York.
3. Atkins, A.G. and Mai, Y.W. (1985) *Elastic and Plastic Fracture*, Ellis

Horwood Limited, London.

4. Chell, G.G. , (Ed.) (1981) Developments in Fracture Mechanics - 2, in *The Mechanics and Mechanisms of Fracture in Metals*, Applied Science Publishers.
5. Panasyuk, V.V., Taplin, D.M.R., Pandey M.C., Andreykiv, O.Y., Ritchie, R.O., Knott, J.F. and Rama Rao, P. (Eds.) (1994) *Advances in Fracture Resistance and Structural Integrity*, Pergamon Press.
6. Suresh, S. (1991) *Fatigue of Materials*, Cambridge University Press.
7. Ditlevsen, O. (1986) Random Fatigue Crack Growth - A First Passage Problem. *Engineering Fracture Mechanics*, Vol. 23, No. 2, pp. 467-477.
8. Bazovsky, I.*Reliability Theory and Practice*, Prentice Hall, Englewood Cliffs.
9. Arsenault, J. E. and Roberts, J. A. (eds.) (1980) *Reliability and Maintainability of Electronic Systems*, Computer Science Press, Potomac.
10. McCormick, N. J. (1981) *Reliability and Risk Analysis, Methods and Nuclear Power Applications*, Academic Press, New York.
11. Jardine, A.K.S. (1973) *Maintenance, Replacement and Reliability*, John Wiley & Sons, New York.
12. Gertsbakh, I. B. (1977) *Preventive Maintenance*, North-Holland Publishing Company, Amsterdam.
13. Madsen, H.O., Krenk, S. and Lind, N.C. (1986) *Methods of Structural Safety*, Prentice Hall, Englewood Cliffs.
14. Melchers, R.E. (1987) *Structural Reliability Analysis and Prediction*, Ellis Horwood Ltd./John Wiley & Sons, Chichester.
15. Thoft-Christensen, P. and Murotsu, Y. (1986) *Applications of Structural Systems Reliability Theory*, Springer Verlag, Berlin.
16 Shinozuka, M. and Yamakazi, F. (1988) Stochastic Finite Element Analysis: an Introduction, in *Stochastic Structural Dynamics*, (eds. S. T. Ariaratnam and G. I. Schuëller), Elsevier Applied Science, London and New York.
17. Sih, G.C. (1973) *Handbook of Stress Intensity Factors*, Institute of Fracture and Solid Mechanics, Leigh University.
18. Cartwright, D.J. and Rooke, D.P. (1976) *Compendium of Stress Intensity Factors*, Her Majesty's Stationery Office, London.
19 Tada, H. (1973) *The Stress Analysis of Cracks Handbook*, Del Research Corporation.
20. Sih, G.C. (ed.) (1973) *Methods Of Analysis and Solutions of Crack Problems*, Nordhoff International Publishers, Leyden.
21. Atluri, S.N. (1986) *Computational Methods in the Mechanics of Fracture*, Vol. 2, North Holland Publishing Company, Amsterdam.
22. Campbell, J.E., Gerberich, W.W. and Underwood, J.H. (1982) *Application of Fracture Mechanics for Solution of Metallic Structural Materials*, American Society for Metals, Metals Park.
23. British Standards Institution. (1991) *Guidance on Methods for Assessing the Acceptability of Flaws in Fusion Welded Structures*. BSI, London. PD 6493.
24. Latzko, D.G.H. (1979) *Post-Yield Fracture Mechanics*, Applied Science Publishers.

30 HIGH STRAIN RATE PHENOMENA IN METAL FORMING: NEW COURSE IN ENGINEERS EDUCATION

M. FOREJT and J. KRECJI
Faculty of Engineering, Technical University Brno
J. BUCHAR
Institute of Physics, Mendel University, Brno, Czech Republic

Abstract
New course has been conceived to introduce the problems of high strain rate processes into the curriculum of mechanical engineering students. The emphasis is on the initial structure influence and its changes during the deformation. The course comprises lectures dealing with measurement of dynamical mechanical properties, structure data, and specific characteristics of various processes such as "conventional" high speed forming, explosive hardening, explosive forming etc. Laboratories contain two parts: first, the measurement of dynamic yield stress using Taylor Anvil Test and analysis of experimental results, and second, the structure studies of specimens obtained both in the Laboratory and in the field tests that employ other loading modes (shaped charges, explosive loading etc.) and evaluation of these experimental data. Spall strength is measured on field experiment specimens.
Keywords: education, engineering, high strain rate, forming, Taylor test, spall fracture, mechanical properties, structure

1 Introduction

The need is being felt to improve, especially in the fourth and fifth years of engineer education, the courses focused on forming technologies. This improvement observes three objectives:
(a) the application of up to date computational methods, (b) development and application of measurement techniques (e.g., measurement of stress variations during the die forming, and (c) allowing the students (and lecturers) to get the "feeling" for material and its structure, to understand the failure under complicated conditions

Teaching and Education in Fracture and Fatigue. Edited by H.P. Rossmanith. Published in 1996 by Taylor & Francis, 2 Park Square, Milton Park, Abingdon, Oxon, OX14 4RN. ISBN 0 419 20700 7.

existing in forming process, to understand the effects of high strain rate. The aim of this contribution is to describe attempts made to fulfil (c).

The course "Forming - selected topics" was started two years ago and it is still being tuned. In due course, the students are encouraged to include these topics in their master and PhD theses. The course consists of lectures and laboratories. Lectures contain following topics: The strain rate as one of the parameters affecting the forming precess; The measurement of mechanical properties under high strain rate conditions (Taylor test, Hopkinson Split Pressure Bar) and their relation to the mechanical properties measured under quasistatic and medium strain rate conditions; Initial structures and their development during deformation; Damage and failure under different strain rate conditions.

2 General background

During the forming, the material is forced to respond to various, usually rapidly changing, parameters - external force, external constrains (e.g., changing profile of die), strain rate, temperature, stress waves. The response consists of slip (dislocation substructure built up, twinning, phase transformations), local rise of temperature, strain localization, damage nucleation and growth, failure. Although in different extent, these effects apply to both product and tool.

Generally, small attention is paid to the initial structure and its evolution during the forming. It is not surprising as there is still lack of complex knowledge about the structure evolvement during deformation processes involving increased strain rate, temperature, and, eventually the stress waves. Special topic is explosive forming and hardening where shock waves are operative.

The relation between structure and mechanical properties is relatively well known for the most deformation modes (quasistatic at room temperature, fatigue, creep) and common structural metals and alloys. This relation is seldom described quantitatively; usually we characterize structure by previous history of the material (composition, heat treatment, irradiation doze, prestrain etc.).

The high strain rate deformation is much less understood. The reasons are, among other:

1. Experimental difficulties
- Short time and high loads
- Sophisticated measurement techniques are essential
- It is impossible to stop the experiment in the "middle" of loading and study the process "sequentially"

2. Intrinsic problems
- The specimen does not experience the same stress in whole volume (as opposite to quasistatic loading)
- Temperature is internal variable (adiabacy)
- Stress and/or shock waves travelling through the specimen reflect on internal and outside surfaces
- Complex and changing stress state
- Final structure is dependent on whole loading history, which is usually not known for particular specimen region and cannot be deduced from externally measured characteristics

- The evaluation of results, e.g., stress characteristics depend on the theoretical model chosen. The discussion of different results (equations) from literature is part of the course.

The overview of relevant deformation processes is in Tab.1.

Table 1. Metal forming technologies and relevant strain rates

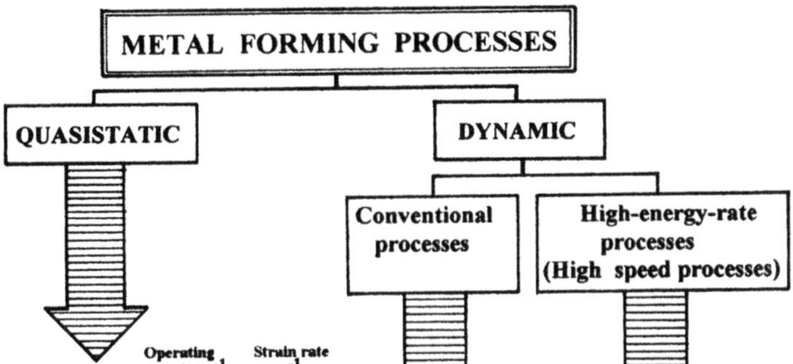

		Operating speed ms^{-1}	Strain rate s^{-1}
Static action	Supperplasticity		10^{-5}-10^{-1}
	Hydraulic machines	0,02-0,25	10 - 10

		Operating speed ms^{-1}	Strain rate s^{-1}
Dynamic action	Crank and screw presses	0,3 - 0,6	4 - 25
	Forging hammers	5 - 8	40 - 160

		Operating speed ms^{-1}	Strain rate s^{-1}
Accelerated solid masses	High speed hammers (pneumatic-mechanical process)	25 - 65	200 - 10^4
	Induced energy (expansion of gases)	10 - 140	
Pressure wave Shock wave	Low-pressure explosive metal forming process	10 - 140	
	High-pressure explosive metal forming process (detonation speed 2000 - 8000 ms^{-1})	250 - 10^3	
Activating energy (impulse generator)	Electrohydraulic forming	300	
	Electromagnetic forming	25 - 250	

3 Laboratories

The experiments performed by students are designed to show all important aspects of high strain rate deformation except the temperature changes (there are no available measurement facilities).

There are only two methods that are experimentally simple and could be used in the laboratories; one when the specimen is loaded from one side and all other surfaces are free (unconstrained), i.e., Taylor Anvil Test (TAT), and second when the change of shape and dimensions is limited, Hopkinson Split Pressure Bar Test (HSPB). In the first case, plastic deformation spreads across the specimen, interacts with elastic waves reflected from free surfaces, but the whole specimen is never stressed to the same level. In the second case, brief periods when the whole specimen is under the influence of the same stress may occur. The presence of free surfaces influences the high strain rate fracture, i.e., spalling. All these effects depend on the geometry and size of the specimens.

The Laboratory (Fig.1) comprises of TAT and HSPB facilities. HSPB contains two maraging bars, the small plate-like specimen is placed between them. Stress wave is induced by the impact of projectile on the free end of the first bar. Capacitance gauges allow measurement of incident, reflected and transmitted (through the specimen) stress waves. The computation then made it possible to determine the stress-strain curve.

For students experiments only TAT is used [1]. TAT is the compressed air gun with barrel 5 m long and 5 cm bore. The specimen impacts, in basic test, on the rigid anvil made from maraging steel (hardness 900 HV10). The compressor could be used up to 4.5 atm. The maximum impact velocities are 250 ms^{-1} (current improvements should allow 300 ms^{-1}). Nozzle velocity is measured by two photodiods.

Two modifications of experiments are used:

1. Rigid anvil - softer specimen. This arrangement allows the demonstration of elastic and plastic waves effects through shape of the specimen and hardness measurements (Figs 2-3). Further, the dynamic yield stress can be evaluated from shape and contraction of specimens (Fig.4). For experiments evaluation, i.e., the measurement of dynamical yield stress, following expression was used:

$$\ln (l_f/l_0) = \rho u^2/2Y \tag{1}$$

where l_f and l_0 are the final length and initial specimen lengths respectively and ρ, u and Y are specimen density, impact velocity and yield stress, respectively [2].

The specimens used in TAT are cylindrical, 25 or 50 mm long, 5 mm in diameter, different cross sections are being prepared. Special problem is the construction of sabots. Students are working on improvements of the sabots (selection of ultralight plastic, optimal shape and their production by injection molding). Until now we used polystyrene sabots machined from plates.

The comparison of results obtained on FCC metals and alloys with different stacking fault energy (SFE) and carbon steel subjected to different heat treatment (tempering temperatures) shows clearly the role of initial structure in high strain rate deformation process.

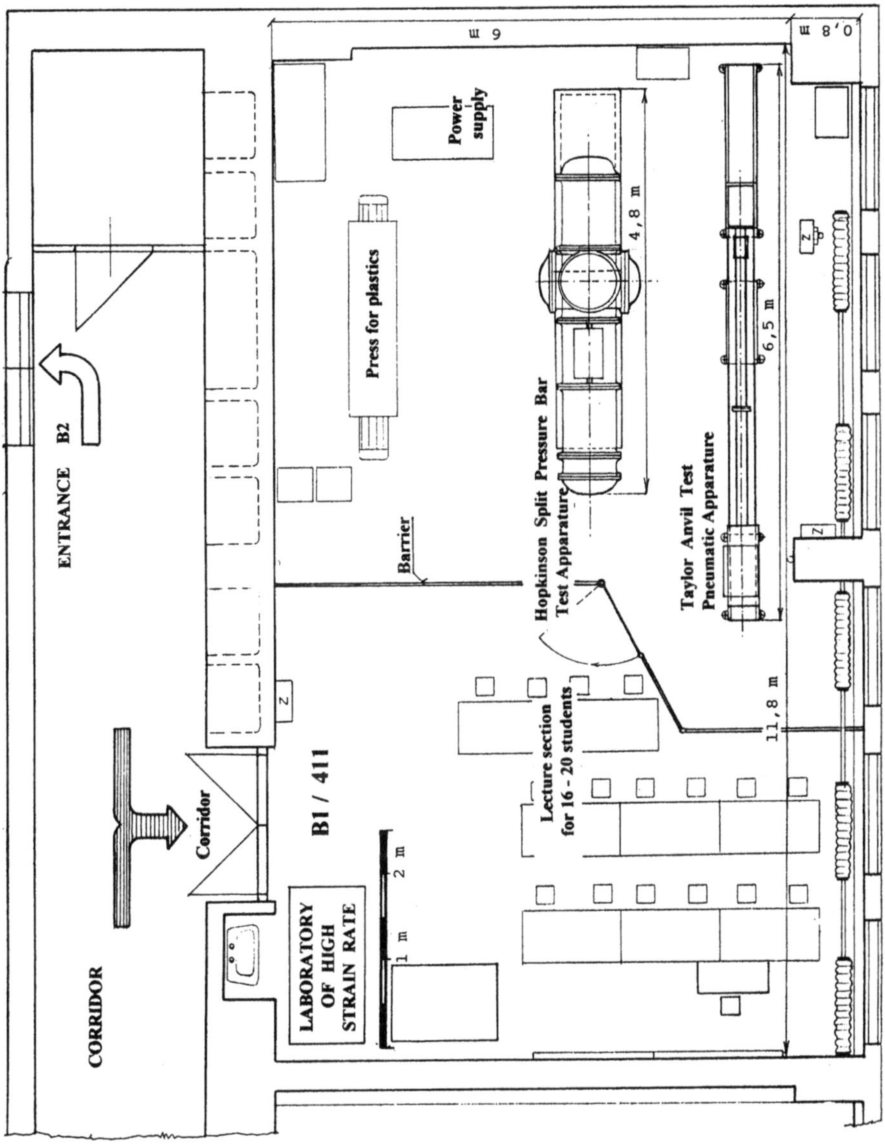

Fig. 1. The layout of the Laboratory of High Deformation Rates

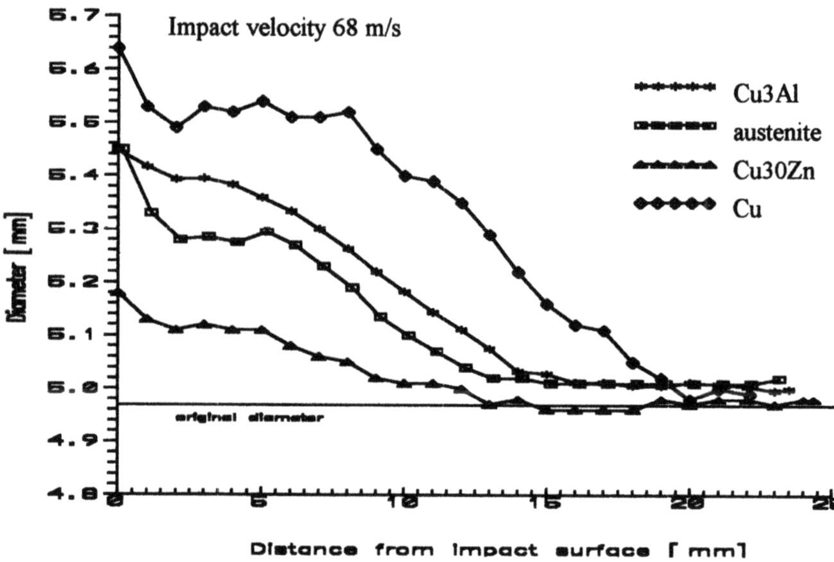

Fig. 2. Taylor test specimens profiles - FCC materials with different SFE

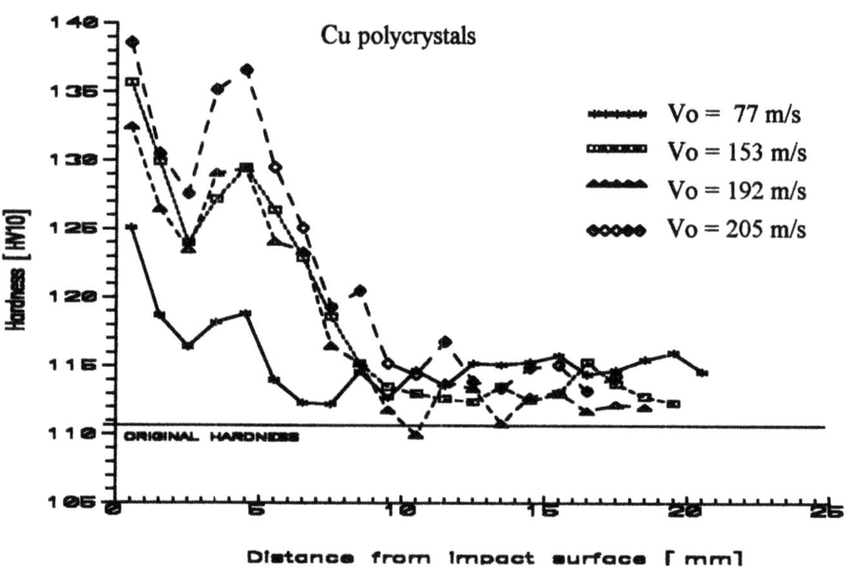

Fig. 3. Hardness HV1 measured along the specimen axis - different impact velocities

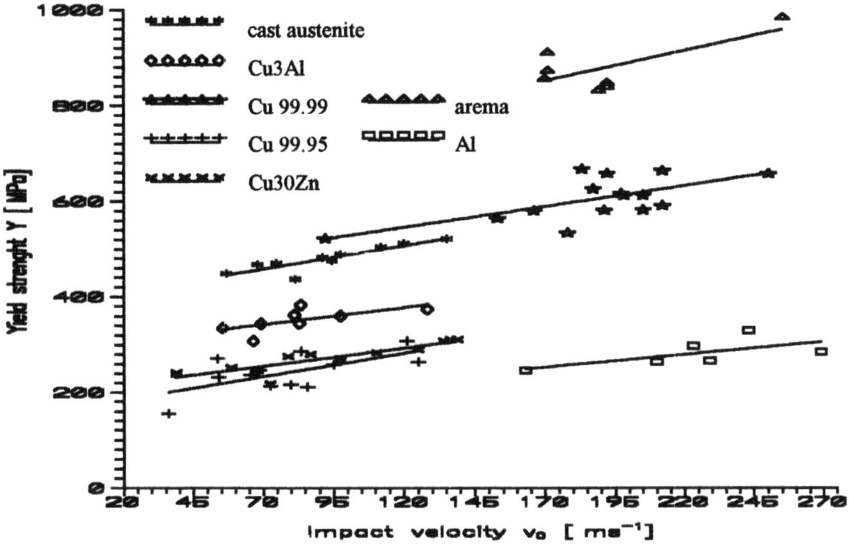

Fig. 4. Dynamic yield stress determined from TAT using Eq.(1)

2. Soft anvil - soft or rigid projectile. This experiments are used to demonstrate the effect of geometrical constraints and the damage evolution. We have used 5 and 10 mm thick copper plates as "anvil" to illustrate the spall behaviour of the material as dependent on the geometry. In 5 mm plates no spall was encountered, while thicker plates exhibit first stages (void nucleation) of high strain rate fracture (Fig.5). Here the importance of material purity is explained on the copper specimens with different volume fraction of oxide inclusions. The profiles and hardness of the specimens are measured (Figs 6-7).

From other experiments (explosive loading), different spall fracture modes are presented. The examples of void and crack failure mechanisms are explained (copper, steel, stainless steel etc), together with spall strength measurement.

These results are confronted with the proposed manufacturing of forming dies by the impact of suitably shaped projectiles.

The results obtained during research tests on HSPB are used to illustrate the stress evolvement in constraint conditions. Similarly the TEM micrographs or eventually the thin foils in TEM obtained from different tests are shown to the students. Dislocation cell size, planar slip in low SFE alloys, dislocation substructure in adiabatic shear bands serves to show behaviour of different materials. Students are asked to evaluate some quantitative parameters such as dislocation cell size (Fig.8). Atlas of microstructures and microanalytical results acquired from explosive cladding, shaped charge jet and solid projectile penetration, explosive loading and other tests, serve to instruct students in cases when direct participation in experiments is not possible. Likewise, the results of failure analysis of forming tools are used. The monographies are available for both students and teachers [3,4,5].

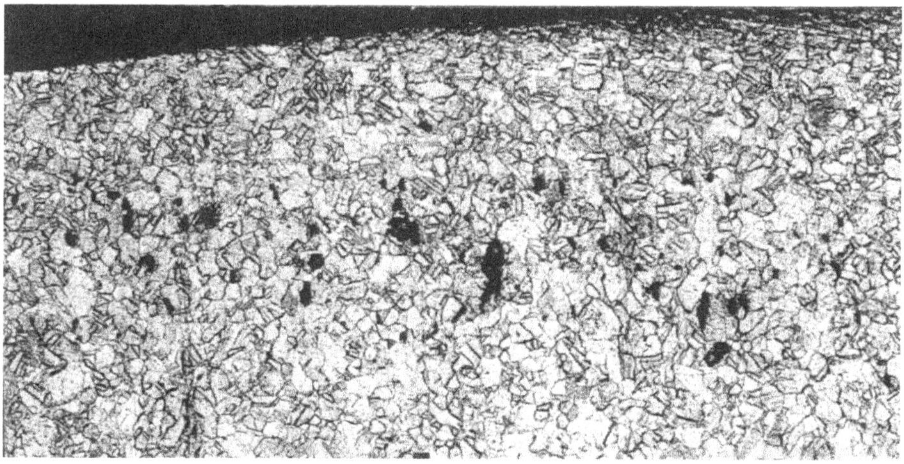

Fig. 5. Metallography of copper target used in TAT. Nucleation of spall fracture (voids) in specimen 10 mm thick. 50 x

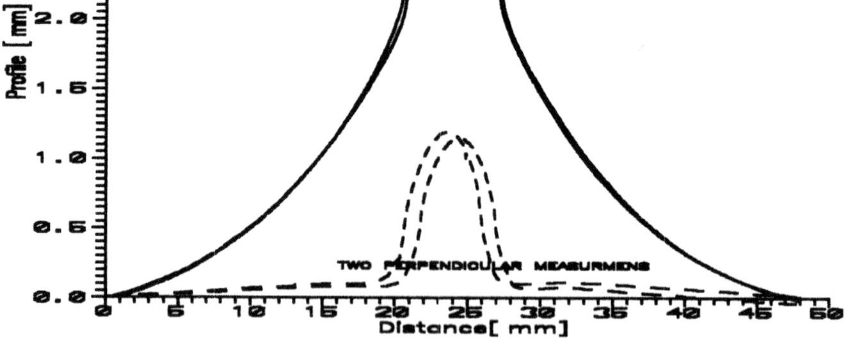

Fig. 6. Profiles of copper discs impacted by copper projectile (front surface)

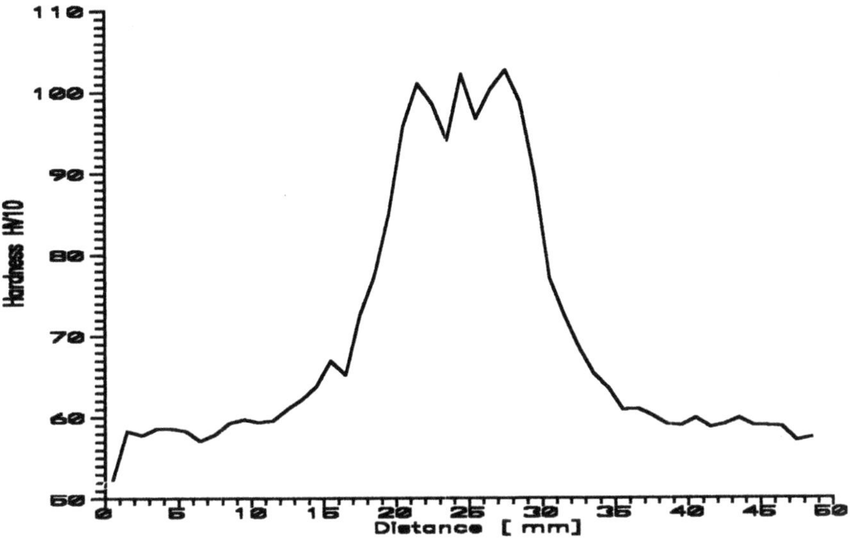

Fig. 7. Copper disc 5 mm, cross section: hardness HV10 on the middle line

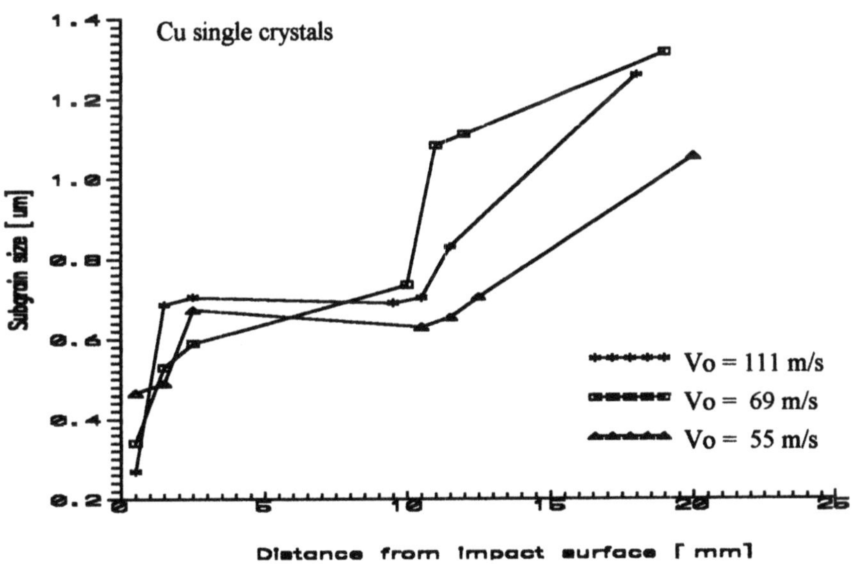

Fig. 8. Dislocation subgrain size along the specimen axis

In addition, the review of literal sources was made easier by developing software that make use of (among other) the COMPENDEX and CURRENT CONTENT and transform them into databases (FORM, HIGHSPEED and IMPACT). The databases are run under software ABSTRACT that allows easy use of databases without special knowledge of database systems [6]. This software is also used by PhD students for preparation of their Thesis.

4 Conclusions

The formability of materials is a complex property that depends on various parameters. Most important are: temperature, strain rate, level of plastic deformation, stress state and structure development. Strain inhomogeneities - sharp strain gradients may cause the strain instability and consequently the damage of material. Insight in high strain rate processes is essential for understanding of metal forming. Many forming technologies, although apparently "quasistatic", use, in fact, such deformation rates that it is necessary to take the strain rate effects into account. These effects are mainly the dependence of strain hardening on deformation rate and stress (shock) waves. Moreover, there are forming technologies using directly the shock waves and related phenomena (e.g., jet formation) like explosive forming.
The relation between material behaviour at high strain rates and its structure is till now obscure.

5 Acknowledgment

Financial support of Czech Ministry of Education and Grant Agency of Czech Republic (Grant No. 106/94/1843) made it possible to concoct this new course for students of engineering technologies, specialization - forming.

6 References

1. Forejt, M., Krejčí, J., Buchar J. (1994) Teorie a technologie tváření vysokými deformačními rychlostmi a energiemi. (Theory and technology of high strain rate and high energy metal forming) Syllabus. Faculty of Mechanical Engineering, TU Brno
2. Woodward, R.L., Burman, N.M. and Baxter, B.J. (1994) An experimental and analytical study of the Taylor impact test. *International Journal of Impact Engineering*, Vol. 15, No. 4. pp. 407-416.
3. Meyers, M. A. (1994) *Dynamic Behaviour of Materials*, John Wiley & Sons, Inc New York.
4. Meyers, M. A, Murr L.E. (Eds) (1981) *Shock Waves and High-Strain-Rate Phenomena in Metals*. Plenum Press. New York.
5. Murr, L.E. (Ed.) (1988) *Shock Waves for Industrial Applications*. Noyes Publications. Park Ridge
6. Krejčová, J., Forejt, M., Krejčí, J. (1994) Vysokorychlostní deformace ve výuce tváření kovů a slitin - Databáze literárních podkladů (High strain rate deformation in metal forming courses - Database system). Faculty of Mechanical Engineering, TU Brno

31 TEACHING HIGH STRAIN FATIGUE IN FINITE ELEMENT ANALYSIS

A.S. MANNING, M.V. BLUNDELL and
J.C. WILLCOX
School of Engineering, Coventry University, Coventry, UK

Abstract

Finite element analysis is used extensively in industry today to solve a wide range of problems and it is necessary to equip engineering graduates with the expertise required to use the technique effectively in industry. The principal requirement for successful finite element modelling and interrogation of results is a thorough understanding of the elements of stress analysis and material behaviour. An understanding of essential finite element theory plus the ability to drive a commercial software system and some knowledge of modelling techniques is also required. This paper addresses the problems of the skills expected of the graduate, the blend of such skills and the teaching of these through simple practical problems using high-strain, low-cycle fatigue.

At Coventry University high-strain, low-cycle fatigue examples are used in the teaching of finite elements as they represent an excellent vehicle in emphasising the care that must be taken in finite element modelling, loading and interrogation of results. Moreover the use of such examples gives students an awareness of the extent and variety of the modes of failure that can occur under different loading systems. Additionally the examples provide a useful introduction to non-linear finite elements. Through the modelling of these simple fatigue load cases and the interrogation of the output however the students learn to recognise the dangers, as well as the power of non-linear finite element analysis within computer aided predictive software.

Keywords: Finite Element Analysis, High Strain Fatigue, Plasticity, Predictive Software.

Teaching and Education in Fracture and Fatigue. Edited by H.P. Rossmanith. Published in 1996
by Taylor & Francis, 2 Park Square, Milton Park, Abingdon, Oxon, OX14 4RN. ISBN 0 419 20700 7.

1 Introduction

In order for the finite element analyst in industry to model practical problems and interrogate results well, the essential prerequisite is to have a thorough knowledge of the subject, be it stress, dynamics or heat transfer. In this paper the problem of fatigue is examined so dynamics and heat transfer is not considered further. In addition to a thorough knowledge of stress analysis the analyst must be able to drive a finite element software system and have some knowledge of finite element theory in order to appreciate limitations of the technique and understand terms used in the software user manuals and hence assist in modelling and interrogation of results.

At Coventry the undergraduates study statics, strength of materials and stress analysis for all of their 3 years, students do not study finite element analysis until part way into their final year when their basic knowledge of stress analysis is considered adequate to be able to model and interrogate results competently.

In the final year undergraduates then study some finite element theory enabling them to appreciate the limitations of the method, and the common expressions used in User Manuals of commercial codes. Such work, combined with the basic knowledge of stress analysis assists in enabling students to perform modelling and results interrogation well. Finally during the final year undergraduates are taught how to drive a software system, the system used at Coventry being ANSYS®. The full set of User Manuals costs about £70, consists of approximately 2500 pages and weighs 15kg, they are intended for the experienced user, not for learning. At Coventry therefore simplified "getting-started" user manuals containing fully documented and annotated worked examples have been developed for student use. Details of the teaching philosophy and method at Coventry is fully described in [1].

Once the basic finite element theory has been completed and the students have a working knowledge of driving the ANSYS® software the students are then given some assignment work and one such assignment in incremental collapse through high-strain low-cycle fatigue is described in this paper.

2 High-Strain, Low-Cycle Fatigue

Power station components such as pressurised pipes and pressure vessels may be subjected to high-strain low-cycle fatigue conditions, particularly under seismic loading. Of concern is permanent hoop ratchetting strain and the ASME design codes must allow for this phenomenon.

If a pipe is subjected to a steady internal pressure causing an elastic hoop stress, plus a cyclic plastic axial loading then a hoop rachetting strain occurs, the strains increasing with each cycle of plastic axial load up to a limiting value.

Experiments carried out to study ratchetting strain and yield surface development are typically carried out on thin tubes subjected to bi-axial stress conditions. The tubes may be subjected to steady internal pressure plus cyclic plastic tensile/compressive a axial loading, or steady end load plus cyclic plastic torsion loading. The tests may be replicated with finite element models, and the model of the internal pressure plus cyclic plastic axial loading is use as a student assignment and the vehicle for teaching and learning at Coventry.

3 Student Assignment - Outline Problem

The physical dimensions of the tube, that is the outer and inner diameters and the tube length are given to the students. In addition the elastic constants, the monotonic uniaxial yield stress and the post-yield or hardening modulus for the tube material, a 304S11 austenetic stainless steel, are also given.

The tube is to be pressurised so that a hoop stress corresponding to ½ uniaxial yield is set up within the tube wall. Axial loading corresponding to ± 1% strain is then to be applied by means of displacement loading and repeated for 20 cycles of plastic strain.

Students are asked to build a finite element model of the tube and then load it according to the prescribed loading above, collect the rachtetting hoop strain data and compare the results with experimental values. Different groups of students compare results so that sensitivity studies of ratchetting data with loading and material properties is obtained.

The finite element solution code to be used is ANSYS® and the non-linear routines in ANSYS® used the von Mises criterion to evaluate onset of yielding, the Prandtl-Reuss flow rules for evaluating increments of plastic strain and a bi-linear kinematic harding law for defining movement of the yield surface.

This plasticity problem provides students with a valuable experience of non-linear finite elements, establishing the correct load steps and number of iterations with each load step. Students also appreciate the dangers of carrying out non-linear finite element work without a good understanding of the subject.

Because the problem involves many load steps it is necessary for students to develop efficient ANSYS® data files involving the parameter and macro facilities. Moreover because of the number of complete cycles of loading involved it is essential for students to perform efficient file handling techniques so that only the required data at the end of each ¼ cycle is extracted from the very large results file and input to another file for subsequent interrogation.

Until recently this assignment would have been possible only on a 'work station' or higher, but hardware development has enabled this work to be carried out on PC machines.

4 Assignment Solution Procedure

The first step in the solution procedure is the construction off the finite element model, because both loading and geometry are axisymmetric the ANSYS® "STIF82", a second order 8-noded isoparametric quadrilateral axi-symmetric element is used. A typical mesh is shown in Fig. 1 which also shows the internal pressure, axial restraint at one end of the tube and axial loading at the other end .

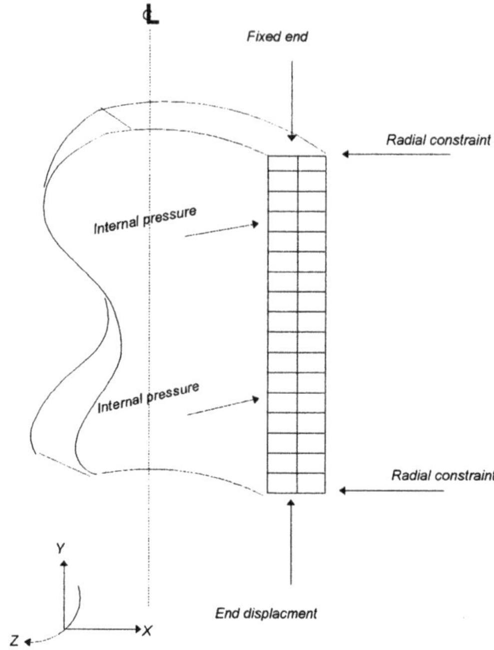

Fig. 1 The Finite Element Model Showing Restraints and Loading

The uniaxial hysteresis loop for a bi-linear kinematic hardening material is shown in Fig. 2. The monotonic uniaxial yield point σ_y must be input together with Poissons ratio υ, Youngs Modulus E and the hardening modulus E_h.

The internal pressure to be applied to the model is evaluated. Typically the pressure should be such that the hoop stress equals ½ uniaxial yield and may be evaluated using thin or thick cylinder theory, depending on the accuracy required.

The first load step of axial load should be such as to just cause yielding in the tube. The hoop and radial stresses are known and the required axial stress is determined using the von Mises yield criterion and solving the resulting quadratic. This derived axial stress is translated into displacement loading using stress-strain and strain-displacement equations.

Subsequent displacement load steps in the plastic region should be equal to or less than the first load step. At each load step convergence is typically achieved in about 5-6 iterations using a convergence ratio of 0.00001. Load step values and convergence ratio settings both influence results.

In Fig. 2 :-

0-A elastic (single load step),
A-B plastic (number of load steps),
B-C elastic (single load step),
C-E plastic (number of load steps),
E-F elastic (single load step),
F-A plastic (number of load steps).

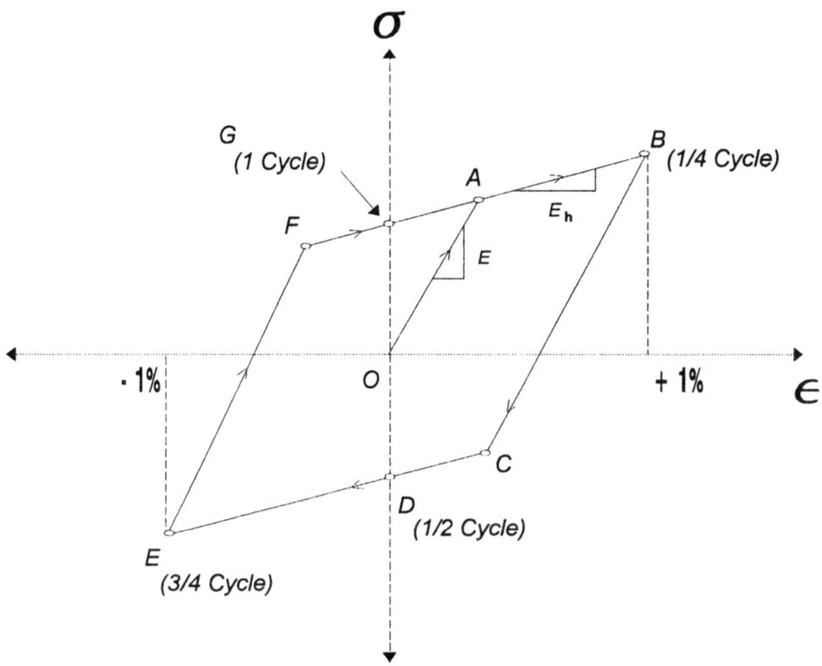

Fig.2 The cyclic hysteresis loop

Because of the number of load steps over 20 cycles of plastic strain it is important that students attempt to write an efficient ANSYS® data file using the parameters and the macro facilities within the preprocessor. Essentially the axial displacement is assigned a letter rather than a numerical value. The different numerical values of displacement being achieved by using a macro, rather like a Fortran 'do-loop'. This use of parameters also enables different groups of students to carry out sensitivity studies, into the effect upon ratchetting strains of different loads and material values

Since rachetting strain data is required at the end of every ¼ cycle over 20 cycles and output file storage is limited it is essential that students extract the required data at the ¼ cycle points and output this to a separate file before proceeding with the next load steps.

5 Results Interrogation and Conclusions

Considerable emphasis is placed upon finite element results interrogation at Coventry, its importance in model validation in general and in particular in non-linear problems such as this assignment is repeatedly highlighted. In this relatively simple exercise much model validation can be performed and serves to illustrate to students the dangers of incorrect non-linear modelling in more complex problems.

For the first elastic load step the derived values of hoop, radial, axial and von Mises equivalent stress can be readily checked against theory. The subsequent ratchetting strain data can also be checked against previously derived experimental and theoretical values.

Fig. 3 illustrates a typical development of plastic hoop ratchetting strain. Similar development graphs can be obtained for other material and loading parameters. Fig. 4 shows a typical plastic hoop strain accumulation curve.

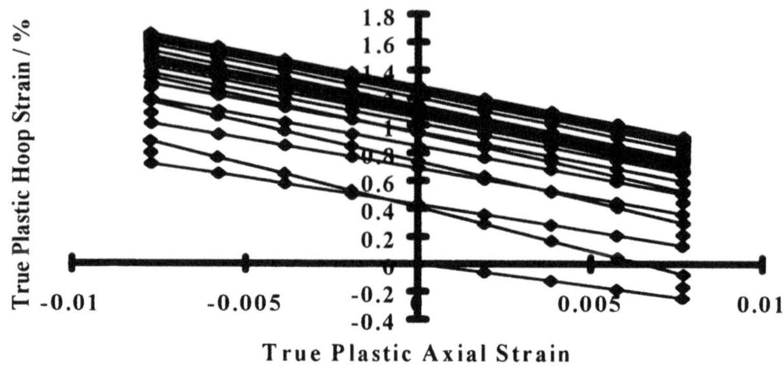

Fig.3 Development of Ratchetting Strain

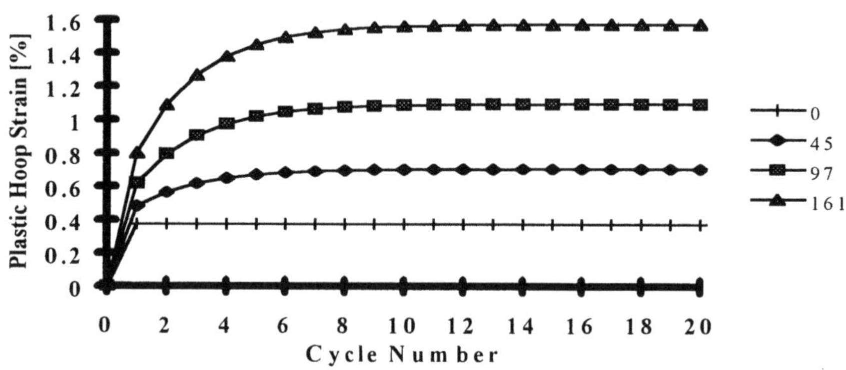

Fig. 4 Hoop Strain Accumulation

This assignment provides students with valuable experience in many aspect of finite element modelling of a non-linear problem, and an awareness of incremental collapse as a failure mechanism.

6 References

1. A.S.Manning, M.V.Blundell and J.C.Willcox (1994) *The Teaching of Finite Element Analysis to Large Groups*, Teaching Science For Technology At Tertiary Level, Stockholm, Sweden.

32 ASSESSMENT METHODOLOGY OF ELEMENTS AND CONSTRUCTIONS RELIABILITY CRITERIA FOR TRANSPORT MACHINES AND EQUIPMENT

M. KOPECKY and F. PESLOVA,
University of Transport and Communications, Zilina, Slovak Republic

Abstract
Modern technology has advenced the development of many new materials and products which in turn have created the need for new and advanced test methods.

The material must be thoroughly investigated for mechanical properties such as fatigue life or maximum strength to permit efficient and economical, as well as reliability and safe design of component structures.

The increase in the reliability level is emphasized also in transport machines and equipment.

In some transportation machinery and equipment, or their elements, as e.q. transmission groups, gear transmissions, also in tower cranes and others, the problem of strength reliability is due to the present-day regulations conditioned by a fatigue process and by knowledge of a fatigue curve.

Keywords: Fatigue, reliability, service-life, strength reliability, random loading, fatigue curve.

1 Introduction

A considerable part of dynamically loaded components in mechanical engineering and in transport is loaded with time-variable strength.

In the case of the means of transport, competitiveness leads to weight reduction savings and extreme operational situations cause high overloading.

Teaching and Education in Fracture and Fatigue. Edited by H.P. Rossmanith. Published in 1996 by Taylor & Francis, 2 Park Square, Milton Park, Abingdon, Oxon, OX14 4RN. ISBN 0 419 20700 7.

A demand to guarantee strength reliability is more important predominantly in case when a failure-free operation of a constructional element can influence the safety of human lives or when an eventual failure can bring about considerable economic losses.

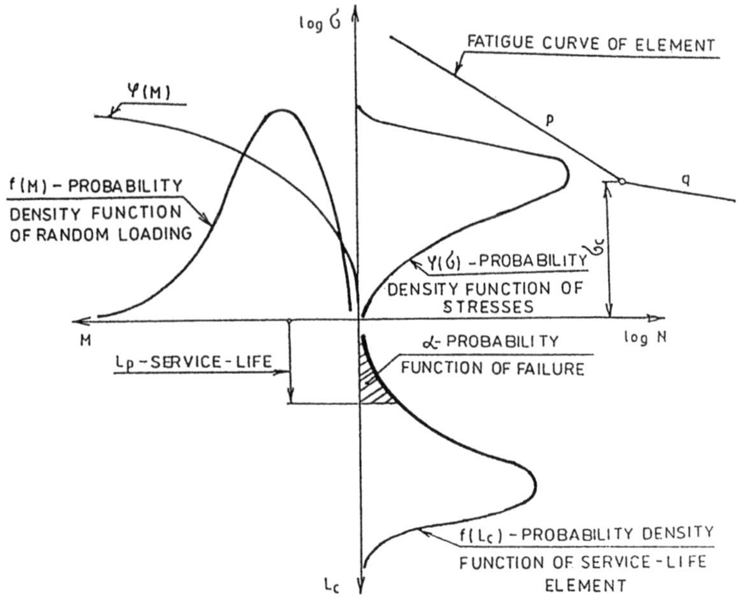

Fig. 1. The dependence of a component service life on loading

The problem of fatigue strength and service-life, as the most important phenomena of strength reliability under those conditions, is connected more or less with a certain degree of uncertainity.

It is probable that the most significant cause of this unfortunate situation is the fact that so far there has not been a single universal theoreticallly and experimentally proved fatigue theory which would consider all the factors that have an influence on the phenomenon. Apart from this, there is not a generally accepted methodology of fatigue tests. The variability of fatigue mechanisms in the Wohler curve belongs among the questions so far not clearly specified, as show in Fig. 1.

It has to be said, that insufficient information about the values of the fatigue curve can lead to errors sometimes higher than 100%. Also, because of economic and others reasons, it is impossible to carry out sufficient amount of fatigue tests on a finished product in order to get reliable information about the fatigue curve. It is further connected with a choice of a suitable cummulation of a fatigue failure which can also lead to different results.

Random operational loading creates a stochastic process of excitation forces. A successful reproduction of the response of this random loading depends on the technical facilities.

With regard to a probability character of the fatigue curve, where its exponents **p**, **q**, have a character of a random quantity, a total service life is also a random quantity characterized by a probability density. The probability that a failure might occur before a required service life **Lp** ends is given by a hatched surface in Fig. 1.

2 The ways to reach the solution goals

A general solution comprises three basic tasks:

2.1 First basic task

To express, by means of failure dependence on damage intensity, reductions in a fatigue curve direction value of the basic material for a constructional element or a complete unit of either a transport machine or equipment, as shown in Fig. 2.

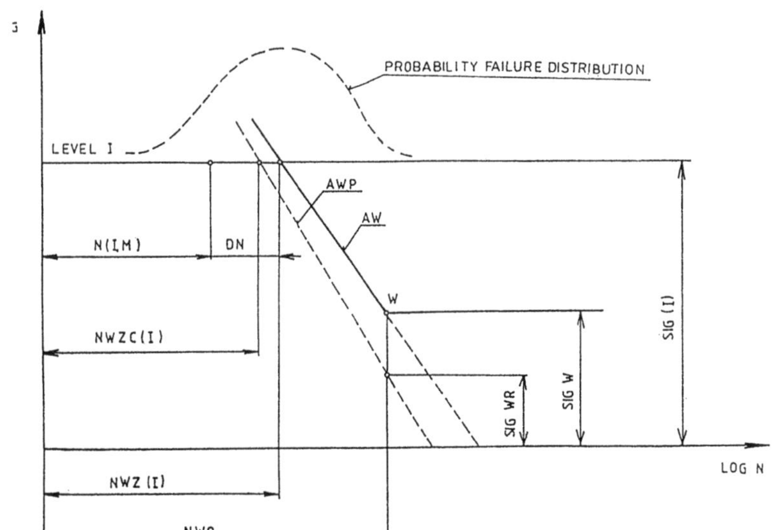

Fig. 2. An estimation of a fatigue curve and its reduction

Insufficient knowledge of a direction value of a fatigue curve is to be expressed by means of an assessment and reduction according to a failure dependence on a damage intensity.

In spite of the fact that a shape of the fatigue curve is idealized to a great extent it was chosen because of its simplicity and possible application of a majority of calculation procedures.

A course of the damage intensity for different loading levels can be a criterion for elimination of insignificant loading levels and for a proper choice of loading levels.

Determining the total failure intensity we can get a basic orientation concerning loading and possible failures in particular elements.

The method is based on a maximum use of computers.The parameters of the fatigue curve reduced in this way are: **NWO**, **AWP**, **SIGWR**.

2.2 Second basic task

An analysis of stochastic loading for transport machines and equipment by means of statistic characteristics. To define an equality amplitude and a frequency of harmonic excitation to a spectrum of random loading by means of functions of statistic moments of power spectral density, as shown in Fig. 3.

An identification of dynamic properties in an operational random excitation and an analysis of a signal with the help of statistical characteristics of stochastic excitation.

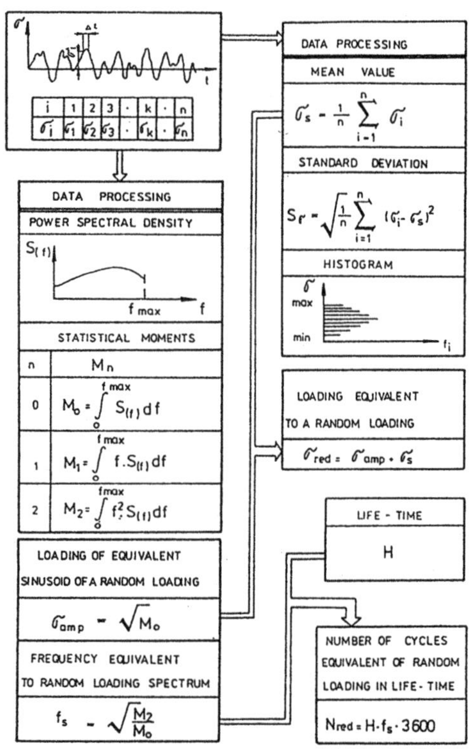

Fig. 3. Scheme of method

2.3 Third basic task

To work out a method of reliability assessment to transport elements and equipment by means of a characteristic curve of strength reliability with the maximum use of computer technology, as shown in Fig. 4.

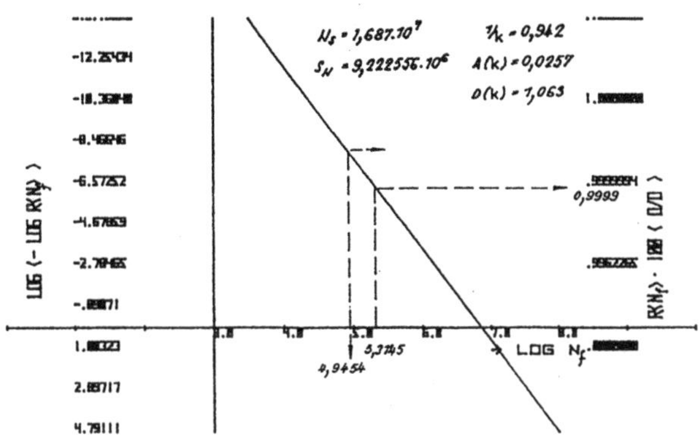

Fig. 4. Statistical curve of the longevity

An expression of a dependence between a stochastic loading of a given constructional element and its reliability by means of an extention with a new magnitude which will express a numerical guarantee in a form of probability.

The dependence between an operating process of a constructional component and its service life $\mathbf{N}_f$ must be expended a variable $\mathbf{R}(N_f)$ which offers a numerical guarantee in a probability form, as shown in Fig. 4.

3 Conclusions

This comprehensive program of general validity has many aspects based on elastomechanics. Its basic part, however, will be solution based on a number of probabilities and on mathematical statistic methods. Only these methods in close synthesis with other technical and physical solutions and creativity can give answers to the above questions.

The synthesis can have many alternatives determined with a character of used calculation methods. A probably continuous application can enjoy those methods which are, from a mathematical point of view, simple, although their simplification has been acquired from a consequence of using certain approximations and limitations.

The application of this methology shorten knowledge of the time to failure of transport machines components and contribute to the safety and economy of mechanical systems.

The results of its application would be presented to transport facility elements.

4 References

1. Kopecky, M. (1992) Non-destructive techniques for measuring random loading signals and their analysis. *VDI-Berichte,* Düsseldorf, Germany, Nr.940, pp. 33-39.
2. Kopecky, M. (1993) Some results of research on the reliability machine component for transportation. *In: 9th Inter.Scien.Conf. UTC,* Zilina, Slovakia, Vol.3/1, pp. 177-182.
3. Peslova, F. (1994) The failure of cost irons with vermiculare graphite alloyded by aluminium. *In: Conf. of Solidification of Metals and Alloys,* Czenstochova, Poland, pp. 37-45.
4. Weibull, W. (1951) A statistical distibution function of wide applicability. *Journal of App.Mechanics,* No.3.

33 THE TEACHING OF FAILURE ANALYSIS AND ACCIDENT RECONSTRUCTIONS: DETAILED COURSE PROGRAM

A.A. JOHNSON
University of Louisville, Dept of Mechanical Engineering, Louisville, KY, USA

This is an extended version of the paper on pages 81–88 and includes fuller versions of the case studies.

Abstract
A course on failure analysis and accident reconstruction for graduate engineers has been developed at the University of Louisville. So far it has been taught three times and each time has been well received by students. The core of the course is a series of case histories selected from the instructor's consulting files. Each student submits a formal written report on each case and collaborates with another student in presenting one case orally to the class. Formal instruction is provided on topics such as fatigue, brittle fracture and stress corrosion cracking which frequently crop up in failure analysis. An attorney who has a degree in engineering teaches product liability law, contract law, civil procedures and rules of evidence. The students study videotapes of a major jury trial to familiarize themselves with courtroom procedures.
Keywords: Accident reconstruction, failure analysis, graduate course, product liability

1 Introduction

Several years ago the Chairman of the Department of Mechanical Engineering at the University of Louisville asked the writer to develop a graduate course on failure analysis and accident reconstruction. He agreed to do so because he had had many years of experience in these areas. At the outset, however, it was apparent that this was a project with formidable problems. Failure analysis and accident reconstruction are not well defined techniques such that one can teach a student a series of steps applicable to every case which he or she takes on. Rather, each case is different and requires a different plan of attack. Experience and a capacity for creative thinking are of great importance. Also, failures and accidents frequently lead to litigation and a

Teaching and Education in Fracture and Fatigue. Edited by H.P. Rossmanith. Published in 1996 by Taylor & Francis, 2 Park Square, Milton Park, Abingdon, Oxon, OX14 4RN. ISBN 0 419 20700 7.

failure analysis or accident reconstruction must be conducted from the onset with the possibility in mind that it may have to be defended in court.

2 Philosophy of the course

With these problems in mind the writer moved ahead with the development of the course. He set two broad objectives for it. First, it would give students an understanding of the diversity of failure analyses and accident reconstructions and of the techniques employed in them. Secondly, it would give students some understanding of state and federal law relating to product liability and of the role of an expert witness in litigation.

It was decided at the outset to base the core instruction in failure analysis and accident reconstruction on a series of case histories taken from the writer's consulting files. It was hoped that by using this approach the students would learn to appreciate that there is no one correct way to conduct a failure analysis or accident reconstruction and that two equally competent investigators can, in good faith, reach different conclusions in a given case.

Because of the importance of ferrous metallurgy and mechanical properties in failure analysis, formal reviews of undergraduate work in these areas were scheduled. New material aimed at deepening the student's knowledge in areas such as fatigue, brittle fracture and stress corrosion cracking, which are of particular importance in failure analysis, were added. Course notes summarizing some of these topics were prepared for distribution to the students. Also incorporated in the schedule were formal lectures on topics which are specific to accident reconstruction. These included accident vehicle examination and the interpretation of tire marks.

As he developed the course, the writer was fortunate to have the assistance of Mr. Gerald Stovall, a practicing attorney who holds degrees in both law and engineering. Mr. Stovall developed lectures and course notes dealing with product liability law, contract law, civil procedures and rules of evidence. Also, a complete videotape of a week-long product liability jury trial was obtained so that students could be shown the structure of a trial and the role of expert witnesses in it.

3 The case histories

The writer's consulting practice, from which the case histories have been drawn, is diverse. He has been involved in a wide variety of cases going back many years. His clients for failure analysis and accident reconstruction cases have included:

(i) Industrial corporations experiencing failures of manufacturing equipment.
(ii) Attorneys representing industrial corporations being sued because of death or injury caused by an alleged product failure.
(iii) Attorneys representing plaintiffs who have been injured or bereaved by an alleged product failure.
(iv) Insurance adjusters trying to asses the cause of accidents in which

policyholders have been killed or injured.

(v) Attorneys representing insurance companies attempting to subrogate claims by pinning liability on other insurance companies or industrial corporations.

(vi) Attorneys representing defendants in criminal proceedings following accidents.

(vii) Prosecuting attorneys preparing cases against criminal defendants following accidents.

The case histories chosen for use when the course was first taught were as follows:

Case History #1

A flatbed tractor-trailer bearing fifteen junked and flattened cars went out of control as it was traveling through downtown Louisville, Kentucky, on an interstate highway. It grazed against a concrete median barrier and the load of junked cars came off the flatbed. Twelve of them went over the median barrier and into the oncoming traffic. There were three fatalities and two people were seriously injured. The driver of the tractor-trailer pled guilty to a string of criminal charges but was treated leniently receiving only probation and a small fine.

This case gives the students an opportunity to analyze a complex case early in the course. They are expected to interpret tire marks on the road surface, the pattern of damage on the concrete median, the final configuration of the accident vehicles, many witness statements, and a plethora of laboratory results. The essential elements of the reconstruction are:

i). The two stacks of junked cars were intended to be secured by six wire ropes tightened using six tighteners mounted along the left side of the flatbed. The two nearest the tractor were not tightened adequately and failed by friction and wear long before the tractor-trailer reached Louisville. Figure (1) shows the fracture surface of a wire taken from one of these two broken ropes. EDAX measurements showed that it was a steel wire coated with an aluminum-silicon alloy which melted and ran onto the fracture surface.

ii). When the tractor-trailer entered a curve in downtown Louisville the load shifted and the right wheels of the tractor left the ground causing the driver to lose control of the vehicle. The left side of the flatbed ground against the concrete median generating a cloud of concrete dust and sparks and causing substantial heating of the wire rope tighteners. Damage to the mounting plates of one of the tigheners may be seen in Figure (2).

iii). The remaining four wire ropes failed by elevated temperature tensile overload. Centrifugal force then took the junked cars off the flatbed.

A more complete account of this case was presented at the Fifth International Conference on Structural Failure, Product Liability and Technical Insurance held in Vienna, Austria, in July 1995 [1].

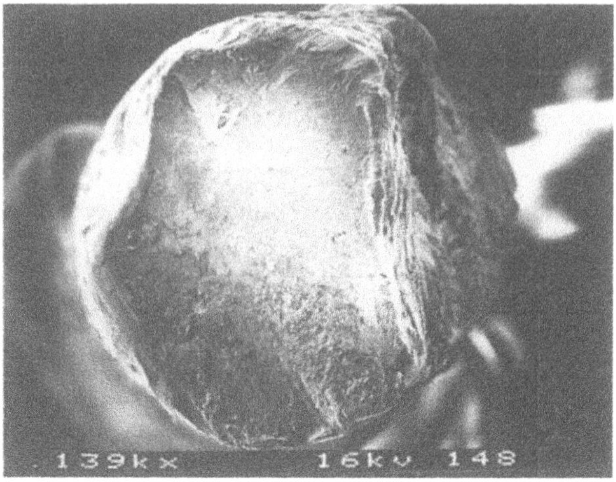

Fig. 1 The fracture surface of a wire from a rope which failed by friction and wear. (Case history #1)

Fig. 2 A damaged wire rope tightener on the left side of the flatbed. (Case history #1)

Case History #2

The second case history is one which does not involve litigation. The exit channel for the hot gases from the final stage of a gas turbine generator was constructed from plain carbon steel. An inner liner, which came into contact with the hot gases, was made from 304 stainless steel. The liner was held in place with cold-drawn 304 stainless steel bolts which were welded to the plain carbon steel channel. The bolts were secured to the liner by nuts which were tack welded to the bolts. There was a 304 stainless steel washer on each bolt on both the inside and outside of the liner. The space between the channel and the liner was filled with an insulating material.

After a few weeks of service the bolts began to fail. Some failures occurred where the bolts were welded to the plain carbon steel channel. Others occurred at the stainless steel liner where tack welds broke and washers suffered excessive wear and cracking. Unfortunately the fracture surfaces were in almost every case corroded so that no fracture markings were visible. The one relatively well preserved fracture surface found is shown in Figure (3). It exhibits some rather poorly defined features which might be "oyster shell" markings characteristic of fatigue or perhaps corrosion fatigue.

This is of course a case in which the students are forced to delve quite deeply into the phenomenon of "sensitization". The relatively cool exit channel end of the 304 stainless steel bolts became sensitized when they were welded to the channel. The relatively hot liner end of the bolts became sensitized because the hot gases passing through the space within the liner are at about the optimum temperature (~600°C) to cause sensitization. Figure (4) is a scanning electron microscope image showing the sensitization at the cold end of one of the bolts. The $M_{23}C_6$ carbide has precipitated on grain boundaries, twin boundaries, slip lines, and perhaps also martensite needles. A detailed interpretation of the microstructure is difficult or perhaps impossible.

Large vibrating stresses were generated by the turbulent 200 mph gas stream within the space defined by the liner. Some gas must necessarily have penetrated into the space between the channel and the liner to produce an acidic condensate. Thus failure probably occurred by corrosion fatigue in the sensitized regions of the bolts.

Case History #3

A tractor-trailer trailer traveling at a moderate speed suffered a spring hanger failure. Examination of the failed hanger revealed that it had been repair welded. Metallographic examinations showed that the hanger was made of malleable cast iron and that it had been repaired using a cast iron welding rod or electrode. On the inside surface of the hanger the repair welding bead missed the fracture over part of its length. In Figure (5), which shows a section through one half of the broken repair weld, the top surface in the photograph is the inside surface of the hanger. The weld bead has missed the fracture and thus served no useful purpose. Fracture occurred through the weld bead on the outside surface of the hanger.

This case gives the students an opportunity to study the microstructures created when malleable cast iron is welded and to learn why it is not usually advisable to

Fig. 3 The fracture surface of a bolt which failed near the weld which secured it to the plain carbon steel exit channel. (Case history #2)

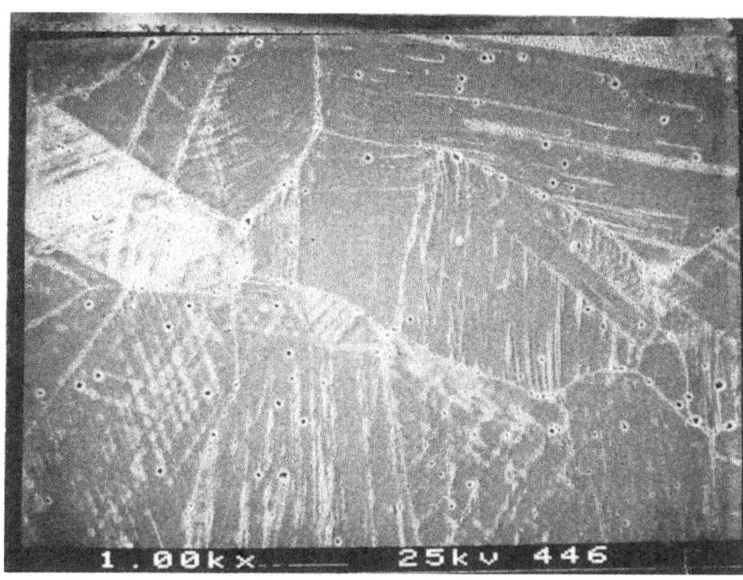

Fig. 4 The microstructure of a sensitized region of a bolt near the weld which secured it to the plain carbon steel exit channel. (Case history #2)

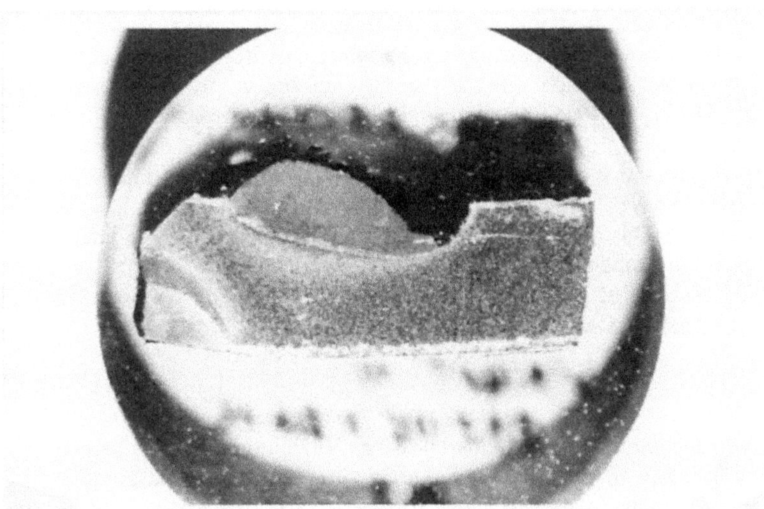

Fig. 5 A metallographic section through one half of the broken repair weld. (Case history #3)

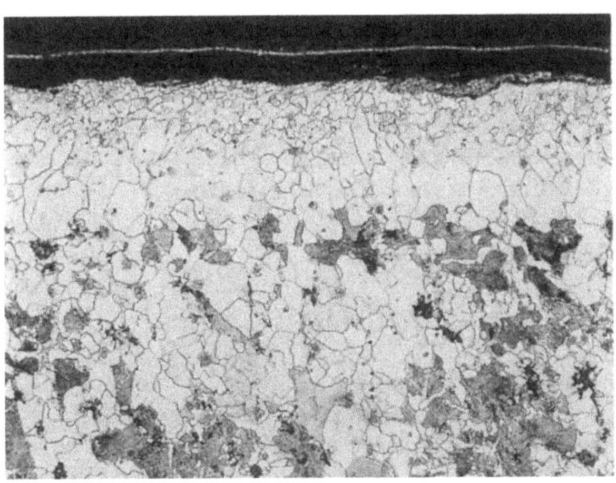

Fig. 6 The decarburized surface layer of the spring hanger (X86). (Case history #3)

attempt to weld it. Malleable cast iron is of course made by first chill casting the product to create white cast iron and then heat treating the casting above the eutectoid temperature to cause the cementite in the white cast iron to break down into graphite and austenite. On slow cooling the austenite breaks down into more graphite and ferrite. During the heat treatment, which is usually carried out in air, the surface is decarburized. This produces a surface layer of ferrite (little or no carbon) and a transition layer containing pearlite (reduced carbon). The ferrite surface layer and part of the transition layer, as seen in a specimen taken from the spring hanger, are shown in Figure (6). This decarburized layer has nothing to do with the failure but its presence teaches the students something about decarburization.

A more detailed account of this case can be found in Volume II of the ASM Handbook of Failure Analysis Case Histories [2].

Case History #4

A total hip prosthesis made of the cobalt alloy known as Vitallium failed after several years of service. It broke in the shaft and the fractures exhibited transgranular fracture with only five grains in each fracture surface. Thus, the casting had an extremely large grain size. Unfortunately, the Hall-Petch parameters for Vitallium have not been measured and so the effect of the large grain size on the mechanical behavior of the prosthesis could not be assessed. The fracture appeared to have started at a surface indentation which was probably caused by the impingement of a tool when the prosthesis was inserted. The patient's X-rays showed that the prosthesis came loose before failure occurred and burnishing observed on the surface of the broken prosthesis was apparently caused by the sliding of the prosthesis within the methyl methacrylate cement which was intended to hold it in place. This was a weak case and the plaintiff failed to establish at trial that the prosthesis was defective. An interesting feature of the case was the observation through Debye-Scherrer X-ray diffractometry that the prosthesis had a hexagonal close-packed crystal structure. As-cast Vitallium has a face-centered cubic structure. The fracture surfaces exhibited mainly cleavage, though part of each was burnished by contact. The patterns on the cleavage facets were characteristic of a hexagonal close-packed metal (Figure 7). A brief account of this case has been given elsewhere [3].

Case History #5

A Chevrolet Blazer ran out of gas and was being pushed by three youths. Another vehicle crashed into the rear of the Blazer trapping one of the youths between the two vehicles and killing him. The accident occurred at night and so it became important to know whether the lights of the two vehicles were on at the time of the accident. This was determined by examining the filaments of lamps not destroyed in the accident.

Lamp filaments are made of potassium doped tungsten which exhibits a ductile-brittle transition. In any kind of test, the transition is above ambient temperature. Thus, when a filament which is not lit receives a mechanical shock, it undergoes

Fig. 7 Cleavage observed on one of the fracture surfaces of the prosthesis. (Case history #4)

Fig. 8 The Nissan Sentra after the accident. (Case history #6)

brittle fracture, if it fractures at all. On the other hand a filament which is lit is at a temperature of about 2500°C which is far above the ductile brittle transition temperature. A mechanical shock causes a helical tungsten filament at this temperature to distend. It may also break.

Lamp filament analyses of this kind are routine in accident reconstructions. In this case it was found that both the tail lights of the Blazer and the headlamps of the other vehicle were lit. A more detailed account of this case, with appropriate photographs, has been published elsewhere [2].

Case History #6

A 1984 Nissan Sentra was stationary at the tail end of a line of traffic on an interstate highway near Cincinnati, Ohio. A Chevrolet van traveling at high speed collided with it causing its gas tank to explode. Both occupants of the Sentra were killed and, according to the coroner, they died of smoke inhalation rather than from injuries caused by the collision. The condition of the Sentra after the accident is shown in Figure (8).

Examination of the gas tank revealed a 16.5 inch (42cm) long tear through one of the heat affected zones of the tank's circumferential weld (Figure 9). Metallographic sections through the weld both at and remote from the tear were prepared. Figure (10) shows the section taken remote from the weld before it was polished and etched. Microhardness measurements on this section and the one through the tear showed that the heat affected zones were significantly softer than either the welds or the unaltered sheet metal from which the tank had been constructed. These results played a central role in the litigation which followed the accident. At issue was whether the relative softness of the heat affected zones constituted a product defect. Work on gas tank welds from several other automobiles made by various manufacturers suggested that other manufacturers have usually found a way of making the heat affected zones harder then the unaltered sheet metal. The case was eventually settled out of court.

Case History #7

Cable television lines in the United States are usually strung on the same poles as those used for electric power distribution and telephone lines. An accident occurred in Louisville, Kentucky, when a lineman was tightening a steel multiwire strand, which had been strung between two southern pine utility poles, prior to attaching a cable television line to it. The lineman had climbed part of the way up one of the poles and was using a strand tightener to tighten the strand when the pole broke above him. The top part of the pole bearing a 7200 Volt transformer, a distribution line, a ground line, and a down guy, began to fall. Almost immediately a second fracture occurred at ground level causing the lower part of the pole to topple over. The lineman was attached to the lower part by a belt and fell with it. The line started a grass fire which caused the lineman to be severely burned. He was also electrocuted and suffered brain damage as a result. The broken pole and the

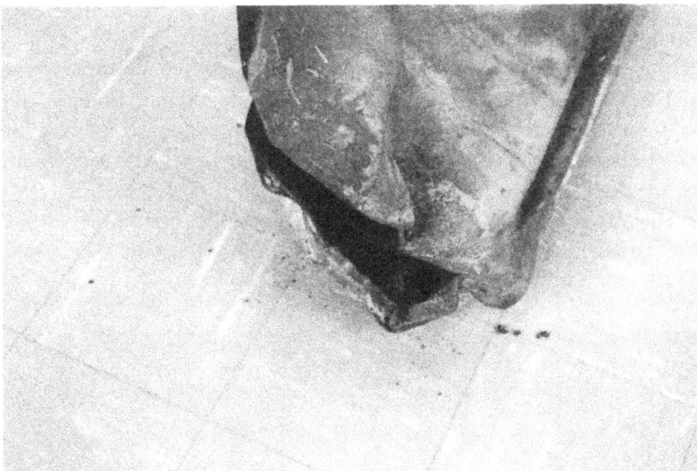

Fig. 9 The 42 cm tear in the heat affected zone of the circumferential weld of the Sentra's gas tank. (Case history #6)

Fig. 10 A metallographic section through the gas tank weld before polishing and etching (Case history #6)

transformer, etc, are shown in Figure (11). The pole was found from the utility company's records to be about forty years old. Forty years is about the half-life of a southern pine utility pole in Kentucky.

the upper fracture was found to be quite complex and to involve a number of stress concentrators. There was a transformer bolt hole, an unused bolt hole, a knot, a woodpecker hole, a machined flat, a region of decay, and a number of weathering checks. All of these played a role in the fracture. The lower fracture exhibited extensive decay. A closer view of the upper fracture is shown in Figure (12).

An analysis of the stresses acting on the pole showed that a sound pole would not have collapsed. The accident was due to the deterioration of the pole which had not been detected at its last biennial inspection.

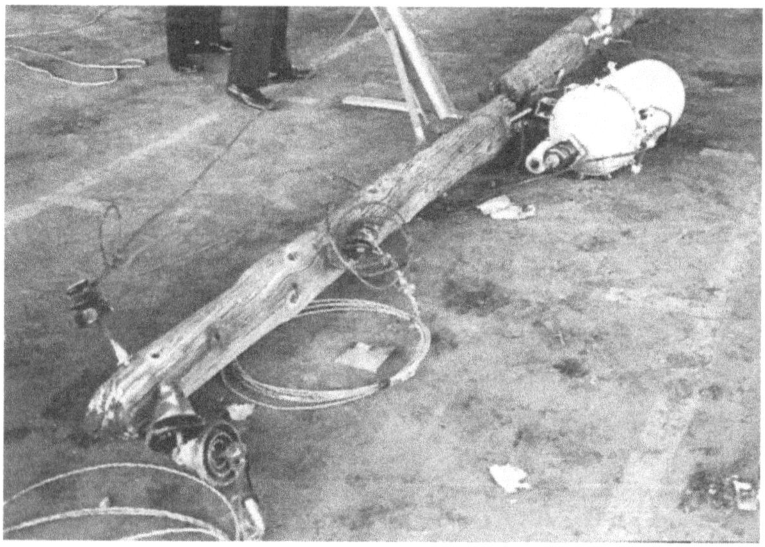

Fig. 11 The broken utility pole after the accident. (Case history #7)

Case History #8

A 7200 volt electric power line strung between two wooden utility poles failed. It came in contact with a wire fence and started a fire along it. A young man who tried to put out the fire came in contact with the live line and was killed. The line was made of a central core of hard drawn steel wire surrounded by seven spirally would wires of hard drawn aluminum. This type of line is known in the industry as swanate ACSR (ACSR stands for "aluminum conductor, steel reinforced"). A metallographic section through a piece of swanate ACSR is shown in Figure (13). The ends of a length of ACSR are usually held in a "dead end shoe" such as the one shown in Figure (14). A piece of swanate ACSR mounted in a dead end shoe is shown in Figure (15).

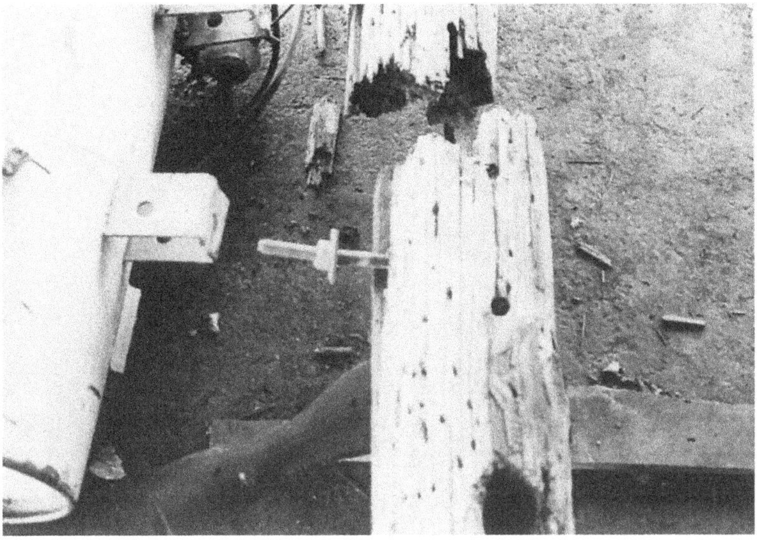

Fig. 12 A closer view of the utility pole's upper fracture. (Case history #7)

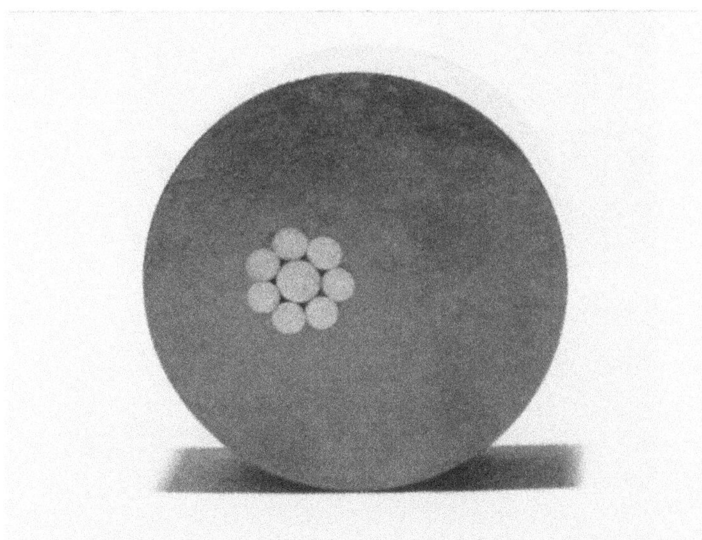

Fig. 13 A metallographic section through a piece of swanate ACSR. (Case history #8)

Fig. 14 A dead end shoe such as is used to mount the ends of a length of ACSR. (Case history #8)

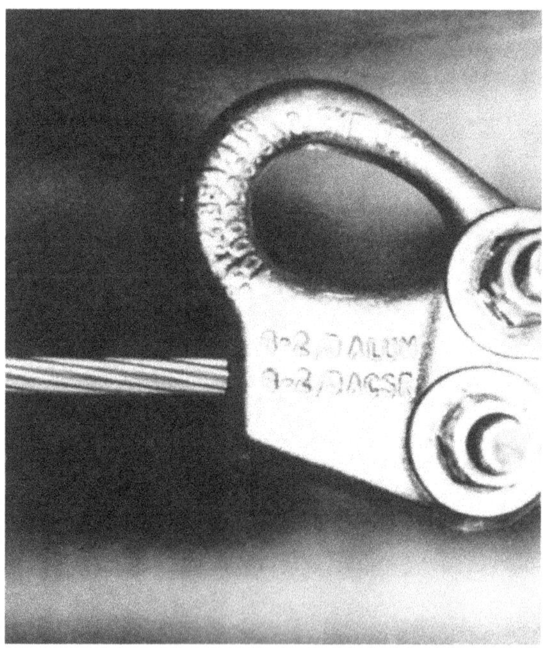

Fig. 15 A piece of swanate ACSR mounted in a dead end shoe. (Case history #8)

The failure occurred where the ACSR was gripped in a dead end shoe. Examination of the fracture surface of the central steel core showed that it had failed by fatigue. It exhibited "oyster shell" markings. Some of the aluminum strands failed by fatigue and some by tensile overload. The accident occurred on a farm in an exposed location with little to break up laminar wind flow. These conditions are conducive to wind induced (aeolian) vibration which is the leading cause of power line failures. It was therefore concluded that aeolian vibration caused the accident. It was essentially a design problem. The failure would not have occurred if the span length had been smaller or dampers had been installed on the line.

Case History #9

Screen failures occurred in grinders used for converting blocks of Hycar artificial rubber into granules. The screens were made from spheroidized 8620 low alloy steel. The grinder consisted essentially of a cylindrical chamber with an axial shaft bearing blades. The blocks rotated in the chamber and struck the screens on each rotation. Impact fatigue cracks developed and eventually, usually after about three months, one or more pieces of the screens broke off. If the grinder was not switched off immediately its blades were damaged and sometimes its shaft was bent. A broken screen is shown , after removal of two small pieces for metallography and chemical analysis, in Figure (16). Some impact fatigue cracks which have jumped from hole to hole are shown in Figure (17).

One solution to this problem might be the choice of a material for the screens which has better impact fatigue properties than the spheroidized 8620 steel. Unfortunately very little is known about the effects of steel chemistry or heat treatment on impact fatigue properties. The most extensive work on impact fatigue is still that of Stanton and Bairstow which was published in 1908 [4]. A reanalysis of their work yielded separate equations for the high and low cycle behavior of a range of irons and plain carbon steels. For failure after about 2500 impacts the energy absorbed per impact, E_i, and the number of impacts to failure are related by the equation: -

$$E_i = E_o + E_k N_f^{-q} \qquad (1)$$

where E_o is the impact fatigue limit, E_k is the impact fatigue parameter and p is the impact fatigue exponent which has a value of 0.6. For failure after less than about 2500 impacts, E_i and N_f are related by the equation:

$$E_i = E_o + \exp\left(\frac{q - D}{C}\right) N_f^{-q} \qquad (2)$$

where C and D are constants and q is known as the low cycle impact fatigue exponent.

Although the development of these equations was an interesting outcome of this failure analysis they were not of much help in resolving the problem. The screen failures were occurring in the high cycle regime and so equation (1) was applicable.

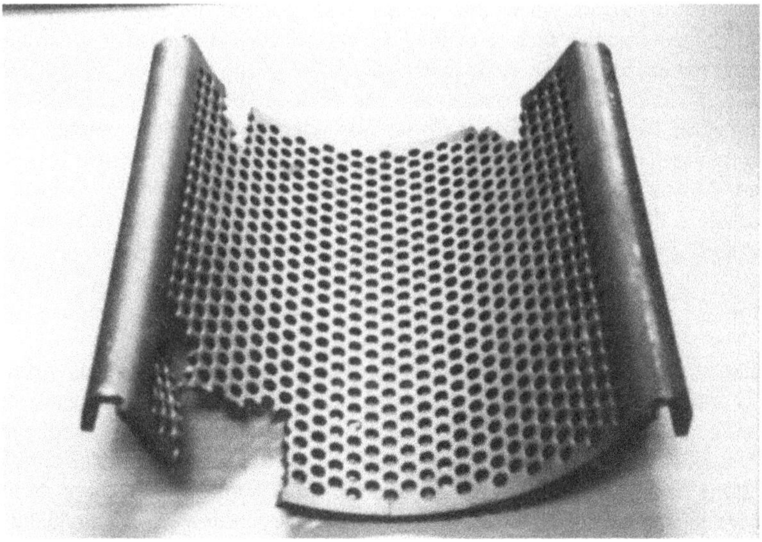

Fig. 16 A grinder screen which failed by impact fatigue. (Case history #9)

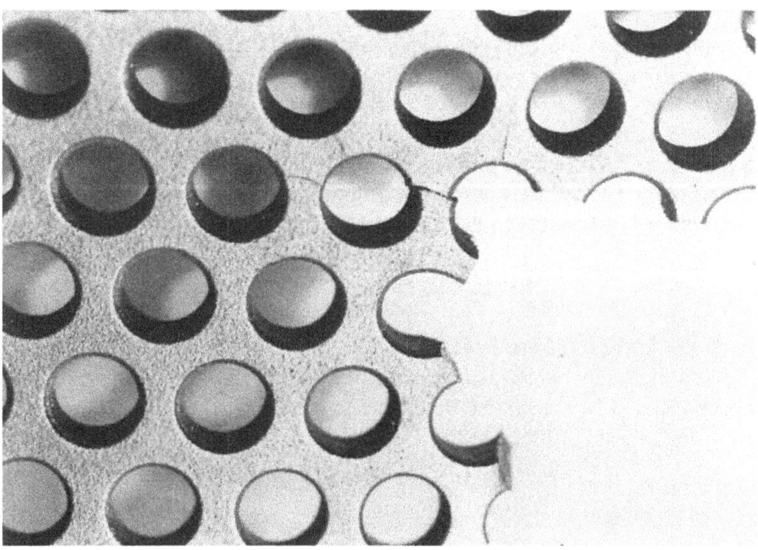

Fig. 17 Impact fatigue cracks in the failed screen. (Case history #9)

The chemistry of the eight steels studied by Stanton and Bairstow was found to influence only E_0 in this equation. The effect of silicon was dominant and it was found that E_0 increased linearly with the square root of the silicon content. However, the silicon contents of Stanton and Bairstow's materials were very low, ranging up to 0.04%, making extrapolations to the higher silicon contents often found in modern steels uncertain.

It was found that this problem was not resolvable by choosing a different material or heat treatment because the information needed to make an informed choice simply does not exist. The problem was resolved by monitoring the grinder's vibration spectrum and changing the screen when a change in the spectrum indicated the development of impact fatigue cracks. Details of this case, and of the analysis which led to equations (1) and (2), have been published elsewhere [5 - 7].

Case History #10

The last case history of the course is one where there was an accident but no failure. It is of course important for an accident reconstructionist to be able to establish that there was no failure so as to avoid unnecessary and inappropriate litigation. A Ford Mustang T-top was being driven at high speed on an interstate highway near Carrolton, Kentucky. The driver lost control of it and it went off the pavement, across the shoulder and into a grassy embankment. It turned over and slid upside down along the gravel shoulder. The driver was seriously injured, losing the sight of both eyes, and the only passenger was less severely injured.

After the accident it was discovered that the tie rod on the front left of the vehicle was broken. The victims' attorney wanted to take the view that the tie rod had broken and caused the vehicle to go out of control. Also, the "A" pillar on the left side, ie, the pillar which supports the left side of the windshield, had bent into an almost horizontal configuration. This had had the effect of forcing the driver backward so that, when the vehicle turned over her face was ground into the gravel shoulder of the highway. Again, the victims' attorney wanted to conclude that the "A" pillar was defective.

Reconstruction of the accident showed that the driver had permitted the left wheels of her vehicle to wander onto the gravel shoulder. When she applied the brakes the resulting uneven traction caused the vehicle to go into yaw. It rotated as it traveled to the right across the highway and shoulder and into the embankment. Its rotation caused the left side of the vehicle to impact the embankment thus fracturing the tie rod on that side. As the vehicle turned over the "A" pillar on the left side violently impacted the embankment. It was this impact that caused its deformation.

The students learned from this case that it is sometimes necessary to tell a client bluntly that an accident has been caused by human error rather than a product defect.

As the course evolved, the American Society for Metals International published the two volume Handbook of Failure Analysis Case Histories [2]. The Handbook uses a standard format for its case histories and that format was adapted for use in the course. Each student was given a copy of one of the two case histories published by the writer in the Handbook [2] as a model. Case histories were limited to 1000 words each plus figures and photographs. Xerox copies of figures and photographs were allowed.

Students were paired and each pair made an oral presentation on one of the case histories to the rest of the class. Since the class size was set at about 20 there were enough students to present all ten case histories during the semester. A series of 35 mm slides was prepared for each case history and these were made available to the students.

4 The course schedule

In the terminology of American higher education, the course was designed to be 3 credit hours at the 600 level. This means that the class is supposed to meet for 3 50 minute periods each week and that all students taking it should have at least an undergraduate engineering degree. It was decided to teach the course in the early evening so that students employed locally full-time could participate. The class therefore met once a week from 5:30 until 8:00 p.m. for fifteen weeks.

The week by week schedule the first time the course was taught was as follows:

Week #1:	5:30 - 6:30	Lecture - Ferrous metallurgy I
	6:30 - 7:00	Introduction of case history #1
	7:00 - 7:30	Introduction of trial videotape
	7:30 - 8:00	Trial videotape
Week #2	5:30 - 6:45	Lecture - Ferrous metallurgy II
	6:45 - 7:00	Introduction of case history #2
	7:00 - 7:30	Student presentation of case history #1
	7:30 - 8:00	Trial videotape
Week #3	5:30 - 6:45	Lecture - Dislocations and Mechanical Properties I
	6:45 - 7:00	Introduction of case history #3
	7:00 - 7:35	Student presentation of case history #2
	7:30 - 8:00	Trial videotape
Week #4	5:30 - 6:15	Lecture - Dislocations and Mechanical Properties II
	6:15 - 6:30	Introduction of case history #4
	6:30 - 7:30	Lecture -Rules of civil procedure I
	7:30 - 8:00	Trial videotape
Week #5	5:30 - 6:00	Lecture - Recovery, recrystallization and grain size effects
	6:00 - 6:30	Student presentation of case history #3
	6:30 - 7:30	Lecture -Rules of civil procedure II
	7:30 - 8:00	Trial videotape

Week #6	5:30 - 6:00	Lecture - Brittle fracture
	6:00 - 6:30	Student presentation of case history #4
	6:30 - 7:30	Lecture -Rules of evidence I
	7:30 - 8:00	Trial videotape
Week #7	5:30 - 6:15	Review for first test
	6:15 - 6:30	Introduction of case history #5
	6:30 - 7:30	Lecture -Rules of evidence II
	7:30 - 8:00	Trial videotape
Week #8	5:30 - 6:30	First test
	6:30 - 7:00	Introduction of case histories #6 and #7
	7:00 - 7:30	Student presentation of case history #5
	7:30 - 8:00	Trial videotape
Week #9	5:30 - 6:15	Lecture - Fatigue I
	6:15 - 7:00	Lecture - Product liability and contract law
	7:00 - 7:30	Student presentation of case history #6
	7:30 - 8:00	Trial videotape
Week #10	5:30 - 6:45	Lecture - Fatigue II
	6:45 - 7:00	Introduction of case history #8
	7:00 - 7:30	Student presentation of case history #7
	7:30 - 8:00	Trial videotape
Week #11	5:30 - 6:15	Lecture - Fractography
	6:15 - 6:45	Lecture - Environmental effects on mechanical properties
	6:45 - 7:00	Introduction of case history #9
	7:00 - 7:30	Student presentation of case history #8
	7:30 - 8:00	Trial videotape
Week #12	5:30 - 6:45	Lecture - Metallurgy of welding
	6:45 - 7:00	Introduction of case history #10
	7:00 - 7:30	Student presentation of case history #9
	7:30 - 8:00	Trial videotape
Week #13	5:30 - 7:00	Lecture -Vehicular accident reconstruction I
	7:00 - 7:30	Student presentation of case history #10
	7:30 - 8:00	Trial videotape
Week #14	5:30 - 6:30	Lecture - Vehicular accident reconstruction II
	6:30 - 7:30	Lecture - Creep
	7:30 - 8:00	Trial videotape
Week #15	5:30 - 8:00	Review for second test

5 The grading system

The American system usually requires that in a graduate course each student be given a grade of A, B, C, or F. This was the case for this course. Since a graduate student must maintain a B average, a C, as well as an F, is a failing grade. Provided a student does a creditable job, therefore, he or she expects to receive a B. Good or excellent performance is rewarded with an A.

For an instructor to decide which students get A's and which get B's he needs some quantitative measure of each student's performance. In the present case each student was assigned a percentage score based on the 10 case histories submitted, the oral presentation, and performance in the two tests. The maximum scores possible in each area were as follows:

Case histories	50%
Oral presentation	10%
First test	20%
Second test	20%

6 Student evaluation of the course

Being very democratic, American Universities usually encourage students to evaluate their instructors. At the University of Louisville standard questionnaires are filled out by each student at the end of each course. This course has now been taught three times, once each in 1992, 1993, and 1994. There are, therefore, three sets of student evaluations available. These are shown in Table I which shows average scores in each of seven areas.

Table 1. The results of student evaluations of the course

	1992	1993	1994
(i) Visual presentation in class	4.55	4.13	4.13
(ii) Answers to questions asked in class	4.52	4.13	4.00
(iii) Effectiveness of classroom activity	4.15	3.81	3.88
(iv) Effectiveness of homework problems	4.33	4.00	4.00
(v) Effectiveness of assigned reading materials	4.30	3.67	4.00
(vi) Instructor's ability to stimulate interest	4.70	3.94	4.13
(vii) Instructor's overall effectiveness	4.57	4.07	4.38

1=poor, 2=fair, 3=average, 4=good, and 5=excellent

These scores compare very favorably with scores achieved by instructors in other engineering courses at the University of Louisville.

7 Resources for the course

Most of the case history files contain print-outs from the searches of computer data bases made as the cases were pursued. Students therefore become very familiar with such searches even though they are not expected to carry them out themselves. The files also contain technical papers and reports obtained by copying material identified in the computer searches. Some of this material will have been found on the shelves of campus libraries. Other parts of it will have been obtained from other libraries through the efforts of library staff. Thus, the students are not expected to do a great deal of independent library work.

Reference works to which the students are introduced include the new two volume ASM Handbook of Failure Analysis Case Histories referred to above [2] and three volumes of the 17 volume ASM Metals Handbook [8-10]. In the field of accident reconstruction the students learn to use the Traffic Accident Investigation Manual by J. Stannard Baker [11].

8 References

1. Johnson, A.A. (1996), Multiple fatalities caused by loss of a tractor-trailer's load, Proc. Fifth Intl. Conf. on Structural Failure, Product Liability and Technical Insurance, to be published.
2. American Society for Metals International (1992 and 1993), Handbook of Failure Analysis Case Histories, Vols. I and II, Materials Park, Ohio, USA.
3. Johnson, A.A., Johnson, D.N., and von Fraunhofer, J.A. (1986), Macroscopy in Failure Analysis. Chapter 2, Microscopy, Fractography and Failure Analysis, Intl Metallographic Society.
4. Stanton, T.E. and Bairstow, L. (1908), The resistance of materials to impact, Proc. Inst. Mech. Engs. pp. 889-919.
5. Johnson, A.A. and Keller, D.J. (1982), Screen failures in a grinder used in the manufacture of artificial rubber, Proc. Intl Symposium for Testing and Failure Analysis, San Jose, California, pp 226-230.
6. Johnson, A.A. and Keller, D.N. (1981), The impact fatigue properties of pearlitic plain carbon steels, Fatigue Engng. Mater. Struc., Vol. 4, pp 279-285.
7. Johnson, D.N. and Johnson, A.A. (1985), The low cycle impact fatigue properties of pearlitic plain carbon steels, Fatigue Fract. Engng. Mater., Struct., Vol. 8, pp 287-294.
8. American Society for Metals International (1978), Metals Handbook, Ninth Edition, Vol. I, Properties and Selection: Irons and Steels, Materials Park, Ohio, USA.
9. American Society for Metals International (1986), Metals Handbook, Ninth Edition, Vol. XI, Failure Analysis and Prevention, Materials Park, Ohio, USA.
10. American Society for Metals International (1987), Metals Handbook, Ninth Edition, Vol. XI, Fractography, Materials Park, Ohio, USA.
11. Baker, J.S. (1976), Traffic Accident Investigation Manual, The Traffic Institute, Northwestern University, Evanston, Illinois, USA.

Author Index

Subject Index

For Product Safety Concerns and Information please contact our EU representative GPSR@taylorandfrancis.com
Taylor & Francis Verlag GmbH, Kaufingerstraße 24, 80331 München, Germany

www.ingramcontent.com/pod-product-compliance
Ingram Content Group UK Ltd.
Pitfield, Milton Keynes, MK11 3LW, UK
UKHW022004180926
475701UK00003B/189
9780367865603